Paleogene Larger Rotaliid Foraminifera from the Western and Central Neotethys

Lukas Hottinger

Paleogene Larger Rotaliid Foraminifera from the Western and Central Neotethys

Edited by Davide Bassi

Springer

Lukas Hottinger (deceased)

Editor
Davide Bassi
University of Ferrara
Department of Physics and Earth Sciences
Ferrara
Italy

ISBN 978-3-319-02852-1 ISBN 978-3-319-02853-8 (eBook)
DOI 10.1007/978-3-319-02853-8
Springer Cham Heidelberg New York Dordrecht London

Library of Congress Control Number: 2014946591

Printed on acid-free paper

Springer is part of Springer Science+Business Media (www.springer.com)

Foreword

Lukas Hottinger's present monograph, entitled *Paleogene larger rotaliid foraminifera from the western and central Neotethys*, including plates and illustrations, has been a long-term project. It was initiated after the publication of Edith Müller-Merz's PhD thesis in 1980. Including and critically discussing a vast amount of research in the field of rotaliid foraminifera, the manuscript had almost reached the stage of publication in August 2011 (see Lukas Hottinger's "Acknowledgments"). However, the author's long lasting serious illness was stronger when he left us on 4 September 2011. It has been his wish that Davide Bassi would take the lead to still make publication possible. Bassi and Philipp Hottinger sorted most of the author's notes, plates and original photographs and included them to the most recent version of the manuscript. Bassi added short abstracts at the beginning of the chapters. Esmeralda Caus (Barcelona, Spain), Johannes Pignatti (Rome, Italy) and Josep Serra-Kiel (Barcelona, Spain) checked the systematic descriptions and double-checked figures and plates. Together with Bassi, they also checked the designations of the holotypes and the diagnoses. Philipp Hottinger provided minor language revisions of the final version of the text.

We are very thankful for the scientific support by Lukas Hottinger's colleagues Esmeralda Caus, Johannes Pignatti and Josep Serra-Kiel. Without their help Lukas' monograph would not have reached its present compact form.

Ferrara, Italy — Davide Bassi
Basel, Switzerland — Philipp Hottinger
Ferrara and Basel, April 2013

Acknowledgements

This book is dedicated to the memory of Yvette Tambareau (Toulouse, 1938–2008) who contributed to this work a number of materials that are exceptional in quality and/or provenience.

Intermittent work on the present volume was started as a sequel of Edith Müller-Merz's thesis published in 1980. During all these years, a great number of colleagues and friends have supported this work by contributing material for the systematic investigation of the Rotaliids. They are too many to be listed here but they are mentioned in the legends of the plates or figures in the text. This may help to appreciate their respective merits. To all of them I am deeply indebted and sincerely grateful. Particular thanks are due also to all those who, for three decades, have guided excursions during workshops or congresses that gave opportunities to collect material.

Considerable technical support was given by Michael Bietenholz who sectioned a significant number of isolated shells, by Pascal Tschudin producing so many thin-sections of different thickness adapted to the particular preservation of the shells in cemented rock and by Saskia Hollaus who printed at standard enlargements so many black and white photographs made under the light microscope. SEM pictures were produced in the SEM laboratory of Basel University directed by R. Guggenheim. Katica Drobne provided a number of important written documents and maps. Many thanks are due also to Nestor Sander for correcting the English language in the first, general part of this book.

Basel, August 2011 Lukas Hottinger

Introduction

During the last century, few foraminiferal taxa have gotten so much attention as *Rotalia trochidiformis* Lamarck, 1804 (see synonymy list of this species below). A revision supported by an ample discussion of this genus and species was published already in 1932 by Davies. There are two reasons for this interest: *Rotalia* is the nominal genus for the foraminiferal family Rotaliidae, a traditional taxon used frequently which plays an important role in the Paleogene biostratigraphy and biogeography of shallow water deposits for petroleum exploration in the Middle East. Moreover, the species *Rotalia trochidiformis* originates from the Lutetian of the Paris Basin where hollow shells are preserved in sediments that can be washed. The material from the Paris Basin was therefore earmarked for analytical studies of general foraminiferal structures supported by scanning electron microscopy (SEM) as carried out by Hansen and Reiss (1971) and by serial sections elaborated by Müller-Merz (1980). Thus, the morphology of *Rotalia trochidiformis* is well known compared to all other members of the family. Many of these have been described under restrictions in observation focussed on the external aspects of the shell as seen in the light microscope. In order to reveal the architecture of the shell, oriented sections of free specimens must be combined with random sections in carbonate rock to reconstruct the three-dimensional structures of the shell from two-dimensional sections because the internal architecture has a diagnostic significance, mainly on the generic level.

The foraminiferal shell of the rotaliids has a rather complex morphology in spite of enveloping only a single cell. The complexity of this architecture is nevertheless high enough to permit the application of study methods developed by comparative anatomy to analyse tissue-bearing organisms. To describe and explain in particular the patterns of the canal system and of the umbilical structures, a number of specialized terms are used. These terms are defined in an illustrated glossary by Hottinger (2006).

The present book supplements the monograph on the Neogene to Recent Rotaliacea (Hottinger 1980 ed.) and follows a similar one on the much smaller group of the Miscellaneidae published recently (Hottinger 2009) that were attributed to the superfamily Nonionacea. The present book presents a revision of most Paleogene rotaliids on the species level. The large-sized taxa were selected that exhibit diagnostic structural features in thin-sections of cemented rock. Special attention was given to the recognition of A and B forms of dimorphic taxa. This was not only to identify the different generations of the same taxon but also to support estimations of the position of the taxon in the K–r gradient of life strategies (Hottinger 1997). Thus, we try to introduce rotaliid representatives as index fossils that can be recognized in random thin-sections of cemented rock.

Accordingly, the illustration of this book is generous. Enlargement standards of 12.5, 25 and 50 magnifications were used to facilitate the comparisons between the different taxa. The selection of taxa is restricted to forms having lived in the Paleocene and the Eocene, where their biostratigraphic significance is much higher than during later epochs. However, some additional rotalid taxa, from the Late Cretaceous or that do not belong to the family Rotaliidae *sensu stricto*, are included in this book in order to demonstrate particular roots of rotaliid phylogenetic lineages in the previous community maturation cycle (see below) or to delimit the taxon Rotaliidae with more precision.

The rotaliid taxa discussed in this book are based on material from the Middle East collected and donated to the Institute of Geology of Basel University by geologists working in the exploration of petroleum or for geological surveys. Collections of borehole material with loose specimens from Qatar collected by Max Chatton complement material of the same region on loan from the Natural History Museum (London). Unfortunately, the detailed biostratigraphical distribution of the taxa from Qatar and some other localities remains confidential, a problem that already had hampered the phylogenetic interpretations in Smout's monograph from 1954. There are no such problems with samples collected during congress excursions in the Salt Range (North-western Pakistan) conducted by A.A. Butt (Lahore University) and the French Pyrenees guided by Y. Tambareau (Toulouse University). As soon as the most important groups of larger foraminifera, and in particular the nummulitids from these places, are revised and tied into a taxonomic and biozonal system without contradictions, we will try to discuss the detailed biostratigraphic background of the taxonomic revisions presented here.

The sections of rotaliid shells illustrated in this book and additional material from the same or additional samples of corresponding regions are deposited in the collections of the Museum of Natural History in Basel (Switzerland), except for material on loan from the Natural History Museum (London) or from P. De Castro's collections in the University of Naples Federico II (Italy) as indicated in the plate legends.

References

Davies LM (1932) The genera *Dictyoconoides* Nuttall, *Lockhartia* nov. and *Rotalia* Lamarck: their type species, generic differences and fundamental distinction from the *Dictyoconus* Group of forms. Trans Roy Soc Edinbourgh 57:397–428, 4 pls

Hansen H-J, Reiss Z (1971) Electron microscopy of Rotaliacean wall strucures. Bull Geol Soc Denmark 20:329–346

Hottinger L (1997) Shallow benthic foraminiferal assemblages as signals for depth of their deposition and their limitations. Bull Soc Géol Fr 168(4):491–505

Hottinger L (2006) Illustrated glossary of terms used in foraminiferal research. Carnets Géol Article 2006, CG2006_M02

Hottinger L (2009) The Paleocene and earliest Eocene foraminiferal family Miscellaneidae: neither nummulites nor rotaliids. Carnets Géol Article 2009/06, CG2009_A06

Lamarck JP (1804) Suite des Mémoires sur les fossiles des environs de Paris. Ann Mus Hist Nat Paris 5:179–188

Müller-Merz E (1980) Strukturanalyse ausgewählter rotaloider Foraminiferen. Schweiz Paläontol Abh 101:5–68, 15 pls

Contents

Part I

The Rotaliid Foraminifera

Rotaliid Shell Architecture and the Palaeodiversity of the Lockhartia Sea

1

Abstract

In the Paleogene the earliest foraminiferal diversity hotspot seems to center in Southern Turkey, in an area of considerable extension around the Lake Van. The centers of Paleogene larger foraminiferal diversity is defined as the Lockhartia communities. This group of complex and large-sized K-strategist rotaliids may serve as index fossil for a so-called Lockhartia Sea, a part of the Paleogene Neotethys, reaching in Asia from Tibet to Pakistan, Afghanistan, Iran and southern Turkey, in Africa from Somalia to Egypt and all over the Arabian Peninsula. The basic structural elements of most rotaliid foraminifera differentiate more than six subfamilies each represented by a single genus that is more than a single phylogenetic lineage. There are three main types of architecture defining the subfamilies Rotaliinae, Lockhartiinae and Kathininae closely evolving together. Counting the taxa per Shallow Benthic Zones reveals common trends in distribution: a rapid increase of taxa in the Paleocene and stability on a very low level from the Cuisian to the beginning of the Bartonian. For the rotaliids, the peak of diversity is observed in SBZ 4, in the Miscellaneidae one zone earlier, in SBZ 3.

1.1 Lamellation Theory (Smout 1954; Hansen 1999)

The study of rotaliid shells had guided Smout (1954) to recognize the basic structural element of most rotaliid foraminifera, the lamellation of the chamber wall (Fig. 1.1). A perforate wall consists of two biomineralized layers, an inner and an outer lamella. They are separated by an organic median layer that reflects the organic template on which the biomineralization of the wall takes place in proximal and distal direction. The presence of a median layer is necessary to produce the pores of the pore-fields for the exchange of small molecules, O_2, CO_2 and others between the living body of the foraminifer and its ambient environment.

In rotaliids, the outer lamella covers not only the wall of the ultimate chamber but also all the free surface of the previously formed shell with each step of the chamberwise growth. Therefore, the number of chambers determines the number of outer lamellas accumulating on the freely exposed surfaces of the shell. In rotaliids, the trochospiral

L. Hottinger, *Paleogene larger rotaliid foraminifera from the western and central Neotethys*, DOI 10.1007/978-3-319-02853-8_1,

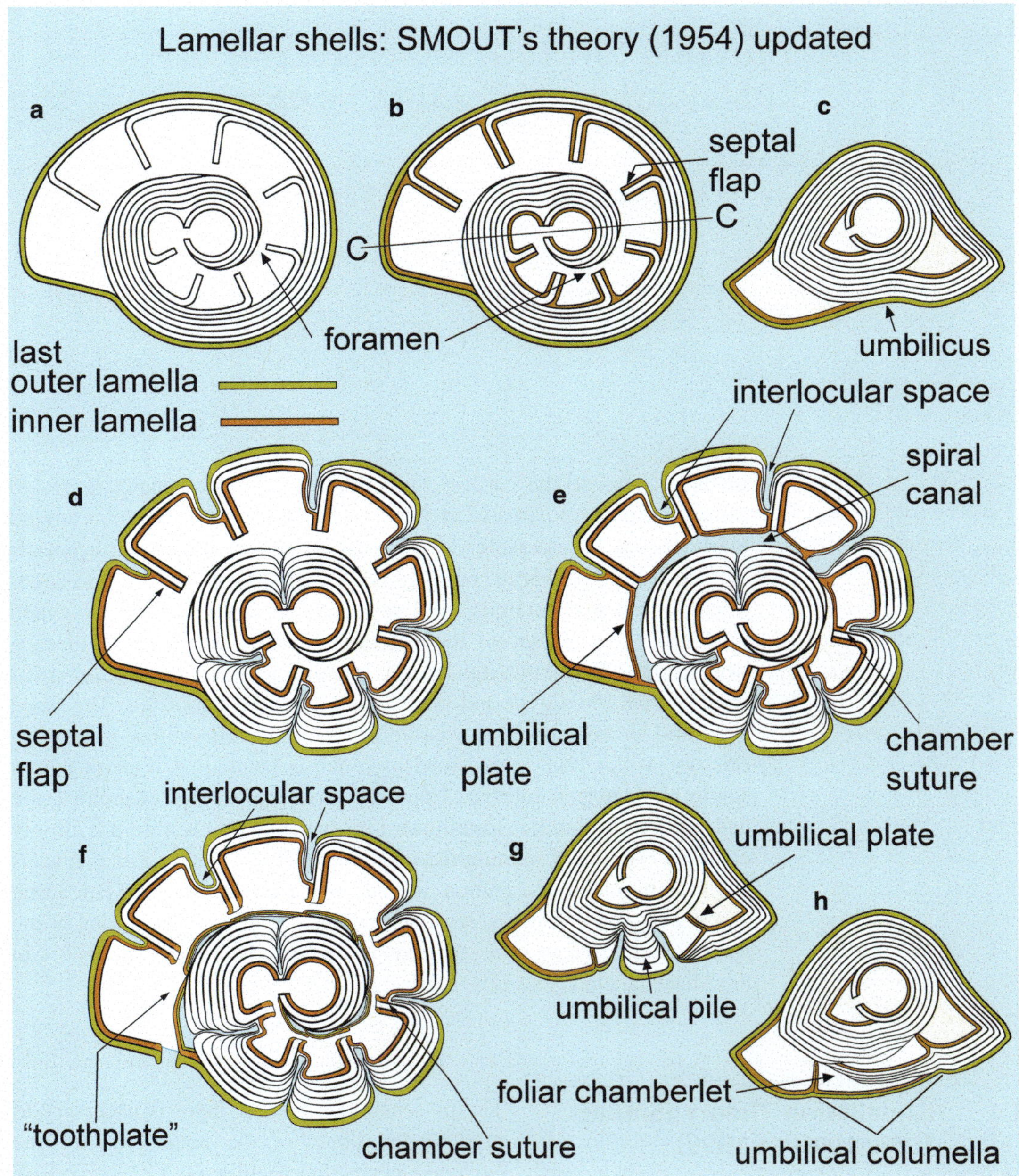

Fig. 1.1 Smout's lamellar theory (1954) updated. The presentation of the different types of lamellation follows the one in Smout's monograph in order to facilitate the comparison. Schematic, not to scale

chamber arrangement enhances the differentiation of dorsal from ventral sides of the shell. Evolute dorsal chamber arrangement exposes the shell surface over more than one whorl and accumulates in the adaxial area an apical umbo. On the ventral side, the chambers are most frequently involute around an umbilicus. Under such circumstances, the number of chambers in the last whorl is reflected by the number of outer lamellae deposited on the chamber wall. However, the thickness of the chamber walls depends not only on the number of lamellae but also on the individual thickness of each lamella. This seems to depend on the water turbulence in the ambient environment.

In the umbilicus, the lamellar structure gets more complicated because the inner and the outer lamellae may form different, independent shapes that produce particular structural elements. The most important of these is the umbilical plate separating the chamber lumen from the umbilical cavity.

Successive chambers may be separated by a simple, tri-lamellar or by a more complex, quadri-lamellar septum that may be visible from the exterior of the shell by the chamber sutures. In rotaliids, the simple septum consists of the primary, bilamellar wall of the previous chamber that is coated by the inner lamella of the ultimate chamber, by the so-called septal flap. The more complex septum is formed by the two primary, bilamellar walls of the successive chambers separated by an interlocular space that is open to the ambient environment. The orifices of the interlocular space forming intraseptal canals are located either on the ventral side of the shell, or all over the ventral and dorsal chamber sutures. In both cases, the interlocular space broadens toward the periphery and receives secondary lamellas from the peripheral surface of the shell. The successive outer lamellas penetrating the narrow interlocular space are extremely thin and fade away towards the interior of the shell. The intraseptal interlocular space is deeply sunk, may be marked on both shoulders by heavy ornamentation and subdivided into a system of tubiform spaces, the intraseptal canal system, that directly communicates with the umbilical cavity and the ambient environment. The original three-layered septum may be restricted to the immediate vicinity of the foramen that in the Rotaliidae always is in interiomarginal position.

In the family Victoriellidae, the intraseptal interlocular space broadens towards the pseudumbilicus by producing a triangular interlocular space between the adaxial ends of subsequent spiral chambers and the previous whorl (Plate 9.11, Fig. 5.3). In this case, the true chamber suture is located at the periphery of the shell and the interlocular space receives secondary lamellation through the pseudumbilicus.

1.2 Elements of Rotaliid Architecture

The most important structural elements of the Rotaliidae are called back to mind in Fig. 1.2. Smaller forms lacking ornamentation on the dorsal side of the shell are closest to *Rotalia sensu stricto*. These forms are distinguished by the aspect of their face dominated by a free-standing central umbilical pile (*Rotorbinella*) or by open slits marking the intraseptal interlocular space, with smooth or feathered shoulders. The adaxial tips of the folia are fused to a columella.

The Lockhartiinae present a peculiar umbilical structure characterized by umbilical cavities that are delimited by successive foliar walls and numerous parallel umbilical piles. This group includes high-spired forms such as *Sakesaria*.

There is the *Kathina* group with their umbilical structure dominated by numerous parallel funnels in a much enlarged columella filling most of the wide umbilicus.

The daviesines are deeply rooted in the Late Cretaceous and characterized by an extremely heavy ornamentation on both sides of the test.

Each of these subfamilies represents more than a single genus that is more than a single phylogenetic lineage. Lockhartiines and kathinines evolve from smaller shells with a simple spiral to much larger shells built by multiple spirals: *Dictyoconoides* and *Dictyokathina* respectively. The daviesines develop a side-line characterized by thin, wide-spired, nearly planispiral shells. The operculini form chambers may have folded septal flaps. In between the folds of advanced species ("*Heterostegina*" *ruida* Schwager, 1863) secondary chamberlets develop like in heterosteginids. Therefore, each of these groups is treated here as a subfamily of the Rotaliidae.

Following Reiss (1963), *Pararotalia* and some of its allies are separated from *Rotalia* and treated as the independent family Pararotaliidae characterised by single, areal foramina linked with a "toothplate" to an

Main rotaliid shell structures

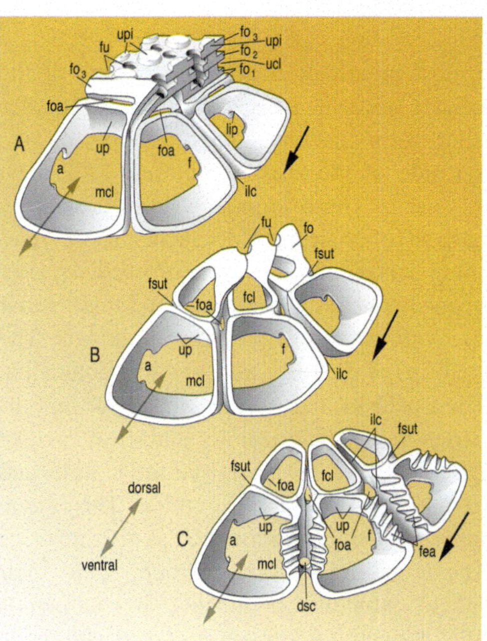

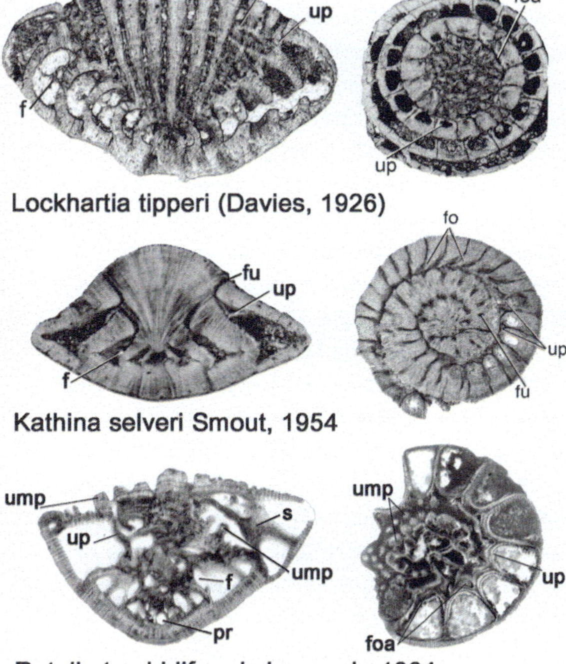

Fig. 1.2 The most important structural features of larger Rotaliidae. Stereographs of the three ultimate chambers of a whorl. The ventral chamber walls are cut away. Previous whorls forming a central umbilical structure are omitted. Schematic, not to scale. Many genera treated in this paper may be defined by combinations of features that are not covered by the three stereographs. The features common to all Rotaliidae are the interiomarginal position of the single foramen, the folium and its separation from the main chamber lumen by an umbilical plate. The *double, gray arrow* indicates the orientation of the models in relation to the dorsal and ventral sides of the shell. The *simple black arrow* indicates the direction of growth. Note the inversion of the orientation of the shells in contrast to the taxonomic part of this paper, where the shells are shown in their life position, the shell face turned towards the substrate. (**a**) Rotaliid shell with superposed foliar walls that are supported by umbilical piles. The umbilical cavities are connected from one level to the next by funnels perpendicular to the base of the conical shell. This is the basic architecture of the Lockhartiinae. (**b**) Rotaliid shell with folia that are fused to a complete cover of the umbilical area of the shell. Funnels perpendicular to this surface connect the foliar chamberlet lumina of previous and of the ultimate whorl with the substrate of the shell in the ambient environment. This is the basic architecture of the Kathininae. (**c**) Rotaliid with an undivided intraseptal interlocular space that opens as a long sutural fissure decorated by feathering. Note the position of the septal suture on the bottom of the interlocular space and the location of a dorsal sutural aperture. Abbreviations: *a* aperture, *fo* folium, *up* umbilical plate, *f* foramen, *foa* foliar aperture, *fu* funnel, *ump* umbilical piles, *pr* proloculus, *s* septum, *fcl* foliar chamberlet lumen, *fea* feathering of the intraseptal interlocular space, *ilc* interlocular space (undivided), *fsut* foliar suture, *mcl* main chamber lumen, *dsc* dorsal sutural canal orifice, *upi* umbilical pile, *ucl* umbilical chamber lumen

umbilical flap that is an extension of the septal flap (see Revets 1993 and discussion below). This is a conservative group starting in Late Cretaceous times and persisting until today. Genera presenting a ventral umbilicus without piles, umbos or columellas in their coiling axis are also excluded from the Rotaliidae as well as all genera with multiple foramina.

1.3 Life Strategies: An Overview

In order to understand the meaning of diversity in rotaliids, we have to introduce, first of all, some basic terms used in biological ecology rather than in micropaleontology, the K- and r-strategy of life in benthic foraminifera.

A K-strategist assigns relatively little material resources to reproduction but cares for his offspring. This is reflected by large shell sizes produced by numerous steps of growth, i.e. many chambers. Their high number reflects relatively long life times that adapt to seasonality in shallow water (Hottinger 1997). Very large, microspheric specimens produce brood-chambers for their offsprings.

The morphology of the K-strategists is restricted to certain stiles of their architecture. Every foraminiferal shell has to respond to basic functions corresponding to an envelope of a single cell living an autonomous life in the marine environment. The growth of the biomineralised cell envelope is performed by addition of new chambers to the previous ones. The successive chambers must communicate between each other by single or multiple foramina in order to guarantee the cohesion of the cell body subdivided into numerous compartments.

Biomineralisation by carbonates is the cheapest way to produce a shell in marine environments. "Cheapest" means in this case the minimal consumption of organic compounds in the cell envelope that are lost to recycling.

Every house needs a door to enter it, windows to light its interior, and a roof to protect the structure from rain. There is a common heritage for all plurilocular (chambered) foraminifera reflecting basic functions of the cell envelope. These are compartmentation of the cells beyond measure into permanent units of production with a suitable volume, protection of the living cell body and mechanical support of the pseudopods extending through apertures into the ambient environment. However, the style of the constructions, position of doors and windows, shape of roofs, ornaments of the walls etc. depends also on the nature of the material used as in houses constructed with wood, bricks or concrete. The arrangement and shape of successive chambers during the ontogeny may determine the shape of the adult shell. Where this shape has an autecological function it must be maintained during ontogeny. In order to keep the shell-shape constant during growth, usually the chamber shape and arrangement has to form a spiral (Hottinger 2006).

Similar to lower plants, the benthic foraminifera have alternating generations that respectively are sexually and asexually produced. Their difference in shell size reflects two different strategies of life in the same species, an enormously successful speciality of larger foraminifera. The communities of larger foraminifera often present two or more species (odd pairs, Hottinger 1999) that differ in size of the same order of magnitude as the A and B generations of a single species. This provides support for the idea to consider size relations as functional.

Size relations and basic functional architecture of the larger foraminifera are repeated several times during earth history. The repeated elements in the morphology of the shell obviously reflect functions that respond to a repetition of the nature of the ambient environment. The larger foraminifera living today permit to interpret the ecological meaning of the basic structures and to extrapolate this into the geologic past that produced analogous morphological elements.

Most K-strategist foraminifera in shallow environments have symbionts that recycle nutrients and thus avoid too much competition for food in environments of low primary production. The symbionts live and reproduce asexually in the host cell. The cell envelope functions in these cases also as greenhouse where the symbionts often are protected by an exoskeleton (Hottinger 2006) from the plasma streaming in the host cell and positioned where there is the easiest exchange of small molecules with the ambient environment through the pores, in the so-called eggholders (Hottinger 1997).

The r-strategy leads to opposed, opportunistic ways of life that reflect rather eutrophic environments. The lifetime of an r-strategist is short and the dimorphism of generations will be poor or absent. K-strategies need a more complex genome that controls a more complex life better adapted to environmental conditions during a long lifetime including seasonality. K-strategists arise from r-strategists by developing the complexity of their genome inherited by an ancestry reaching far back in time.

1.4 Cyclical Community Maturation

A rise of K-strategists is observed in regions of oligotrophy, where the r-strategists cannot compete for the lack of food. Oligotrophy usually needs long periods of climatic stability in the environment and a permanent separation of antiestuarine or gyral current patterns from nutrient sources produced on land by erosion. Such a process is called Global Community Maturation (GCM). This is repeated in successive periods of the Earth history. Each period produces similar K-strategist foraminiferal shells that often have the same size and general outline but are dissimilar in architectural design reflecting their different genetic origin. Some of these may reflect also different grades of organisation, the higher grades usually being more successful than the lower grades. However, a new set of K-strategists affects also the conditions of life in the new community reaching some equilibrium only after a certain time. This is a process similar to seres, communities following each other in a certain area from pioneering plants to trees growing over many years and determining usually the climax of the succession. A sere of plant communities is realised by the immigration and emigration of existing species under the pressure of environmental change. In the marine realm, the immigration of species has been measured by the velocity of Lessepsian migration through the Canal of Suez from the Red Sea into the eastern Mediterranean that reaches 1,000 km in 100 years. This does not depend on the mode of reproduction, the mode of the specific motility nor on the size of the migrant organism.

Community maturation is realized by a simultaneous development of the genomes after the (true) extinction of K-strategists in the preceding period by loss of the corresponding genome. The preparatory time interval from the extinction event to the first K-strategists, that show a recovery in the K-phenotype, is in the order of magnitude of 10 million years.

The area of recovered K-strategist communities spans the globe or at least an ocean like the Tethys in comparatively broad, latitudinal, climatic zones. But there are also longitudinal differences in the composition of the communities that are still difficult to understand. There are centers, hot spots of foraminiferal diversity, that migrate eastward with geologic time. In the Paleogene, the earliest hotspot seems to center in Southern Turkey in an area of considerable extension around the Lake Van. In the Middle and Late Eocene, the hot spot seems to migrate into Iraq and Iran forming a center in Oman. In this region, the presence of high numbers of taxa depends of course also on the very different states of systematic description of the taxa that compose the communities. However, in the course of geologic time, the number of taxa at the generic level has another rate of growth than the number of species. For the moment we will define the centers of Paleogene foraminiferal diversity by a common denominator, the Lockhartia communities. This group of complex and large-sized K-strategist rotaliids may serve as index fossil for a so-called Lockhartia Sea, a part of the Paleogene Neotethys, reaching in Asia from Tibet to Pakistan, Afghanistan, Iran and southern Turkey, in Africa from Somalia to Egypt and all over the Arabian Peninsula. Some important participants of this Lockhartia community are the porcelaneous *Neorhipidionina*, *Somalina*, various archaiasines, *Rhabdorites* and *Neotaberina*. Most of them are new. They accompany *Orbitolites* ("*Opertorbitolites*" included), the only representative of Paleogene discoidal porcelaneous groups in the western Mediterranean. Both rotaliid genera that have multiple spirals, *Dictyokathina* and *Dictyoconoides*, characterize also the Lockhartia Sea as well as *Sakesaria*, a high-spired lockhartiid.

The analysis of the foraminiferal family Rotaliidae permits to study a group of benthics where there are K- and r-strategists with a common heritage. There are three main types of architecture defining the subfamilies Rotaliinae, Lockhartiinae and Kathininae closely evolving together. Counting the taxa per SBZ zones reveals common trends in distribution: a rapid increase of taxa in the Paleocene and stability on a very low level from the Cuisian to the beginning of the Bartonian. Comparing the frequency curve of taxa with the one of miscellaneids (Figure 1.3 a,b), as well as with the curve of nummulitids (Hottinger 2001), the rotaliids have a much lower number of taxa at the species level; many genera are represented in a zone by a single species. Therefore, the number of rotaliid taxa is essentially the same on the generic and the specific level. In nummulitids, the peak of the curve of counts on the generic level is more or less coincident but the highest number of species is observed much later, reflecting competition in favour of the nummulitids.

The phases of community maturation in foraminiferal K-strategists may be interpreted as follows:

- Phase 1: no phenotypic response to the development of the genome for about 10 million years.
- Phase 2: many groups experiment with K-strategies using their respective heritage to develop new ways of life. Only few have success over long periods of time.
- Phase 3: the successful genera develop parallel lineages of species that adapt to different niches within the shallow environment. Different shell sizes in congeneric species reflect different strategies of life ("odd pairs", Hottinger 1999) and support an interpretation of the dimorphism of generations in the life cycle as a differentiation of the life strategy within the same species. This differentiation primarily concerns the seasonality in shallow water to be seen in the development of the vegetation cover.
- Phase 4: with certain equilibrium between the members of a community the number of parallel lineages of species is reduced. In each region, the successful lineages are different. The resulting endemism weakens the communities in competition with immigrants of a higher grade. Heterosteginids and cycloclypeids have folded septal flaps that are of higher grade than *Assilina* and *Nummulites* without folding. The former survive until today by numerous species, the latter go extinct except for a single species in each genus.

The immigration and success of higher graded taxa in a community may reconduct the maturation cycle to earlier phases. This is in analogy with seres that may be interrupted or gradually reconducted by regular destructive events such as destructive seasonal hurricanes that keep the succession cycle in a time frame limited to about a year.

1.5 Diversity in the Lockhartia Sea

This biogeographic term designates an area in the Paleogene Neotethys that is characterized by the presence of taxa classified in the rotaliid subfamily Lockhartiinae (i.e. the genera *Lockhartia*, *Dictyoconoides* and *Sakesaria*; Fig. 1.4). These are associated with a considerable number of other Paleogene foraminiferal genera that have a similar paleogeographic distribution. The generic diversity of the foraminiferal K-strategists in the Paleogene peaks in SBZ 4–5, more or less at the Paleocene–Lower Eocene boundary. For the rotaliids, the peak of diversity is observed in SBZ 4, in the Miscellaneidae one zone earlier, in SBZ 3. Both groups decline or disappear completely, obviously under the pressure of the very successful competition by the nummulitids, the alveolinids and the orbitolitids. The genera of the Lockhartiinae, all of them K-strategists, closely reflect these processes of community maturation in geologic time. However, there is also a geographic peak of foraminiferal diversity, the so-called diversity hot spots, that determine nature and number of the taxa present in an appropriate facies. During the Paleogene, the hot spots of benthic foraminiferal diversity seem to migrate during the Early Eocene from an area around the Lake Van in Southern Turkey to the Middle Eastern realm in the Persian Zagros and along the southern shores of the Arabian Peninsula up to Somalia. The hot spot

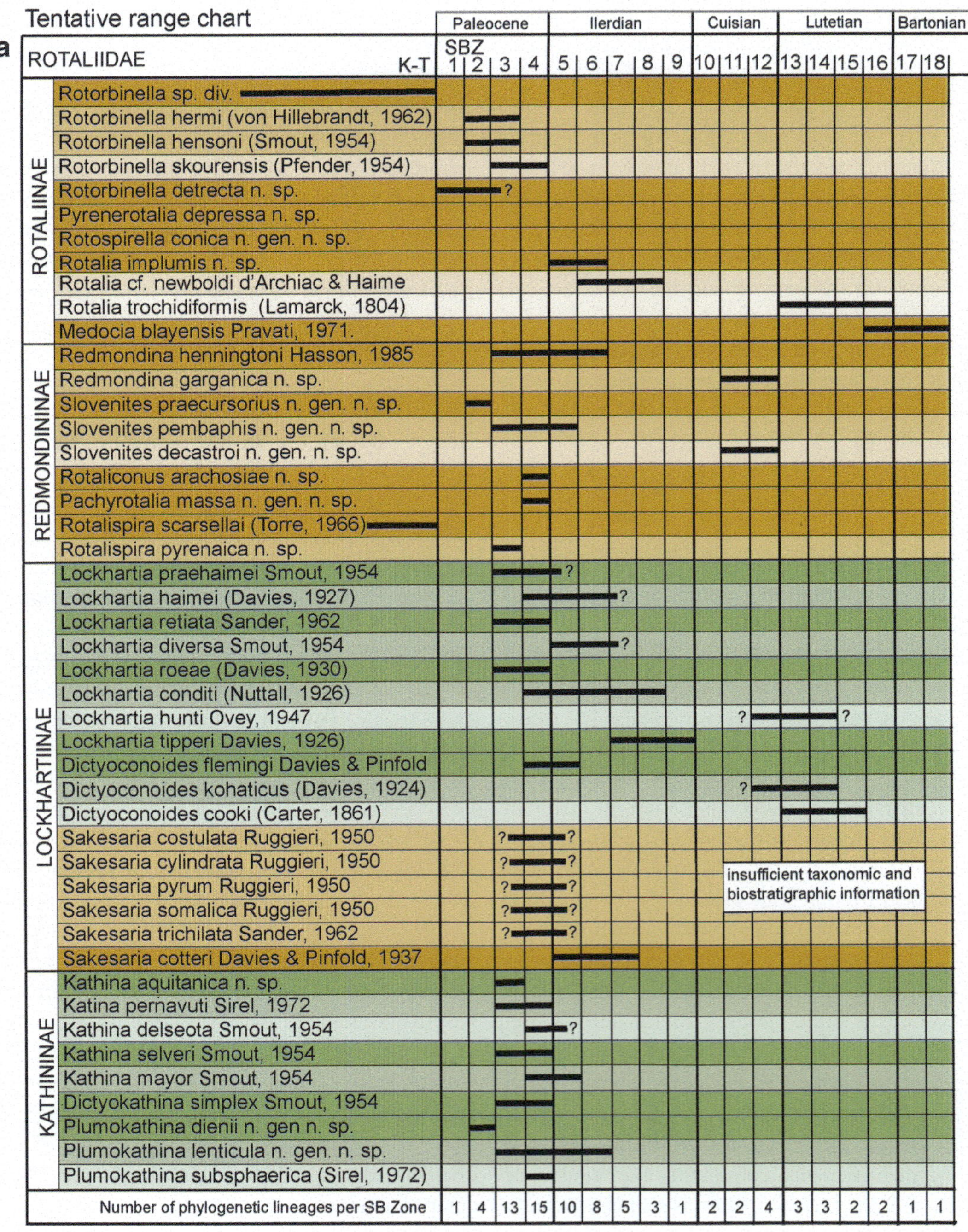

Fig. 1.3 (continued)

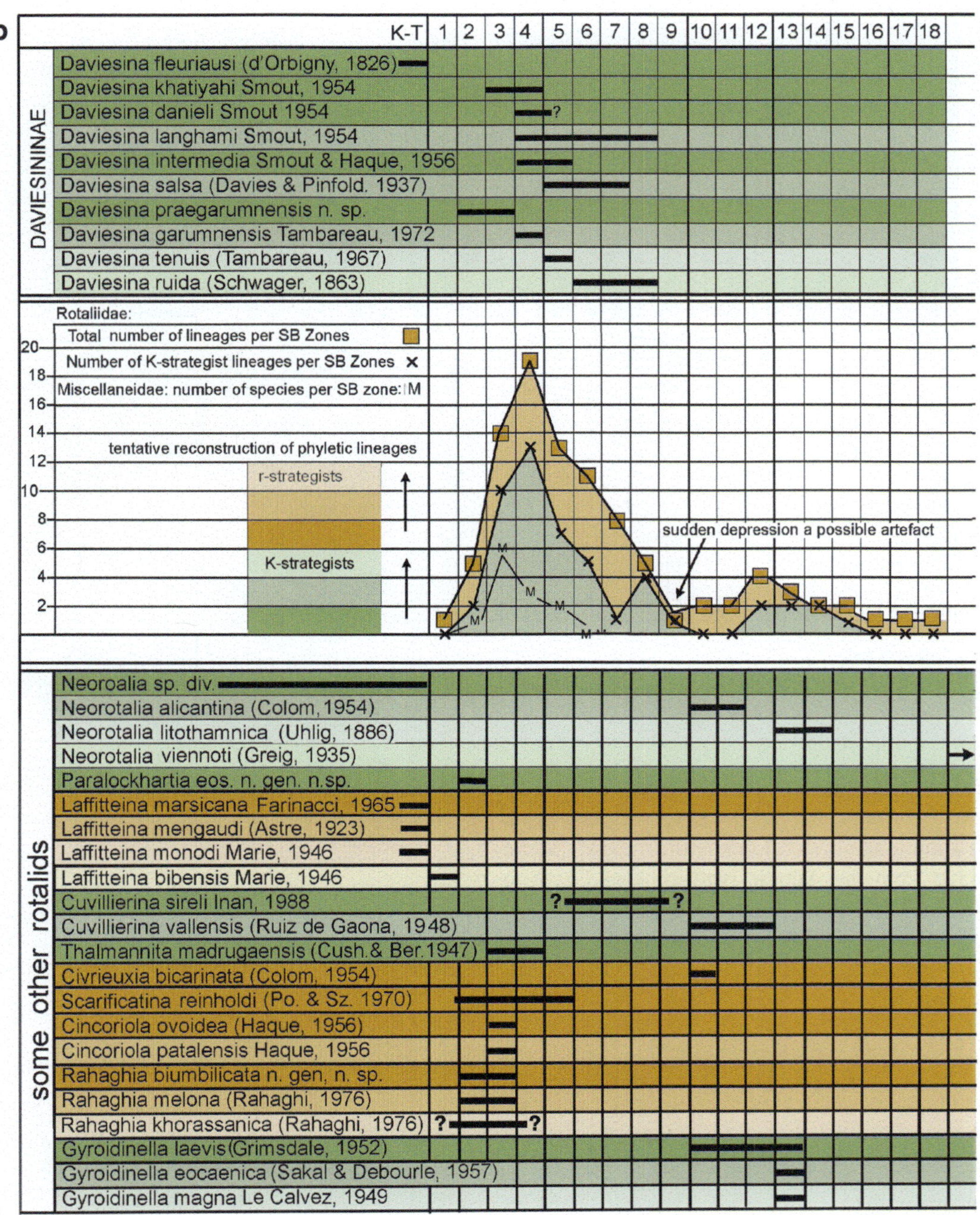

Fig. 1.3 (a–b) Tentative range chart of species classified as Rotaliidae and of selected taxa excluded from this family. Age given as Shallow Benthic Zones proposed by Serra-Kiel et al. (1998). *Taxa coloured green*: K-strategists; *taxa coloured brown*: r-strategists, as explained by Hottinger (1997). In order to avoid difficulties due to the uncertainty of the total specific ranges, the number of phyletic lineages per SB Zones has been counted, rather than the number of species. The sudden drop of all curves in SBZ 9 is an artefact due to the rareness of the facies that correspond to foraminiferal K-strategy in that period of time. Note the asymmetry of the curves indicating more than a single factor that determines the frequencies. For comparison, a curve of numbers per zone of the species attributed to the family Miscellaneidae by Hottinger (2009) is included in the table. The curve shows the peak of diversity one zone earlier than the one of the Rotaliidae, but the asymmetry is very similar. Selected rotalids excluded from the family Rotaliidae discussed in this paper are not included in the counts. They very incompletely represent their systematic and biostratigraphic relationships. Their inclusion in the counts would heavily bias the results

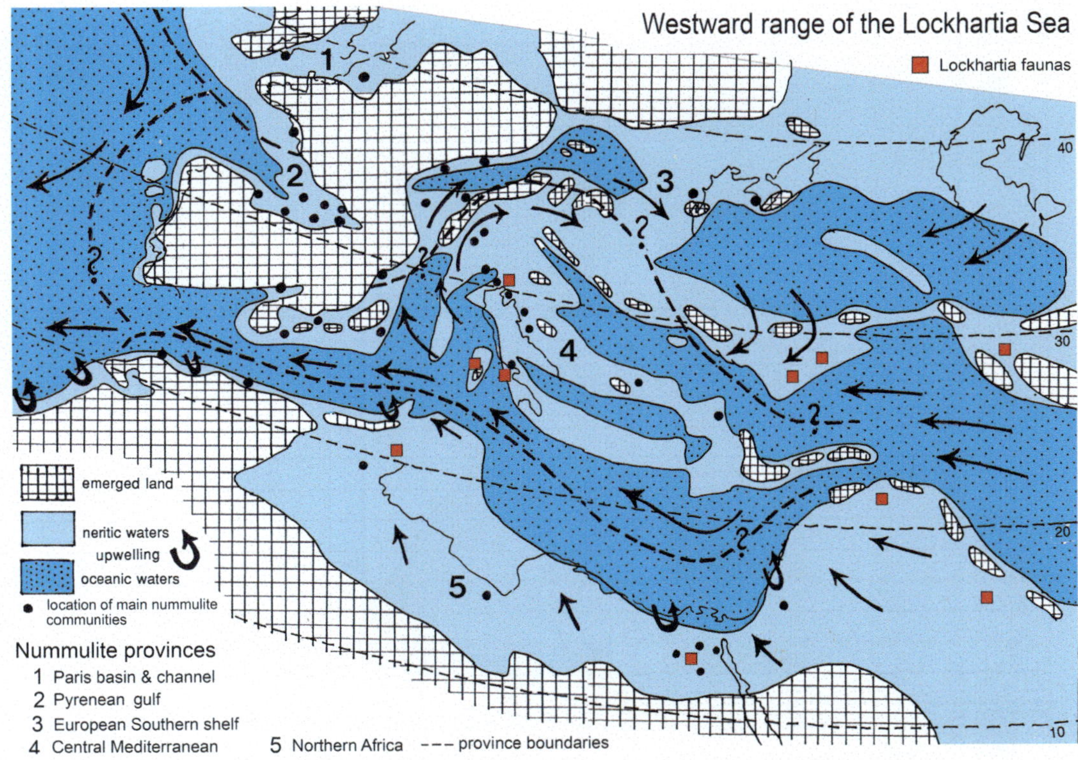

Fig. 1.4 The Lockhartia Sea in Paleocene to Middle Eocene times

area is in the center of the range of geographic distribution of the Lockhartiine taxa. These reach from the Adratic platform and Egypt to Southern India and Tibet.

References

Hansen HJ (1999) Shell construction in modern calcareous Foraminifera. In: Sen Gupta BK (ed) Modern foraminifera. Springer, Dordrecht, pp 57–70

Hottinger L (1997) Shallow benthic foraminiferal assemblages as signals for depth of their deposition and their limitations. Bull Soc Géol Fr 168 (4):491–505

Hottinger L (1999) Odd partnerships, a particular size relation between close species of larger foraminifera, with an emendation of an outstandingly odd partner, *Glomalveolina delicatissima* (Smout, 1954), Middle Eocene. Eclogae geol Helv 92:385–393

Hottinger L (2001) Learning from the past. In: Levi-Montalcini R (ed) Frontiers of life, vol 4, Part 2: Discovery and spoliation of the biosphere. Academic, London/San Diego, pp 449–477

Hottinger L (2006) The "face" of benthic foraminifera. Boll Soc Paleontol Ital 45:75–89

Hottinger L (2009) The Paleocene and earliest Eocene foraminiferal family Miscellaneidae: neither nummulites nor rotaliids. Carnets Géol Article 2009/06, CG2009_A06

Reiss Z (1963) Reclassification of Perforate Foraminifera. Bull Geol Surv Israel 36:1–111

Revets SA (1993) The foraminiferal toothplate, a review. J Micropaleontol 12:155–169

Schwager C (1863) Die Foraminiferen aus den Eocaenablagerungen der lybischen Wüste und Aegyptens. Palaeontographica 30:79–154, 6 pls

Serra-Kiel J, Hottinger H, Caus E, Drobne K, Ferrandez C, Jauhri AK, Less G, Pavlovec R, Pignatti J, Samsó JM, Schaub H, Sirel E, Strougo A, Tambareau Y, Tosquella J, Zakrevskaya E (1998) Larger foraminiferal biostratigraphy of the Tethyan Paleocene and Eocene. Bull Soc Géol Fr 169:281–299

Smout AH (1954) Lower Tertiary foraminifera of the Qatar peninsula. Brit Mus (Nat Hist), 96 pp, 44 figs, 15 pls

2 The System of the Rotaliidae, an Overview

Abstract

Large Paleogene Rotaliidae all represent moderate K-strategists similar to the nummulitids to which they are associated. They arise as Tethys-wide groups during the Paleocene and reach their highest diversity during the earliest Lower Eocene (SBZ 5). They document a Global Community Maturation cycle that ends for the rotaliids by their disappearance after the Middle Eocene while other groups like the nummulitids and the orthophragminiforms continue up to the Eocene-Oligocene boundary. Under the generic definitions the species are grouped in order to reflect what is supposed to represent a phylogenetic lineage in the order of their appearance in geological time. This meets the demands of an easier taxonomic identification and to reflect in the same time a phylogenetic hypothesis.

2.1 Overview

Large Paleogene Rotaliidae belongs to the new subfamilies Daviesininae, Kathininae, Laffitteininae, Cincoriolinae and Lockhartiinae. By their large size, their strong dimorphism and their distribution of canal orifices on the surface of their shells they all represent moderate K-strategists similar to the nummulitids to which they are associated. Their strategy of life obliges them to adapt rapidly to the changing conditions in shallow water. They arise as Tethys-wide groups during the Paleocene and reach their highest diversity during the earliest Lower Eocene (SBZ 5). Later, the representatives of the lockartiines, *Lockhartia, Dictyoconoides* and *Sakesaria,* dominate the rotaliid associations and sometimes the complete shallow benthic communities.

The rotaliid distribution pattern on the time scale reflects the basic rules of global community maturation (GCM) cycles, that is: (a) preparing the metabolism for K-strategy, (b) experimenting with ways of life, (c) revelling in success, (d) gaining size and endemism. Thus, they document a GCM cycle that ends for the rotaliids by their disappearance after the Middle Eocene while other groups like the nummulitids and the orthophragminiforms continue up to the Eocene–Oligocene boundary. In the Middle Eocene already the first K-strategists of the conservative neorotalias appear with a new species that has microspheric specimens with incomplete chamberlet cycles like a

L. Hottinger, *Paleogene larger rotaliid foraminifera from the western and central Neotethys,*
DOI 10.1007/978-3-319-02853-8_2, © Springer International Publishing Switzerland 2014

miogypsinid. These neorotalias will partially substitute the extinct Lockhartiinae.

The biogeographical distribution of the Rotaliidae starts with species that are ranging throughout the western and central Neotethys during the Paleocene. In the earliest Early Eocene, most rotaliids disappear from the eastern shores of the Atlantic, in particular from the Pyrenean Gulf, may be by the numerous nummulitid competitors. During the following periods, the biogeographic range of the larger rotaliids is restricted to the central part of the Neotethys, from Turkey to Tibet in Asia, and from Egypt to Somalia in Africa. Another index fossil with an identical range is the porcelaneous *Somalina*, whereas the archaiasines are newcomers in the Bartonian. This is what we call the Lockhartia Sea. The large carbonate platforms in the South of the Arabic Peninsula (Dhofar in Oman in particular) exhibits in addition numerous porcelaneous genera of which several are new. The relations of the Tethyan Rotaliids with the Caribbean forms on the other side of the Atlantic remain obscure as long as these taxa have not been revised.

The arrangement of the taxa in the text below tries to meet the demands of an easier identification of the taxa and to reflect in the same time a phylogenetic hypothesis. Thus, under the generic definitions the species are grouped in order to reflect what is supposed to represent a phylogenetic lineage in the order of their appearance in geological time.

The taxa selected for this paper characterize a community indicating a hot spot of diversity in the Middle and Near East during the Late Paleocene. This community is geographically restricted to the shallow shelves around the Western Indian Ocean and the Himalayan shores of the Neotethys. The corresponding marine area is called the Lockhartia Sea. The present paper may also serve to enhance the census of this particular community.

2.2 Superfamily Rotaliacea Ehrenberg: Family Rotaliidae Ehrenberg, 1839

The superfamily Rotaliacea needs an entirely new systematic concept, in order to integrate the advances in structural analysis of the architecture of the shells during the last century. Loeblich and Tappan's (1987) definition of the family Rotaliidae obviously is insufficient: "test trochospiral throughout, with radial canals or fissures and intraseptal and subsutural canals". This definition obliges to unite too many benthic genera with different architectural patterns and stratigraphic ranges in a family that loses any significance. Already in 1963, Reiss laid the foundation for a new classification of the perforate foraminifera that integrated Smout's (1954) lamellar theory of foraminiferal shell construction with ideas of J. Hofker sen. (1927, 1951) about the tooth-plates and their relation to teeth in imperforate forms. Taking into account these structural elements in the umbilical region of the shell leads to the recognition of three main groups of rotaliaceans with family rank: (1) the rotaliids, a Late Cretaceous to Paleogene group of bilamellar, trochospiral shells with an umbilical plate, (2) the ammoniids, a Neogene group with an umbilical cover plate, and (3) the pararotaliids, a conservative group reaching from Late Cretaceous to Recent, with a kind of tooth-plate. There are supplemental groups of similar rank, *Laffitteina* for instance. The systematic ambient environment of the Rotaliidae is too poorly known to decide where and how to place the laffitteinas in the system. The present paper is limited to the revision of the rotaliid genera and species from the Paleogene, in particular from the Lockhartia Sea. The Late Cretaceous ancestors are discussed and illustrated by Boix et al. (2009). Moreover, some taxa resembling rotaliids but belonging to other families are also discussed and illustrated here in order to elaborate the definition of the Rotaliidae by showing what is not a rotaliid.

A revised diagnosis of the Rotaliidae is proposed as follows: Bilamellar-perforate shell composed of trochospirally arranged chambers that are connected by a single intercameral foramen in interiomarginal position. The chamber lumen is separated from the umbilical interlocular space by an umbilical plate that is an extension of the septal flap. The umbilical plate delimits with a suture the folium, a triangular umbilical extension of the ventral chamber wall that has an anterior or median-ventral foliar aperture. There is a spiral and an intraseptal interlocular space transformed in various ways into a canal system. The umbilicus is filled with shell material forming free umbilical plugs, numerous umbilical piles or a solid mass perforated by funnels.

Excluded from the rotaliids are trochospiral, bilamellar benthic foraminifera that lack any kind of columella, or with multiple foramina or with a tight system of enveloping canals on both, ventral and dorsal sides of the shell.

The Rotaliidae are subdivided here into subfamilies that have each a separate phylogenetic root in the Late Cretaceous or the Early Paleocene: the Rotaliinae with *Rotorbinella* since the Cenomanian, the Redmondininae with the Paleocene *Redmondina*, the Lockhartiinae with *Rotospirella* (Senonian), the Kathininae with *Kathina aquitanica* (Paleocene) and the Daviesininae with *Daviesina fleuriausi* (Maestrichtian).

A key to the subfamilies and their generic subunits is given below. It is designed to help finding particular taxa discussed in this paper. Note that the key has been formulated for only those taxa discussed in this paper, not for all rotaliid genera mentioned by Loeblich and Tappan (1987). Thus, the key helps to find one's way in the systematic part of this paper but is not an instrument for classifying all described rotaliids.

2.3 Identification Key to the Paleogene Genera of the Superfamily Rotaliacea

Shell walls lamellar-perforate; chamber arrangement trochospiral; foramen single, in interiomarginal position; folium and umbilical plate present; face ventral	ROTALIACEA
Dorsal side smooth; periphery angular to sharp; dorsal side of shell evolute, ventral side involute; umbilicus filled with a single, undivided axial pile or a columellar filling	ROTALIIDAE
Columellar filling of the umbilicus by superposed folia fused at their adaxial tips or by a single, undivided pile. Few funnels may appear in the columellar umbilical fill revealing their composite nature	ROTALIINAE
Umbilicus filled with a single pile in axial position. Folia short, mostly free, unfused. Ventral septal furrows without feathering	*Rotorbinella*
Low trochospiral shells with a sharply angular periphery. Umbilicus covered by heavy, oblique foliar walls that do not fuse in the axis of the shell. No feathering nor an umbilical plug but an incipient columella formed by the foliar "propeller blades"	*Pyrenerotalia*
Hemispherical to conical shells with a coarse, dense perforation. Periphery rounded to angular but lacking any keel or imperforate bands. Sutures flush. The narrow umbilicus is filled with slender parallel piles, a single one per folium and separated by narrow funnels. The thin but extended folia cover more than half of the umbilical cavity	*Rotospirella*
Evolute dorsal surface of the shell smooth. Umbilicus filled with a columellar structure	*Rotalia*

(continued)

produced by the fusion of the adaxial tips of the folia with their secondary lamellation; shoulders of septal furrows feathered or smooth. Few, irregularly directed funnels	
Umbilical silling fused to a compact mass perforated by few, large funnels that are not always parallel to the axis	*Medocia*
Umbilical walls very coarsely perforate. Spiral and cameral sutures flush, unmarked by imperforate shell material. Periphery rounded. Canal system with low numbers of dorsal orifices. Umbilicus empty or filled with parallel radial piles resulting from a thickening of the mostly imperforate foliar walls in successive whorls	**REDMONDININAE**
Spiral chambers dorsally evolute and inflated. Folia short, in the axis of the spiral chamber, free or touching each other to form a ring of imperforate material or small umbilical piles. Intraseptal interlocular space undivided, an open radial slit on the ventral side, a single orifice at the dorsal intersection of the cameral with the spiral suture. There is a dorsal umbo producing a scarcely perforated, often brilliantly white cap on the apex of the shell	*Redmondina*
Broadly conical to lenticular shells with a wide umbilicus filled with umbilical piles that are generated by the thickening of the foliar walls and fused to a solid mass. In between the piles, numerous coarse funnels represent the umbilical interlocular space. They spring from the triangular interlocular space at the confluence of the interseptal space and two successive foliar chamberlets	*Slovenites*
Subspherical to hemispherical shells with coarsely perforated chamber walls covered by an enveloping canal system on the dorsal and the ventral sides of the shell. Some irregularly placed funnels maintain the communication in the umbilical cavity that is obscured by the ventral enveloping canal system	*Rotaliconus*
covering also large parts of the foliar walls	
Subspherical shells with very thick, coarsely perforated walls. The dorsal side including the peripheral part of the shell exhibits an enveloping canal system. Umbilicus filled with radial, heavy piles fused to a solid umbilical mass pierced by funnels. No ventral enveloping canals	*Pachyrotalia*
Dorsal side evolute with various ornamental patterns or smooth, ventral side covered with a dense population of piles of similar, uniform size, without marking the axial realm. The folia cover each large parts of the shell face and support each the site of origin of the free umbilical piles. The folia exhibit an aperture in the foliar suture that is a slit approximately symmetric to the axis of the chamber. Foramen single, in interiomarginal position and of small size. Large shells may be constructed with multiple spirals of main chambers and a more or less distinct dimorphism of generations	**LOCKHARTIINAE**
Hemispherical shells with coarsely perforated chamber walls. Periphery angular to sharp and distinctly keeled. Spiral and cameral sutures imperforate and raised. Umbilicus often with few or without piles but filled with perforate, very long folia superposed like blades of a propeller. There is a large spiral canal	*Rotalispira*
Low-trochospiral shells with an evolute dorsal side and a wide umbilicus filled with numerous umbilical piles. Structural features corresponding to the definition of the subfamily. Dimorphism usually restricted to the nepiont. There are many species grouped according to their dorsal ornamental patterns in four parallel phyletic lines	*Lockhartia*
Low-trochospiral, large sized shells of low-conical to discoidal shape. Chamber arrangement in multiple spirals, at least in the microspheric	*Dictyoconoides*

(continued)

generation. The megalospheric generation has much smaller shells and may lack supplemental spirals. Foramen of small size, a compressed arch in interiomarginal position	
High-trochospiral shells with a dorsal ornamentation of various kinds that characterizes the numerous species described in the literature. The surface of the shell face is restricted in size and much inclined in respect to the axis of the shell. The umbilical architecture is dominated by the large folia supporting umbilical piles. Due to the high-spired shell, the umbilical structures are distorted around the coiling axis	*Sakesaria*
Discoidal to roughly lenticular or conical shells. Chamber arrangement organized in single or multiple spirals. Chambers per whorl numerous and dense. Foramen single, a low arch or slit in interiomarginal position, immediately below the dorsal cover of the spiral chamber. The folia are small, inclined foreward. Their tips are fused to each other and with the umbilical fill produced by the previous chambers to form a solid mass perforated by numerous parallel funnels. Some species may have a massive central umbo. No feathering	**KATHININAE**
Shells with simple spiral chamber arrangement. Umbilical architecture corresponding to the subfamily Kathininae	*Kathina*
Shell discoidal, often distorted and irregular. Spiral multiple in both generations. Umbilical architecture conforms to the Kathininae	*Dictyokathina*
Shell lenticular, with an unkeeled periphery. Chamber arrangement always single-spired. Umbilical plates heavy, admitting on their umbilical side a broad, spiral interlocular space considered as a spiral canal. Dorsal and ventral umbos almost of equal size, smooth. Ventral intraseptal interlocular space undivided, a radial slit with massive feathering visible also in axial sections as vertical striations	*Plumokathina*
Heavily ornate, biumbilical shells with almost identical appearance on both sides of the shell. Dorsal surface ornate with pustules and pseudospines. Ventral surface with similar ornaments alined between radial furrows marking the septal sutures. Chambers dorsally and ventrally involute, almost symmetrical. Foramen asymmetrically V-shaped, bearing an asymmetrical "tooth" in axial section. The foramen is riding on the periphery of the previous whorl. Its asymmetry helps to identify the ventral and the dorsal side of the shell. The umbilical plate separates in both umbilici the chamber lumen from a kind of narrow umbilical canal on both sides of the shell	**DAVIESININAE**
Today the subfamily is not subdivided into several genera in spite of the recognition of four phyla evolving in parallel lineages. The reason for this is to stress the nevertheless close relationship of taxa that were attributed to many different families such as the Miscellaneidae or the Nummulitidae. Thus, coarsely or finely ornamented shells without or with trabecules, or forms with supplementary stolons resulting in folded septal flaps all bear the same generic name. More accurate definitions are revealed on species level	*Daviesina*

2.4 Identification Key to Some Genera Excluded from the Rotaliacea

Trochospiral shells with single, areal foramina and a "toothplate" as umbilical closure of the spiral chamber, with or without a dense enveloping canal system extending over the whole surface of the shell, or trochospiral shells with multiple areal foramina lacking any kind of plates.

Chambers arranged in a single, low trochospiral. Spiral chambers evolute on the dorsal side, involute on the ventral side, keeled and covered with spinose ornaments. Foramen a single, areal slit in the apertural face of the chamber. Presence of a "toothplate"	PARAROTALIINAE
Adaxial, umbilical piles separated from the more peripheral realm by a deep circular slit that represents the major part of the umbilical cavity	*Pararotalia*
Furrow between adaxial umbilical piles and more peripheral ventral ornaments covered by an extension in adaxial direction of the umbilical flap	*Neorotalia*
Ventral, bilamellar chamber walls extending as flying cover over a large part of the umbilical cavity. They are supported by radial piles like in *Lockhartia*	*Paralockhartia*
Chambers arranged in a single, almost plane spiral, involute on both sides, and covered on both sides by a narrow system of enveloping canals. Adaxial realms with umbos pierced by numerous funnels. Foramen a single, narrow slit extending obliquely over the adaxial apertural face. An umbilical plate separates the chamber lumen from the umbilical cavity	LAFFITTEININAE
Almost planispiral-involute shells with a rounded periphery. Enveloping canal system initiated by a double row of intraseptal canal orifices, similar on both sides of the test	*Laffitteina*
Lenticular shells with almost planispiral chambers that are involute on both sides. Periphery sharp but unkeeled. Foramen an areal, comma-shaped slit. The enveloping canal system is derived from heavy feathering of the ventral septal sutures. Both umbos perforated by numerous radial funnels opening to the ambient environment as polygonal network	*Cuvillierina*

2.5 Identification Key for Some Rotalids with Single Foramina and Lacking Umbilical Plugs or Umbos

Small-sized, almost planispiral-evolute shells with a single foramen. Umbilical area flat or concave filled with long spikes. Ornamentation consisting of heavy parallel costae separately disposed over the lateral surface of each chamber in oblique direction in respect to the radius of the spiral	*Thalmannita*
Small-sized, trochospiral shell with a flat dorsal side and a concave umbilical side with an open umbilicus. Periphery truncate, with a double carina. Single areal foramen. Interlocular space wide open in the septa and with a large, spiked orifice opening to the ambient environment in the septum between the two peripheral carinas	*Civrieuxia*

2.6 Identification Key for Some Rotaliid Shells with Multiple Areal Foramina

Small-sized shells near to planispiral, with an inflated rounded periphery. Foramina multiple, each with a circular peristome. Umbilicus covered by extensions of the ventral wall of the spiral chambers. These support	*Scarificatina*

(continued)

an abaxial ornamentation of radial rows of minute pustules. The axial area of the umbilicus is dominated by one or several parallel, straight ridges that are independent of the number or shape of the spiral chambers. Each ridge separates a double line of tubular openings	
Trochospiral, lamellar-perforate shells with sets of multiple areal foramina with peristomes. They cover large, interiomarginal areas of the septal face. No true umbilicus nor canal system	**CINCORIOLINAE**
Hemispherical, dorsally evolute and ventrally involute shells with smooth walls that present a uniform, dense perforation. Periphery rounded or sharply angular. Septa dorsally straight, bent backwards and ventrally extending over the axial area, covered by numerous, closely spaced foramina that extend over the whole face of the shell. There is no true umbilicus nor umbilical plates	*Cincoriola*
Lenticular to subspherical shells with involute spiral chambers. Septa extending over the axial area of the shell, with multiple foramina bearing heavy, circular peristomes. Septa with intraseptal canal systems opening into both, dorsal and ventral umbilicus	*Rahaghia*

2.7 Identification Key for Some Rotaliid Shells of the Family Victoriellidae

Dorsally evolute, ventrally involute, subspherical to conical shells with a false umbilicus filled with partially fused pustules growing on the shoulder of the umbilical depression. Interlocular space opening dorsally in a single orifice. Foramen a single, interiomarginal arch	*Gyroidinella*

References

Boix C, Villalonga R, Caus E, Hottinger L (2009) Late Cretaceous Rotaliids (Foraminiferida) from the Western Tethys. Neues Jb Geol Paläontol Abh 253(3):197–227

Loeblich AR, Tappan H (1987) Foraminiferal genera and their classification. Van Nostrand Reinhold, New York, 1, 970 pp; 2, 212 pp, 847 pls

Reiss Z (1963) Reclassification of Perforate Foraminifera. Bull Geol Surv Israel 36:1–111

sen Hofker J (1927) The foraminifera of the Siboga expedition. Tinoporidae, Rotaliidae, Nummulitidae, Amphisteginidae. Brill, Leiden, 78 pp

sen Hofker J (1951) The toothplate Foraminifera. Arch Neeraland Zool 8(4):353–372

Smout AH (1954) Lower Tertiary foraminifera of the Qatar peninsula. Brit Mus (Nat Hist), 96 pp, 44 figs, 15 pls

Part II

Systematic Palaeontology: Family Rotaliidae

Subfamily Rotaliinae Ehrenberg, 1839 — 3

Abstract

The subfamily Rotaliinae is defined by the absence of dorsal ornaments and of umbilical spaces subdivided by free-standing piles. This group embraces five genera (*Rotorbinella*, *Pyrenorotalia*, *Rotospirella* n. gen., *Rotalia*, *Medocia*). Ten species (*R. skourensis*, *R. hermi*, *R. hensoni*, *R. detrecta* n. sp., *P. depressa* n. sp., *R. conica*, *R. implumis* n. sp., *R.* cf. *newboldi*, *R. trochidiformis*, *M. blayensis*) are described and illustrated.

The subfamily Rotaliinae (superfamily Rotaliacea, family Rotaliidae Ehrenberg, 1839) is erected to accommodate the genus *Rotalia* and its immediate relatives *Rotorbinella* and *Medocia*. The group of taxa is defined by the absence of dorsal ornaments and of umbilical spaces subdivided by free-standing piles. The adaxial tips of the folia are imperforate and separated from the chamber wall by the foliar suture marked by notch in the sutural depression. The foliar tip fuses in the axial area of the shell. The foliar suture reflects the position of the umbilical plate that separates the main chamber lumen from the foliar cavity. The latter is connected with its neighbour by the foliar aperture opening the communication between the umbilical and intraseptal interlocular spaces. The result is a spiral communication called spiral canal.

In this subfamily, we include also similar forms but with a free-standing umbilical plug: *Rotorbinella*. In the Late Cretaceous, we also find Rotaliines with simple structures like in *Rotorbinella*. The latter might represent ancestral forms of Cenomanian age, while other, Santonian forms are specialized to a degree (Boix et al. 2009) which forbids such an interpretation. They would represent rather K-strategists of a previous community maturation cycle (Hottinger 2001). The taxa of this group are defined by the morphology of the shell's face.

Another branch of this subfamily exhibits few, simple funnels in the columellar umbilical filling, such as in *Medocia* Parvati, 1971. Its dorsal outer surface is smooth as in *Rotalia sensu stricto*.

3.1 *Rotorbinella* Bandy, 1944

Type species: *Rotorbinella colliculus* Bandy, 1944

Diagnosis: Rotaliidae with a single, undivided umbilical pile (see also Revets 2001). Ventral intraseptal interlocular space undivided. Shoulders framing the intraseptal space without

L. Hottinger, *Paleogene larger rotaliid foraminifera from the western and central Neotethys*,
DOI 10.1007/978-3-319-02853-8_3, © Springer International Publishing Switzerland 2014

feathering. Umbilical interlocular space comparatively wide, forming a spiral canal. Species of *Rotorbinella* represent the phylogenetic root of the rotaliid family reaching back to Cenomanian ages. They exhibit the simplest architecture within the family. They are no K-strategists by the lack of an adult dimorphism of generations, the absence of particular ornaments on their face and by their small overall shell size remaining below 1 mm in equatorial diameter. Accordingly they are often found on or in soft substrates rich in clay minerals.

Rotorbinella skourensis (Pfender in Moret, 1938); Plate 3.1, Figs. 1–18.

1938 *Rotalia skourensis* —Pfender in Moret, p. 62, text-fig. 8.

1991 *Rotalia* sp. 1 —Wan, p. 10, pl. 1, figs. 15–17.

1991 *Rotalia skourensis* —Pfender. Herbig, p. 54, pl. 5, fig. 1.

2009 *Rotalia trochidiformis* Lamarck—Afzal et al., pl. 1, fig. 12.

Remarks: *Rotorbinella skourensis* was originally described by Pfender from Maroccan beds that are dated as Paleocene by indirect, sedimentological arguments (Herbig 1991). The same shells have appeared also in the first fossiliferous beds of the Salt Range section at Dhak Pass, northern Pakistan. Their occurrence in the Hangu Formation confirms the Paleocene age that was attributed so far to this species. They also occur in the Himalayan Paleocene (Wan 1991). *Rotorbinella skourensis* differs from *R. hensoni* by its higher ventral convexity supported by a single thickening on each foliar wall (foliar pile). The dorsal chamber septa are arcuate and strongly bent backwards.

Rotorbinella hermi (von Hillebrandt, 1962); Fig. 3.1A–E; Plate 3.2, Figs. 14–20.

1962 *Rotalia hermi* —von Hillebrandt, p. 116, pl. 10, figs. 13–14; pl. 15, figs. 16–17.

Remarks: *R. hermi* is one of the few well described *Rotorbinella* from the Paleocene of the Northern realm of the Alps. It is defined in this paper by a thickening of the anterior shoulder of the chamber, extending adaxially over the whole foliar wall and forming there a single rounded projection that is separated from the central umbilical pile by a deep furrow. The notch indicating the position of the plate suture is indistinct. The ventral sutures are falciform and inclined backwards, the dorsal sutures are straight and much inclined. The ratio equatorial to axial diameter varies from 1.8 to 2.4. There are 10–12 chambers in the last whorl. No dimorphism is observed.

Rotorbinella hensoni (Smout, 1954); Figs. 3.2, 3.3A–F and 3.4I; Plate 3.2, Figs. 1–13.

1954 *Rotalia hensoni* —Smout, p. 45, pl. 15, fig. 8.

1972 *Rotalia perovalis* (Terquem) pars. —Samuel et al., pl. 37, figs. 1, 3.

1973 *Discorbis perovalis* (Terquem) —Ferrer et al., p. 38, pl. 1, figs. 12–14.

1976 *Rotalia saxorum* d'Orbigny—Ho et al., p. 47, pl. 25, figs. 9a–c.

1998 *Rotorbinella hensoni* (Smout) —Haynes and Nwabufo-Ene, p. 60–62, pl. 5, figs. 1–2.

2006 *Rotorbinella* sp. —Hottinger, p. 86, pl. 2, figs. 11–16.

Remarks: The taxon *Rotorbinella hensoni* is a basket to receive Paleocene to Ilerdian Rotorbinellas lacking features of specific character. The shells are smooth on both sides and exhibit almost radial, slightly curved ventral chamber sutures lacking any feathering. The umbilical pile forms a solid, undivided plug. There is a faint notch indicating the position of the foliar suture corresponding to the umbilical plate. The latter is positioned comparatively far from the umbilical pile and creates a broad spiral canal around the umbilical pile. The folia are large but do not overlap, nor do they fuse to form a columella. The intraseptal interlocular space is restricted to the ventral adaxial part of the septum where it creates a single, undivided room communicating with the umbilical space around the umbilical plug.

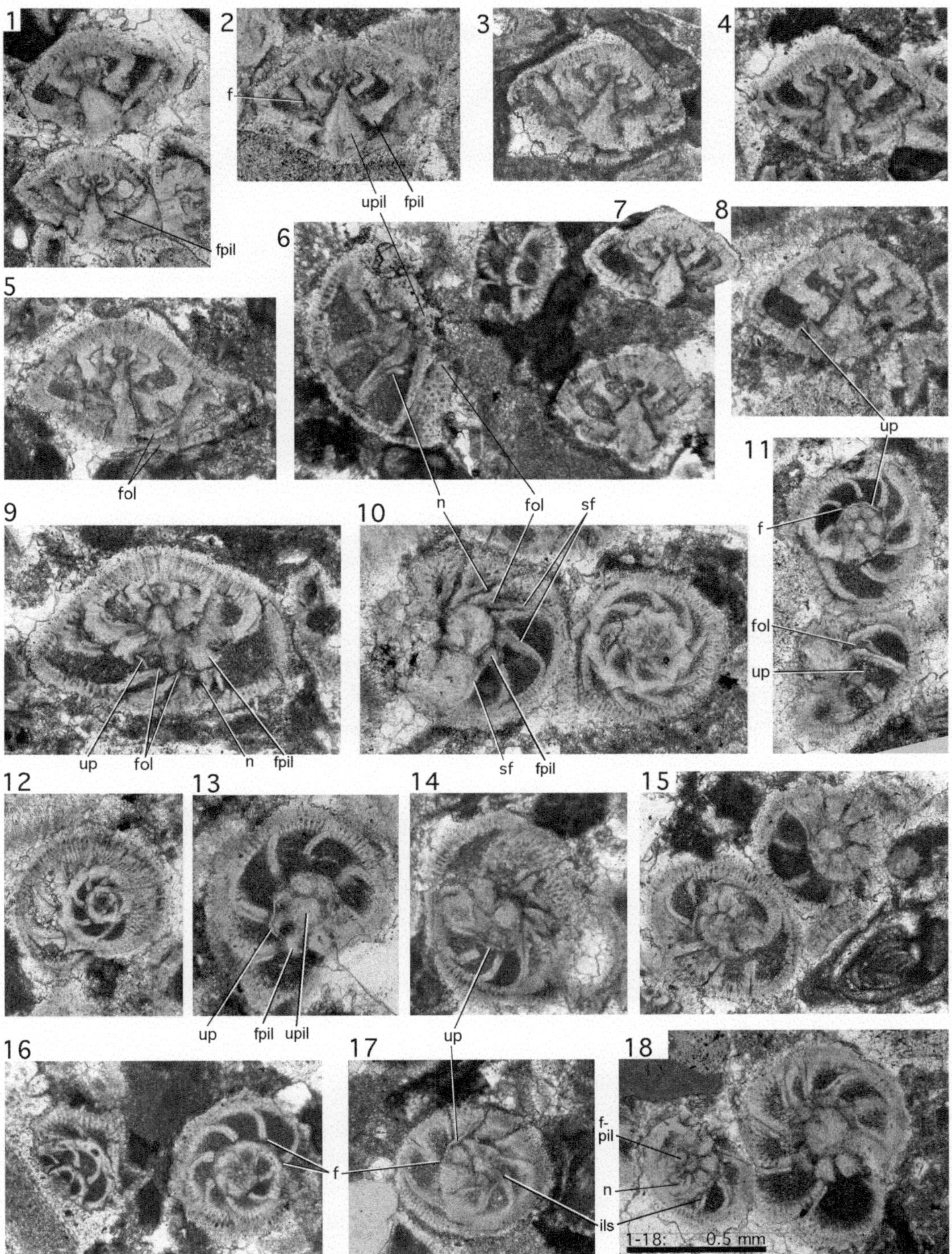

Plate 3.1 *Rotorbinella skourensis* (Pfender in Moret, 1938), topotypes from north of Tinerhir, section SA 11 in Herbig (1991), southern Morocco, Paleocene. (**1–8**) Axial sections showing the rotorbinellid, single umbilical plug. (**6**, **9**, **17**) Oblique sections. (**10–16**, **18**) Sections perpendicular to the axis of coiling. Note the pile-forming projections of the foliar walls that characterise the species. Abbreviations: *fpil* foliar pile, *f* foramen, *upil* umbilical pile, *up* umbilical plate, *fol* folia, *n* notch, *sf* septal flap, *ils* interlocular space

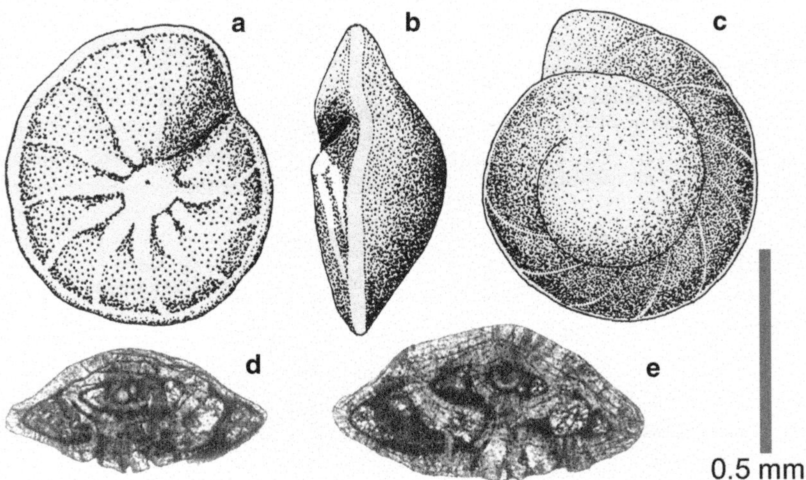

Fig. 3.1 *Rotorbinella hermi* (von Hillebrandt, 1962). (**A–E**) Selection of original illustrations by von Hillebrandt at standard enlargements. (**A–C**) Drawings illustrating the umbilical, peripheral and spiral views. (**D–E**) Oblique-centered sections in transparent light, missing the umbilical plug in the coiling axis (From Wasserfall Graben, Lattengebirge, Oberbayern, Southern Germany; Lower Paleocene)

Rotorbinella detrecta n. sp.; Fig. 3.4A–H.

Holotype: Specimen figured in Fig. 3.4A.

Type locality and type level: Pic d'Orhi, sample 10.3, collected by Y. Tambareau. Paleocene (SBZ 2).

Derivation of name: *detrectus* (Lat.) means reduced (in size).

Diagnosis: This new species has all the characters of the genus *Rotorbinella* but is of much smaller size as compared to *Rotorbinella hensoni*. Compare with Fig. 3.4. The single umbilical pile stands free, separated from the foliar chamberlets by a circular, deep, open furrow. The spiral of *R. detrecta* is wider open and exposes the seven chambers of the last whorl to comprimation due to the settling of the encasing sediment by the loss of water during early diagenesis. The megalosphere reaches about half the size of the one in *R. hensoni*.

Remarks: *Rotorbinella detrecta* was discovered by Yvette Tambareau in limestones of SBZ 2 age in the Aquitaine where it is associated with *Pyrenerotalia depressa* n. sp. and *Slovenites precursorius* n. sp.

3.2 *Pyrenerotalia* Boix et al., 2009

Type species: *Pyrenerotalia longifolia* Boix et al., 2009

Pyrenerotalia depressa n. sp.; Fig. 3.5A–I.

Holotype: Specimen figured on Fig. 3.5A.

Type locality and type level: Pic d'Orhi, sample 10.3, collected by Y. Tambareau. Paleocene (SBZ 2).

Derivation of name: *depressa* (Lat.), flattened in morphology.

Diagnosis: Medium sized, rotaliid shell exhibiting large, stout, much inclined folia similar to *Pyrenerotalia longifolia* from the Late Cretaceous. The oblique folia are overlapping in the abaxial part of the umbilicus like a ship's propeller and bear on their abaxial section, besides the foliar suture, a single pile of lamellae. The adaxial tips of the folia fuse to form a slender, small-sized umbilical plug. The interlocular spaces in the septa are wide open and subdivided into a row of canals. In oblique sections of the

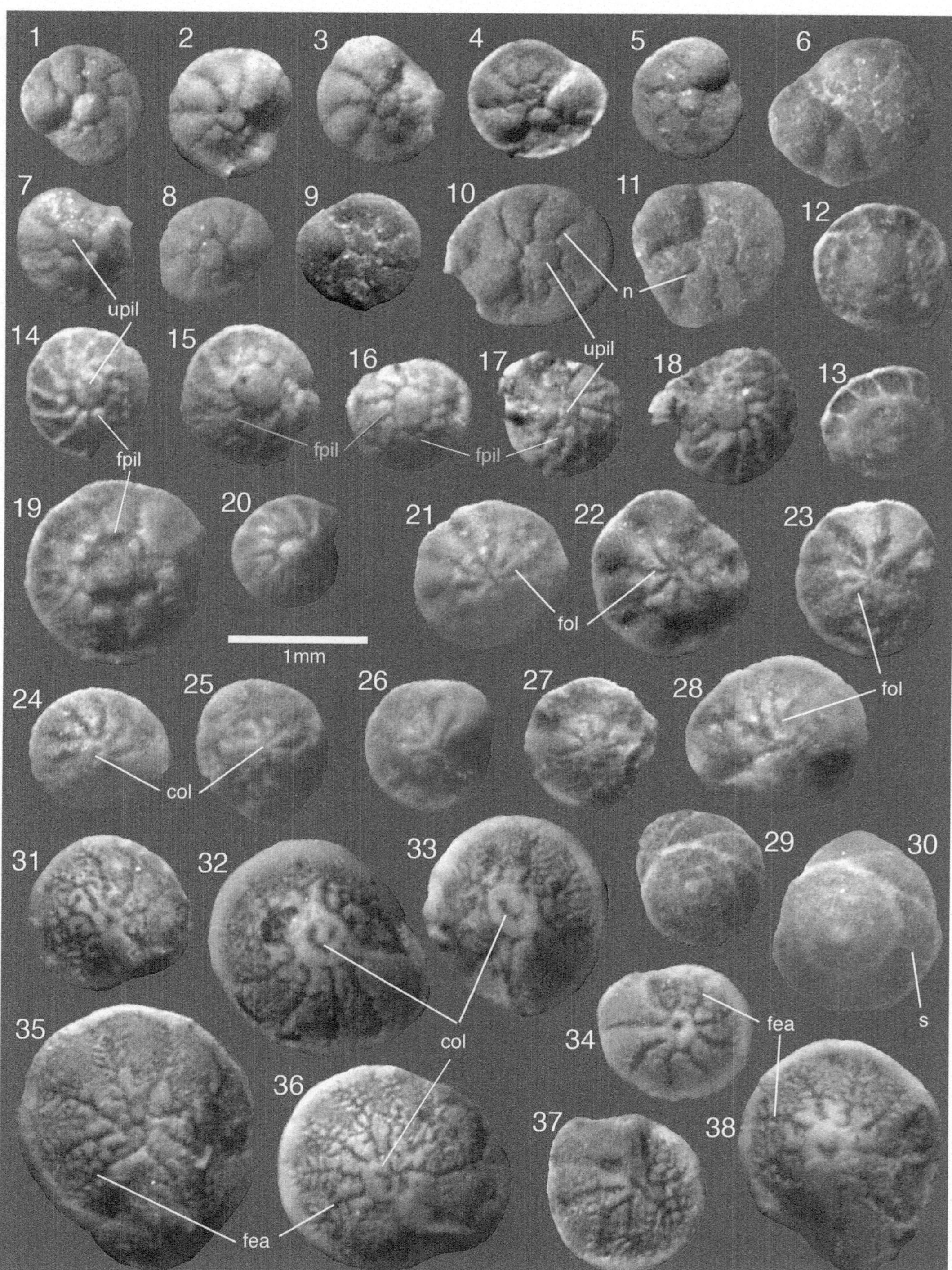

Plate 3.2 (**1–13**) *Rotorbinella hensoni* (Smout, 1954); sample 73907A, Couiza in the Aude Valley, Aquitaine, Southern France; sample 73907 from Locality 1 in Schaub (1981, p. 48, Fig. 46). (**1–11**) External, ventral view of shell face showing unfeathered chamber sutures and heavy, free standing umbilical plug. (**12–13**) Dorsal view of shell. (**14–20**) *Rotorbinella hermi* (von Hillebrandt, 1962); external view of isolated specimens under light microscope; note the large, undivided umbilical pile and the heavily limbate ventral chamber sutures. (**21–30**) *Rotalia implumis* n. sp.;

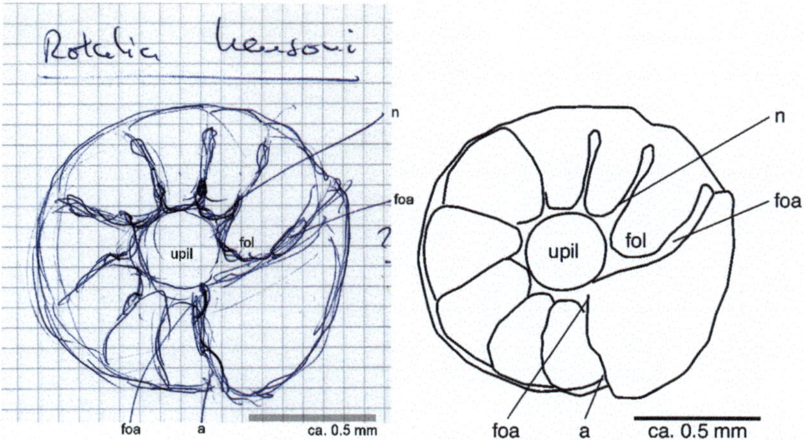

Fig. 3.2 *Rotorbinella hensoni* (Smout, 1954). Lukas Hottinger's hand sketch approximately at standard enlargement of the holotype's face deposited in the Natural History Museum (London). The specimen is marked as from "Paleocene of Qatar" according to Smout's (1954, pl. 15, Fig. 8) monograph. Abbreviations: *a* aperture, *fol* folia, *foa* foliar aperture, *n* notch, *upil* umbilical pile

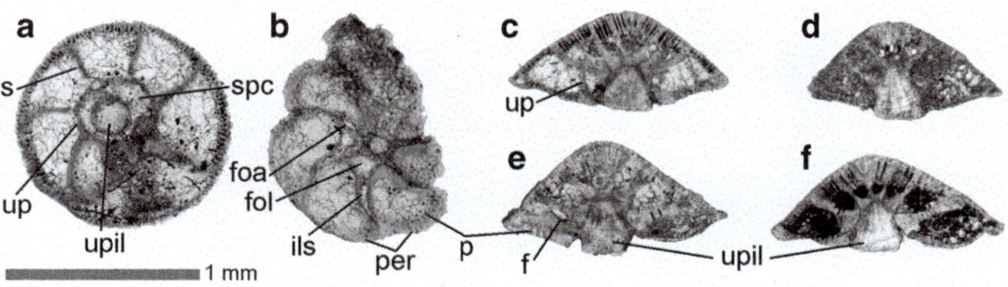

Fig. 3.3 *Rotorbinella hensoni* (Smout, 1954). (**A–B**) Sections perpendicular to the coiling axis. (**C–F**) Axial sections. (**B**, **D**) Sample 73907, Couiza, Aquitaine, SW France, lower Ilerdian (SBZ 6). (**A**, **C**, **N**, **F**) From Le Quillet, Petites Pyrénées, France, Paleocene (SBZ 4). Abbreviations: *f* foramen, *s* septum, *p* pores, *up* umbilical plate, *upil* umbilical pile, *spc* spiral canal, *foa* foliar aperture, *fol* folia, *ils* intraseptal interlocular space, *per* periphery

shell, they appear (Fig. 3.5B) as row of foramina or stolons but they are positioned along the median plane of the septum without interrupting its outer walls. The interlocular space forms an additional spiral canal below the dorsal whorl suture. There seems to be a single dorsal orifice of that canal system. All these structural details are found also in the Cretaceous type species of *Pyrenerotalia*.

The position and numbers of the intercameral apertures in *Pyrenerotalia depressa* are not entirely clear. Some intersections (Fig. 3.5A–B) seem to reveal an interiomarginal row of foramina. However, their shaping is so irregular that I

Plate 3.2 (continued) all specimens from the sample 95116 at Dhak Pass, Salt Range, Pakistan; Hangu Formation, first foraminifera-bearing bed, Paleocene (SBZ 3). (**21–28**) Ventral views of shell face. Note the straight, unfeathered and deep ventral cameral sutures and the fusion of the foliar tips into a columella. (**29–30**) Dorsal view of shell showing strong inclination of the septa. (**31–38**) *Rotalia* cf. *newboldi* d'Archiac and Haime, 1853; sample Sp 562, Pyrenean Basin, Northern Spain, Tremp section in Schaub (1981, p. 52, Fig. 51); middle Ilerdian (SBZ 8). Ventral view of shell face showing incipient feathering of the interlocular space marking the ventral cameral septa and the often annular fusion of the foliar tips into a columella. Abbreviations: *fpil* foliar pile, *upil* umbilical pile, *fol* folia, *n* notch, *s* septum, *col* columella, *fea* feathering of interlocular space

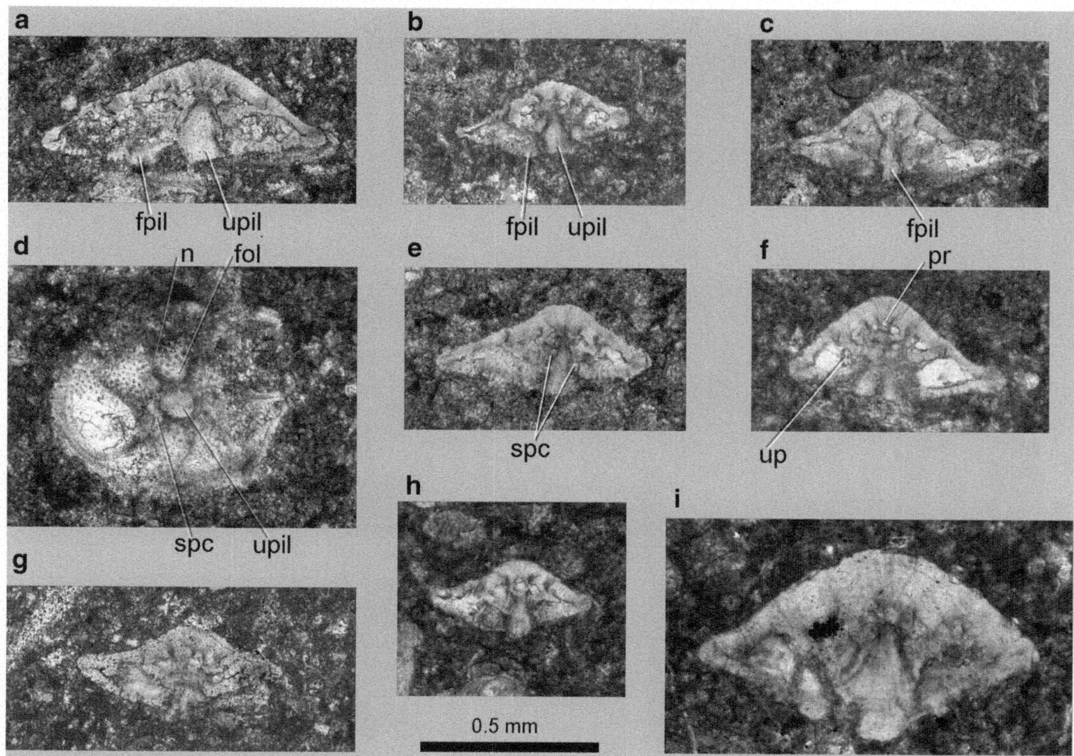

Fig. 3.4 *Rotorbinella detrecta* n. sp. (**A–H**) and *Rotorbinella hensoni* (Smout, 1954) (**I**). (**A–H**) Axial and sub axial sections showing free-standing umbilical plug. Note the partial compression of the chambers in the last whorl that is due to the compaction of the encasing, fine-grained sediment during early diagenesis. (**D**) Section perpendicular to coiling axis, tangential to the shell's face. (**I**) Axial section, for comparison, from the same locality as *R. detrecta*: LAP collected by Y. Tambareau; Paleocene (SBZ 2). Abbreviations: *fpil* foliar pile, *up* umbilical plate, *upil* umbilical pile, *spc* spiral canal, *fol* folia, *pr* proloculus, *n* notch

am sceptical about such an interpretation. I would rather consider them as some kind of loopholes connecting the chamber lumen with the intraseptal interlocular space.

There are about 12 almost isometric chambers in the dorsal side of an adult whorl. The ratio of equatorial to axial diameter is distinctly larger than the one of the Cretaceous species. The magalosphere reaches a diameter of about 0.075 mm.

Remarks: The material of this species is limited to a few random sections but it has nevertheless been given a name because the species is an additional marker of SBZ 2 associated to *Slovenites praecursorius* n. sp. and *Plumokathina dienii* n. sp. (Fig. 3.5J).

3.3 *Rotospirella* n. gen

Type species: *Lockhartia conica* Smout, 1954

Diagnosis: Conical shells with a flattened face covered by very long but extremely delicate folia, superposed like blades of a propeller. At each foliar suture a long, slender pile builds up to fill the adaxial part of the umbilicus without any mutual fusion. Foramen single, in interiomarginal position, loop shaped or compressed, possibly with a peristome.

Rotospirella conica (Smout, 1954); Fig. 3.6A–E.

1954 *Lockhartia conica* —Smout, p. 53, pl. 4, figs. 1–3.

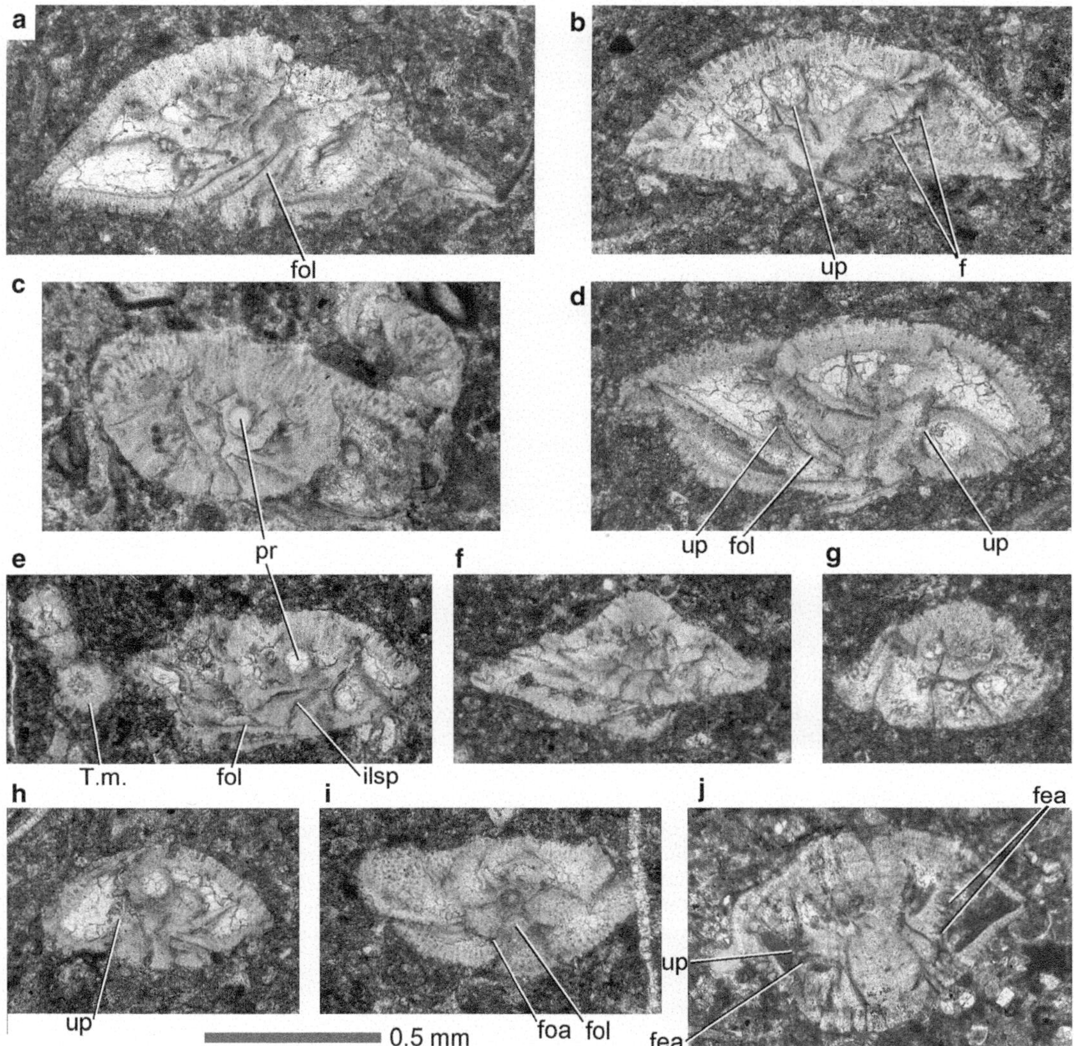

Fig. 3.5 *Pyrenerotalia depressa* n. sp. (**A–I**) and *Plumokathina dienii* n. sp. (**J**). (**A**) Subaxial section showing the extended, obliquely overlapping folia. (**B**, **D**) Transverse sections parallel to the coiling axis. (**C**, **E**) Oblique centered section much inclined in respect to the coiling axis; note the large proloculus, possibly a megalosphere. (**F–I**) Oblique section with different inclinations in respect to the coiling axis. Note in (**I**) the proloculus. Its small diameter is due to the tangential position of the section in respect to the center of the megalosphere. All from LAP collected by Y. Tambareau; Paleocene (SBZ 2). (**J**) Subaxial section showing the marks of the feathering in the septal sutures. From Pic d'Orhi, sample 10.3, collected by Y. Tambareau; Paleocene (SBZ 2); the presence of *P. dienii* in these Pyrenean rotaliid faunas is an important argument to attribute the corresponding beds to SBZ 2. Abbreviations: *fol* folia, *f* foramen, *up* umbilical plate, *pr* proloculus, *ilsp* intraseptal interlocular space, *foa* foliar aperture, *fea* feathering of interlocular space

1991 *Lockhartia conica* Smout —Wan, p. 162, pl. 1, figs. 28–29.

Remarks: The original description of this species by Smout (1954) does not give structural details of the shell but the axial section figured on pl. 4, fig. 3 shows the umbilical structure to be different from the one in Lockhartiids by the lack of regular umbilical "chamberlets". The dorsal

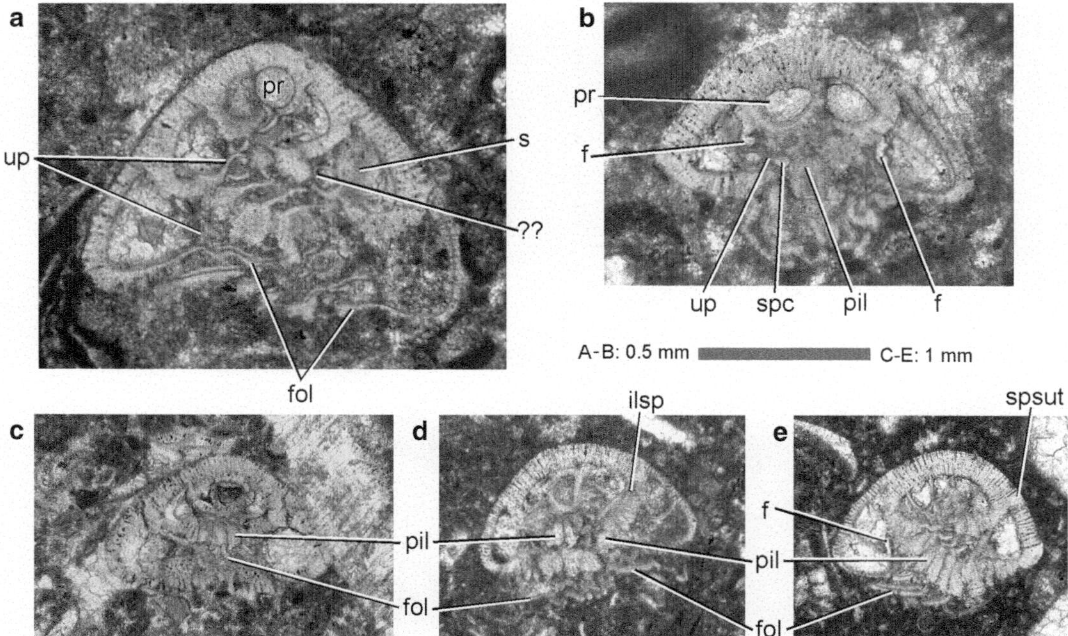

Fig. 3.6 *Rotospirella conica* (Smout, 1954). (**A–E**) Megalospheric specimens; sections with inclinations of about 20°–40° in respect to coiling axis. Note the umbilical piles obliquely superposed like thickened propeller blades that only by occasion support the extended foliar walls; from sample Kar 12, collected by J. Braud, Kuh-e-Kargan, Kermanshah, Zagros, Iran. Abbreviations: *up* umbilical plate, *fol* folia, *s* septum, *pr* proloculus, *f* foramen, *spc* spiral canal, *pil* pile, *ilsp* intraseptal interlocular space, *spsut* spiral suture

surface of the shell is smooth, the sutures flush. There are about 10 chambers in adult whorls. Megalosphere 0.012 mm in diameter.

Rotospirella conica n. sp. is distinguished from *Pyrenerotalia depressa* n. sp. by a much higher conical shell shape, by delicate, thin foliar walls and by numerous but free, unfused, slender umbilical piles.

3.4 *Rotalia* Lamarck, 1804

Type species: *Rotalites trochidiformis* Lamarck, 1804

Rotalia implumis n. sp.; Fig. 3.7A–I; Plate 3.2, Figs. 21–30.

Holotype: Specimen figured in Fig. 3.7G.

Type locality and type level: Bed 15116 in the Dhak Pass section, Salt Range, Pakistan; Hangu Formation; Paleocene (SBZ 3).

Derivation of name: *implumis* (Lat.), featherless.

Diagnosis: A rotaliine shell with a totally smooth dorsal side of the test. There are 10–12 chambers in the last whorl with much inclined and poorly curved septa below the dorsal chamber walls. No dimorphism has been recognized. On the ventral side of the test, the chambers are delimited by radial slits representing the interlocular septal space. Only in the extreme peripheral portion the slits are slightly inclined backward. The folia distinctly appear as narrow, radial, clear triangles by their lack of pores. Their adaxial tips are fused by a sudden backward kink, hiding a broad columellar structure that corresponds to the earlier whorls. There is no or only a faint, shallow feathering on the abaxial part of the shoulders of the interlocular space. The equatorial to axial diameter ratio varies from 1.8 to 2.3. There are 9–10 chambers in the last whorl.

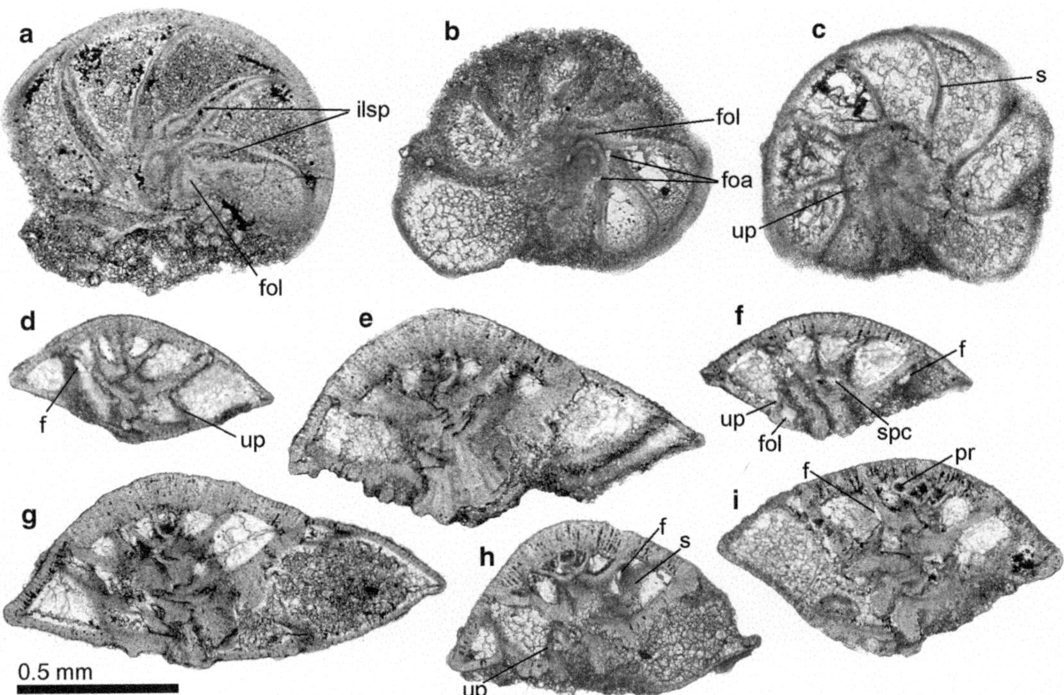

Fig. 3.7 *Rotalia implumis* n. sp.; sample 95116, Dhak Pass, Salt Range, Pakistan; first beds of the Hangu Formation bearing foraminifera; Paleocene (SBZ 3). (**A–C**) Sections more or less perpendicular to the coiling axis. (**D**, **F**) Oblique, uncentered sections. (**E**, **G–I**) Approximately axial, centered sections. (**G**) Holotype; note the columellar superposition of the successive foliar walls. Abbreviations: *f* foramen, *s* septum, *ilsp* intraseptal interlocular space, *pr* proloculus, *fol* folia, *foa* foliar aperture, *up* umbilical plate, *spc* spiral canal

Remarks: The columellar structure filling the umbilicus of *R. implumis* justifies its attribution to the genus *Rotalia sensu stricto*.

Rotalia* cf. *newboldi d'Archiac and Haime, 1853; Plate 3.2, Figs. 31–38; Plate 3.3, Figs. 1–17; Plate 3.4, Figs. 5–13.

1853 ?*Rotalia newboldi*—d'Archiac and Haime, p. 347, pl. 36, figs. 17 a–d.

1927 *Dictyoconoides newboldi* (d'Archiac and Haime)—Davies, p. 279, pl. 22, figs. 1–4.

1931 *Dictyoconoides newboldi* (d'Archiac and Haime) —Nuttall and Brighton, p. 57, pl. 4, figs. 1–3.

1932 *Dictyoconoides newboldi* (d'Archiac and Haime)—Davies, p. 408.

1937 *Lockhartia newboldi* (d'Archiac and Haime) —Davies and Pinfold, p. 46, pl. 5, fig. 23.

1953 *Rotalia trochidiformis* Lamarck —Gill, p. 840, pl. 90, figs. 1–12.

1954 *Rotalia trochidiformis* Lamarck pars. —Smout, p. 43, pl. 1, fig. 1; non pl. 1, figs. 2–6.

1973 *Rotalia trochidiformis* (Lamarck) —Ferrer et al., p. 44, pl. 2, fig. 5–7; ?pl. 9, figs. 1–3.

1993 *Rotalia trochidiformis* (Lamarck)—Weiss, p. 252, pl. 3, fig. 3

2008 *Rotalia trochidiformis* Lamarck —BouDagher-Fadel, p. 362, pl. 6.25, fig. 5.

Remarks: The so-called *Rotalia trochidiformis* identified by Ferrer et al. (1973) in the Ilerdian of the Pyrenees have no cover of prominent cylindrical papillae on their face, but otherwise they are rather similar to the Middle Eocene *R. trochidiformis*. Specimens with feathered shoulders of the septal interlocular space but lacking papillae are also frequent in the Ilerdian of the Salt Range (Pakistan).

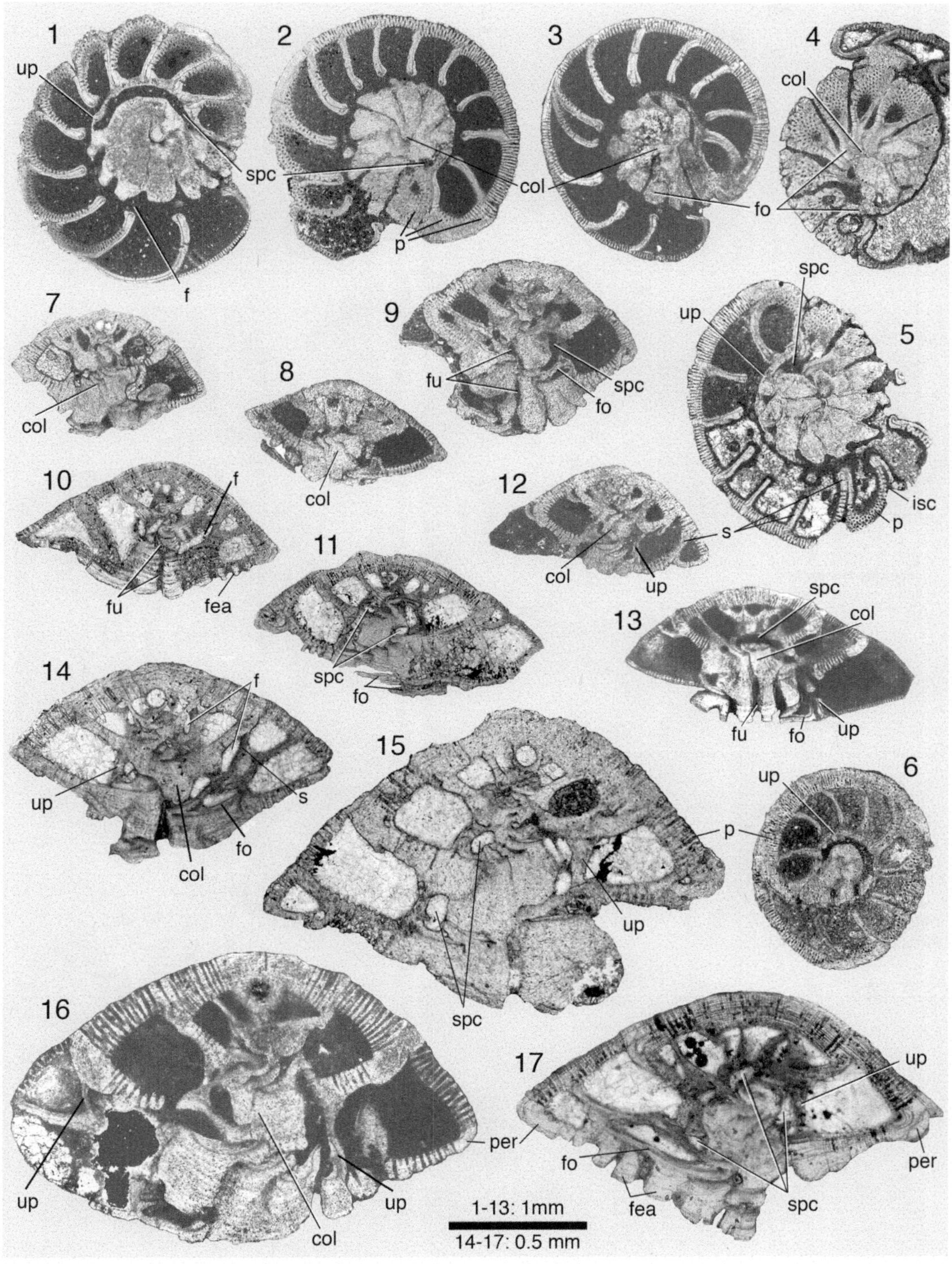

Plate 3.3 *Rotalia* cf. *newboldi* d'Archiac and Haime, 1853; (**1–5**, **7–9**, **12**, **16**) from sample E 327a; (**13–15**) from sample Sp 891; Aren section, Tremp, Northern Spain, see Schaub (1981); late Ilerdian (SBZ 9); (**10–11**, **17**) from sample Sp 688, Tremp section, Northern Spain, see Schaub (1981). (**1–6**) Sections perpendicular to the axis of shell coiling. Note the columella formed by the adaxial tips of the foliar walls and the large spiral canal. (**7–9**, **13**, **17**) Oblique sections. (**10–12**, **14–16**) Axial sections. Abbreviations: *col* columella, *f* foramen, *fea* feathering of interlocular space, *fo* folium, *fu* funnel, *isc* intraseptal interlocular space, *p* pores, *s* septum, *spc* spiral canal, *up* umbilical plate, *per* imperforate periphery

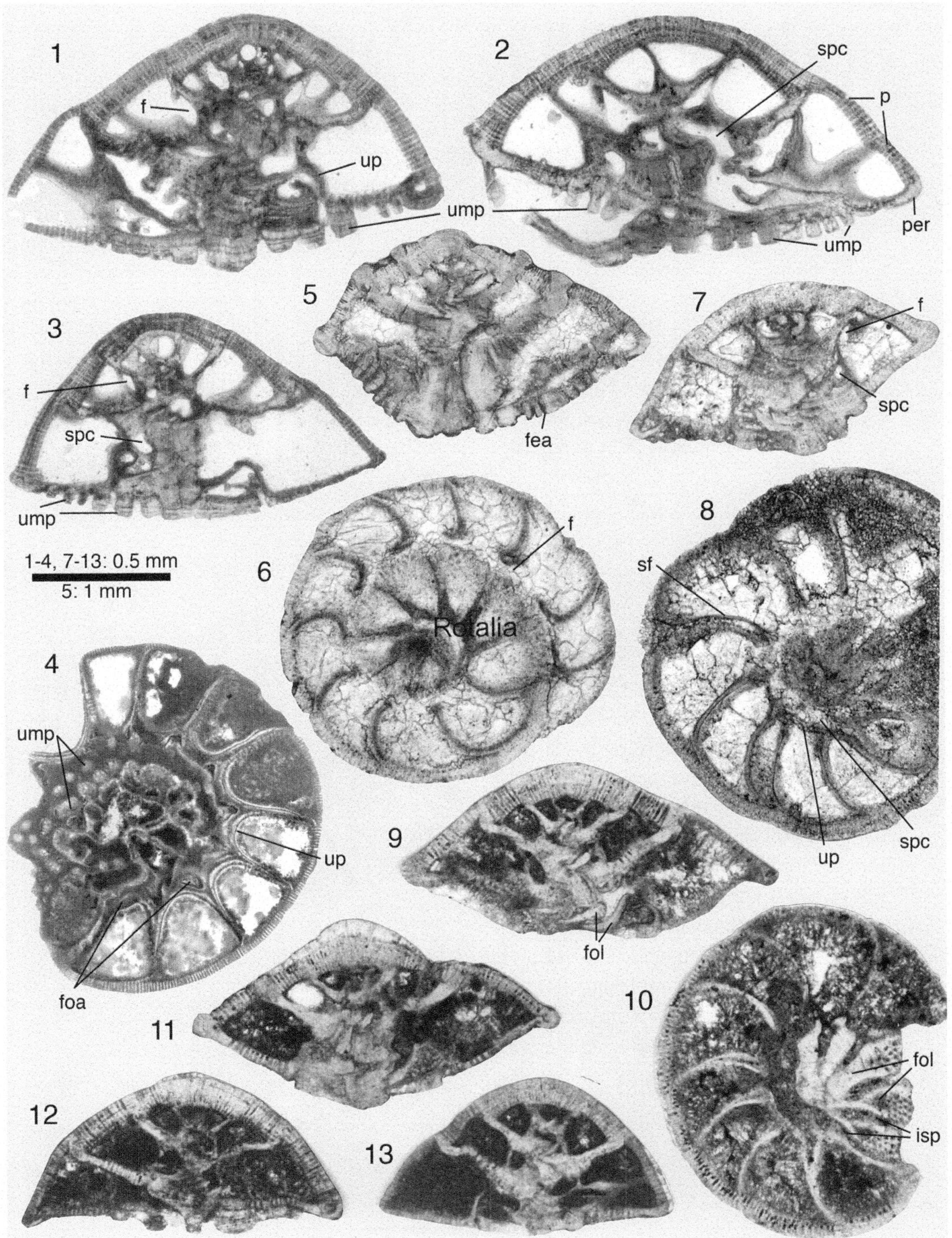

Plate 3.4 *Rotalia trochidiformis* (Lamarck, 1804); (**1–3**) from Ferme de l'Orme, Paris Basin and (**4**) from Bois Goüet (Bretagne), see Lehmann (1961, pp. 615–617); Middle Eocene, Lutetian. (**1**) Axial section. (**2–3**) Oblique sections. (**4**) Section perpendicular to coiling axis; note the development of heavy papillae forming umbilical papillae on the ventral surface of the chambers. (**5–13**) *Rotalia* cf. *newboldi* d'Archiac and Haime, 1853; (**5–6**) from sample 93545, Nilawahan section, Nammal Fm, Salt Range, Pakistan, middle Ilerdian; (**7**, **9–13**) from sample 93563, Dandot, Salt Range, Patala Fm; (**8**) from sample 95520, Nammal Gorge, Patala Fm., Early Eocene,

Davies and Pinfold (1937) have used the name *R. newboldi* d'Archiac and Haime that had been introduced in 1853 for similar forms from Sind in South-Eastern Pakistan, where Ilerdian shallow water sediments are known to be present. The types of this species seem to be lost. No topotype material is available today. The original figures show numerous, densely grouped, falciform septa on the dorsal side of the shell. I have never observed a similar density of septa anywhere else among the rotaliines but since the drawing from 1853 is rather schematic, I feel entitled to neglect this difference between the original and Davies and Pinfold's illustrations. The use of this species name nevertheless is tentative.

Rotalia trochidiformis (Lamarck, 1804); Plate 3.4, Figs. 1–4.

1804 *Rotalites trochidiformis*—Lamarck, p. 183; lectotype illustrated by Davies 1932, pl. 3, fig. 4.
1862 *Discorbina trochidiformis* (Lamarck) —Carpenter et al., p. 204.
1932 *Rotalia trochidiformis* (Lamarck) —Davies, p. 408, pl. 2, figs. 8–15; pl. 3, figs. 1–13.
1954 *Rotalia trochidiformis* (Lamarck)—Smout, p. 43, pl. 1, figs. 1–6.
1958 *Rotalia trochidiformis* (Lamarck) —Reiss and Merling, p. 13, pl. 1, figs. 1–4, 6; pl. 5, figs. 6–7, 10–11.
1963 *Rotalia trochidiformis* (Lamarck) —Reiss, p. 91, pl. 4, fig. 11; pl. 5, figs. 3, 18.
1971 *Rotalia trochidiformis* (Lamarck) —Hansen and Reiss, p. 339, pl. 19, figs. 1–6; pl. 20, figs. 1–5; pl. 21, figs. 1–6.
1971 *Rotalia trochidiformis* Lamarck—Parvati, p. 5, text-fig. 1, pl.1, figs. 1–5.
1980 *Rotalia trochidiformis* Lamarck—Müller-Merz, p. 34, text-figs, 16–17, pl. 2, figs. 1–4; pl. 9, figs. 1–4; pl. 15, fig. 3.
1982 *Rotalia trochidiformis* (Lamarck) —Lévy et al., p. 34, pl.1, figs. 1–9; pl.2, figs. 1–8.
1984 *Rotalia trochidiformis* (Lamarck) —Lévy et al., p. 382, pl.1, figs. 1–3.
1990 *Rotalia trochidiformis* (Lamarck) emended. Haynes and Whittaker, p. 97, text-figs. 1–3; pl. 3, figs. 1–6; pl.2, figs. 1–3; pl. 3, figs. 1–5.
1992 *Rotalia trochidiformis* Lamarck—Hansen and Revets, p. 167, pl.1, figs. 1–3.

Remarks: The topotypes from the Paris Basin show the characteristic columellar umbilical filling produced by the fused adaxial tips of the folia and the comparatively wide bore of the spiral canal separated from the chamber lumen by a vertical umbilical plate. This Lutetian species has a ventral face covered by heavy cylindrical papillae (Plate 3.4, Figs. 2–4). These cover the whole face, out of reach by the feathered ornaments on the shoulders of the sutural interlocular space. In axial sections, the feathers may appear similar to the pustules but the latter may be easily distinguished in sections tangential to the dorsal face.

3.5 *Medocia* Parvati, 1971

Type species: *Medocia blayensis* Parvati, 1971

Medocia blayensis Parvati, 1971; Fig. 3.8A–I; Fig. 3.9A–F.

1954 *Rotalia trochidiformis* Lamarck pars. —Smout, p. 43, pl. 1, figs. 2–6.
1971 *Medocia blayensis*—Parvati, p. 17, text-fig. 4; pl. 2, figs. 2–4; pl. 3, figs. 1–5; pl. 4, figs. 2–6; pl. 5, fig. 5.

Plate 3.4 (continued) Ilerdian. (**5**) Microspheric specimen, axial section. (**6**) Megalospheric specimen, section perpendicular to the coiling axis of the shell. (**8**) Section perpendicular to axis; note the largeness of the spiral canal and of the interlocular space. (**7**, **9–13**) Axial and (**10**) perpendicular sections; note the flattened ventral surface of the shell in (**12**) and (**13**) that is a consequence of the reduction of lamellar thickness of the shell in low energy environments with soft substrates rich in clay minerals. Abbreviations: *f* foramen, *up* umbilical plate, *ump* umbilical piles, *spc* spiral canal, *p* pores, *per* periphery, *foa* foliar aperture, *fea* feathering of interlocular space, *fol* folia, *sf* septal flap, *isp* intraseptal space

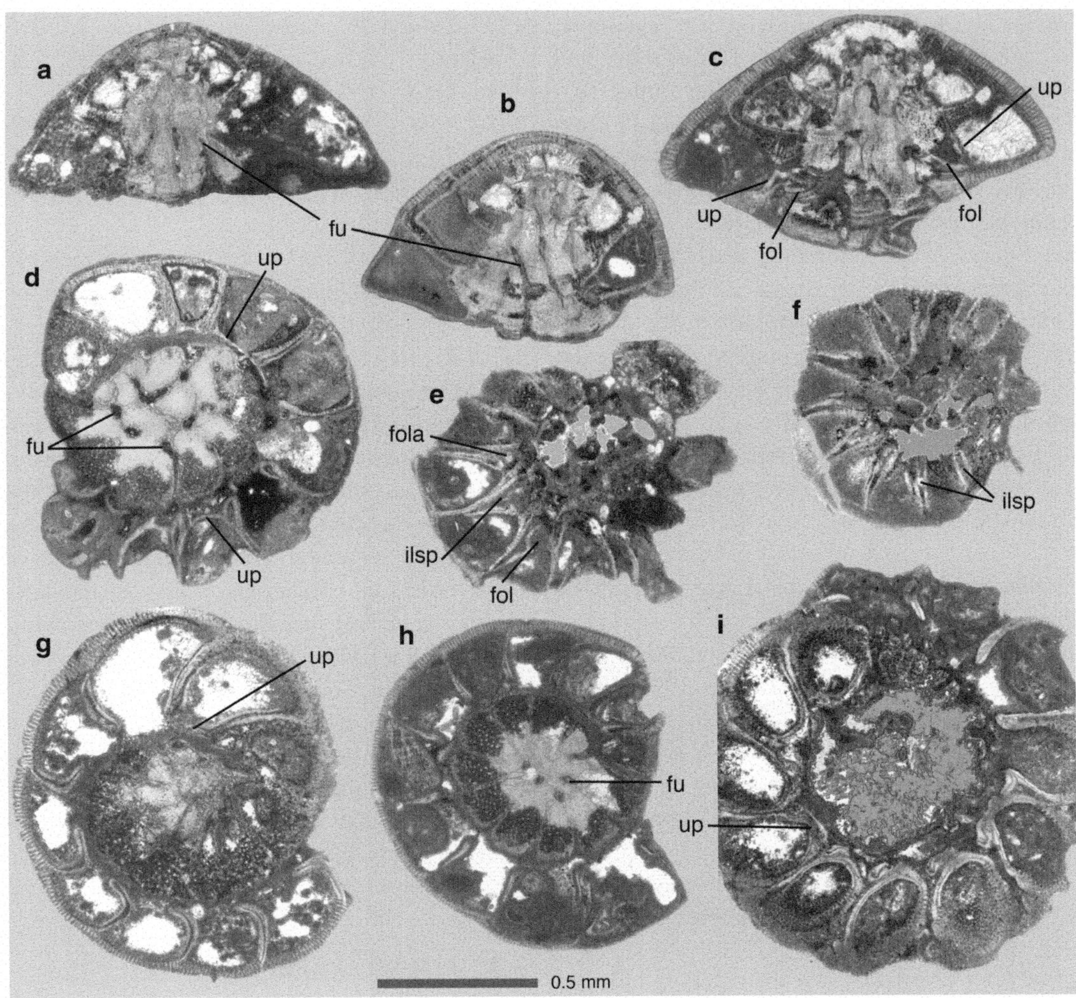

Fig. 3.8 *Medocia blayensis* Parvati, 1971; Cotentin area, Western France, see Lehmann (1961); late Lutetian (SBZ 16). (**A–I**) Megalospheric specimens from the type region, with partially silicified shells. (**A–F**) From Bois Gouët. (**G–I**) From Fréville. Abbreviations: *fu* funnel, *up* umbilical plate, *fol* folia, *foa* foliar aperture, *ilsp* intraseptal interlocular space

1978 *Medocia blayensis* Parvati—Le Calvez and Blondeau, p. 28, pl. 1, figs. 1–2.

2007 *Medocia blayensis* Parvati —Hottinger, p. 15, pl. 9, figs. 2, 5, 8; pl. 10, fig. 3; pl. 12, figs. 1–9; pl. 13, figs. 1–10.

Remarks: *Medocia blayensis* was described by Parvati (1971) in great detail and its description has been updated according to modern terms of rotaliid morphology by Hottinger (2007). It is not repeated here. The illustrations in Hottinger (2007, figs. 12–13) document the wide geographic range of this species throughout the Tethyan realm. As to its age range, the species has been discovered so far in beds of late Lutetian to Bartonian age. A few specimens (Plate 4.7, Figs. 13–14) to be included at least in the genus *Medocia* have been found in association with *Slovenites decastroi* n. sp. on Monte Gargano, Italy, in beds with an age of Early Eocene, Cuisian (SBZ 11–SBZ 13).

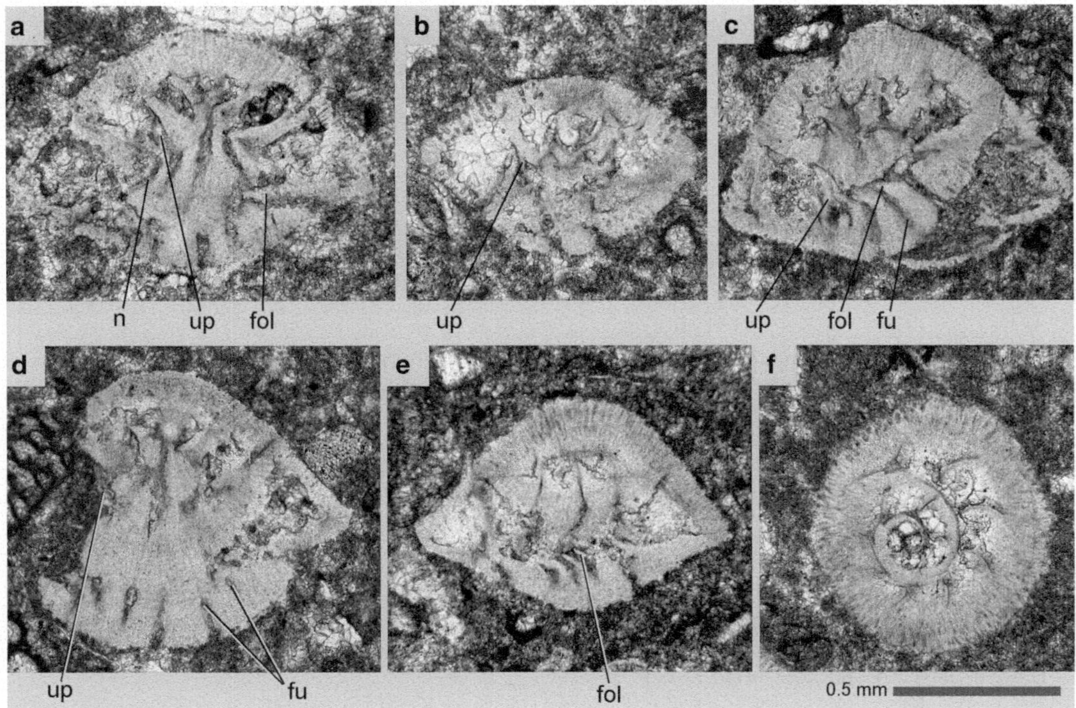

Fig. 3.9 *Medocia blayensis* Parvati, 1971; sample 00006, Oman. Lutetian. (**A–F**) Megalospheric specimens from Oman documenting the presence of this species in the Middle East; note the coarse funnels in the shell's columella, the most important diagnostic feature. (**A–E**) Oblique sections. (**F**) Centered section perpendicular to the coiling axis. Abbreviations: *n* notch, *up* umbilical plate, *fol* folia, *fu* funnel

References

Afzal J, Williams M, Aldbridge R (2009) Revised stratigraphy of the lower Cenozoic succession of the Grater Indus Basin in Pakistan. J Micropaleontol 28:7–23

Archiac A. d', Haime J (1853) Description des animaux fossiles du groupe nummulitique des Indes, précédée d'un résumé géologique et d'une monographie des Nummulites, 373 pp, 36 pls, Gide & Baudry, Paris

Bandy OL (1944) Eocene foraminifera from Cape Blanco, Oregon. J Paleontol 18:366–377

Boix C, Villalonga R, Caus E, Hottinger L (2009) Late Cretaceous Rotaliids (Foraminiferida) from the Western Tethys. Neues Jb Geol Paläontol Abh 253 (3):197–227

BouDagher-Fadel MK (2008) Evolution and geological significance of larger benthic foraminifera, Developments in palaeontology & stratigraphy 21. Springer, Amsterdam, 540 pp

Carpenter WB, Parker WK, Jones R (1862) Introduction to the study of the foraminifera. Ray Society, Robert Hardwicke, London, 319 pp, 22 pls

Davide LM (1927) The Ranikot beds at Thal (North-West Frontier Provinces of India). Quart J Geol Soc 83 (2):260–290

Davies LM (1932) The genera *Dictyoconoides* Nuttall, *Lockhartia* nov. and *Rotalia* Lamarck: their type species, generic differences and fundamental distinction from the *Dictyoconus* Group of forms. Trans Roy Soc Edinbourgh 57:397–428, 4 pls

Davies LM, Pinfold ES (1937) The Eocene Beds of the Punjab Salt Range. Palaeontologia Indica Calcutta 24 (1), 79 pp, 7 pls

Ehrenberg CG (1839) Über die Bildung der Kreidefelsen und des Kreidemegels durch unsichtbare Organismen. Phys Abh k preuss Akad Wiss 1838:59–148, 1 tab, pls. 14, Berlin

Ferrer J, Le Calvez Y, Luterbacher H-P, Premoli-Silva I (1973) Contribution à l'étude des foraminifères Ilerdiens de la region de Tremp (Catalogne), Mém Mus Nat Hist Nat Paris, ser C, 29. Éditions du Muséum, Paris, pp 3–107

Gill WD (1953) Facies and fauna in the Bhadrar beds of the Punjab Salt Range Pakistan. J Paleontol 27:824–844

Hansen H-J, Reiss Z (1971) Electron microscopy of Rotaliacean wall structures. Bull Geol Soc Denmark 20:329–346

Hansen HJ, Revets S (1992) A revision and reclassification of the Discorbidae, Rosalinidae and Rotaliidae. J Foram Res 22:166–180

Haynes JR, Nwabufo-Ene K (1998) Foraminifera from the Paleocene phosphate beds, Sokoto, Nigeria. Rev Esp Micropaleontol 30:51–76

Haynes JR, Whittaker JE (1990) The status of *Rotalia* Lamarck (Foraminifera) and of the Rotaliidae Ehrenberg. J Micropaleontol 9(1):95–106

Herbig H-G (1991) Das Paläogen am Südrand des zentralen Hohen Atlas und im Mittleren Atlas Marokkos. Stratigraphie, Fazies, Paläogeographie und Paläotektonik, Berl Geowiss Abh, Reihe A, 135. Selbstverlag Fachbereich Geowissenschaften, Berlin, 289 pp

Ho Y, Zhang B, Hu L, Shang J (1976) Mesozoic and Cenozoic foraminifera from the Mount Julmo Lungma Region. In: A report of the scientific expedition in the Mount Julmo Lungma 1966–1968. Paleontology (Beijing), spec publ 2:1–76, 36 pls (in Chinese)

Hottinger L (2001) Learning from the past. In: Levi-Montalcini R (ed) Frontiers of life, vol 4, Part 2: Discovery and spoliation of the biosphere. Academic, London/San Diego, pp 449–477

Hottinger L (2006) The "face" of benthic foraminifera. Boll Soc Paleontol Ital 45:75–89

Hottinger L (2007) Revision of the foraminiferal genus *Globoreticulina* Rahaghi, 1978 and of its associated fauna of larger foraminifera from the late Middle Eocene of Iran. Carnets Géol Article 2007/06, CG2007_A06

Lamarck JP (1804) Suite des Mémoires sur les fossiles des environs de Paris. Ann Mus Hist Nat Paris 5:179–188

Le Calvez Y, Blondeau A (1978) La microfaune "Biarritzienne" du Lutétien du Cotentin. Bull Inf Géol Bassin Paris 15(2):21–31

Lehmann R (1961) Strukturanalyse einiger Gattungen der Subfamilie Orbitolitinae. Eclog Geol Helv 54:599–667

Lévy A, Mathieu R, Poignant A, Rosset-Moulinier M, Rouvillois A (1982) Données nouvelles sur *Rotalia trochidiformis* Lamarck (Foraminifera). Emendation du genre *Rotalia* Lamarck. Géol Méditerr 9:33–41

Lévy A, Mathieu R, Poignant A, Rosset-Moulinier M (1984) Une nouvelle conception des familles Discorbidae and Rotaliidae. In: Benthos '83, 2nd international symposium on Benthic Foraminifera, Pau, Bordeaux, pp 381–387

Moret L (1938) Contribution à la paléontologie des couches crétacées et éocènes du versant sud de l'Atlas de Marrakech (Maroc), Notes Mém Serv Min Carte Géol Maroc 49. Imprimerie officielle, Rabat, 77 pp

Müller-Merz E (1980) Strukturanalyse ausgewählter rotaloider Foraminiferen. Schweiz Paläontol Abh 101:5–68, 15 pls

Nuttall WL, Brighton AG (1931) Larger Foraminifera from the Tertiary of Somaliland. Geol Mag 63:49–65, 4 pls

Parvati S (1971) A study of some rotaliid Foraminifera. Proc Kon Ned Akad Wetensch Ser B74(1):1–26, 4 pls

Reiss Z (1963) Reclassification of perforate Foraminifera. Bull Geol Surv Isr 36:1–111

Reiss Z, Merling P (1958) Structure of some Rotaliidea. Bull Geol Surv Isr 21:1–19

Revets SA (2001) The genus *Rotorbinella* Bandy, 1944 and its classification. J Foram Res 31:315–318

Samuel O, Borza K, Köhler E (1972) Microfauna and Lithostratigraphy of the Paleogene and adjacent Cretaceous of the Middle Vah valley (West Carpathians). Geol Ustav Dionysza Stura, Bratislava, 246 pp, 153 pls

Schaub H (1981) Nummulites et Assilines de la Tethys paléogène. Taxinomie, phylogenèse et biostratigraphie. Mém Suisses Paléontol 104:1–236, 116 figs, 18 pls

Smout AH (1954) Lower Tertiary foraminifera of the Qatar peninsula. Brit Mus (Nat Hist), 96 pp, 44 figs, 15 pls

von Hillebrandt A (1962) Das Paleozän und seine Foraminiferenfauna im Becken von Reichenhall und Salzburg. Bayer Akad Wiss Math-Naturwiss Kl 108:9–182

Wan X (1991) Paleocene larger foraminifera from Southern Tibet. Rev Esp Micropaleontol 23:7–28

Weiss W (1993) Age assignments of larger foraminiferal assemblages of Maastrichtian to Eocene age in northern Pakistan. Zitteliana 20:223–252

4 New Subfamily Redmondininae

Abstract

The new subfamily Redmondininae is characterized by coarsely perforated and strikingly thick chamber walls, a reduced umbilical filling but with a canal system that tends to extend onto the dorsal side of the shell. This group encompasses four genera (*Redmondina*, *Slovenites* n. gen., *Rotaliconus*, *Pachyrotalia* n. gen.). Seven species (*R. henningtoni*, *R. garganica* n. sp., *S. praecursorius* n. sp., *S. pembaphis* n. sp., *S. decastroi* n. sp., *R. arachosiae* n. sp., *P. massa* n. sp.), are described and illustrated.

This is a group of rotaliids with coarsely perforated and strikingly thick chamber walls, a reduced umbilical filling but with a canal system that tends to extend onto the dorsal side of the shell. Few and coarse funnels may appear in the umbilical filling. *Redmondina* Hasson, 1985, represents a branch of small-sized shells with a relatively simple morphology, *Slovenites* n. gen. a parallel group with a more complex architecture and a peculiar dorsal ornamentation by perforate pustules. In Pakistan's SBZ 4 (Upper Paleocene), they are associated to *Pachyrotalia massa* with an enveloping canal system invading the dorsal side.

All rotaliines and redmondinines remain limited in size during all the Paleogene; there are never extreme morphological variants like multiple spirals (*Dictyoconoides*, *Dictyokathina*) with their corresponding, extreme dimorphism.

4.1 *Redmondina* Hasson, 1985

Type species: *Redmondina henningtoni* Hasson, 1985

Remarks: The trochospiral shells with very coarse pores are comparatively small. The dorsal ornamentation is scarce and often obscured by the large pores. The periphery of the shell is rounded, without any differentiation of the perforation. The septa are radial, curved and inclined backwards on the dorsal side of the shell. They house a simple canal system with orifices near the junction of the spiral suture with chamber sutures. There seems to be a single radial canal in the septum below the septal suture (Plate 4.3, Fig. 2). The ventral side of the shell is characterized by radial, deeply sunk septa admitting a large, radial interlocular space. The

L. Hottinger, *Paleogene larger rotaliid foraminifera from the western and central Neotethys*,
DOI 10.1007/978-3-319-02853-8_4, © Springer International Publishing Switzerland 2014

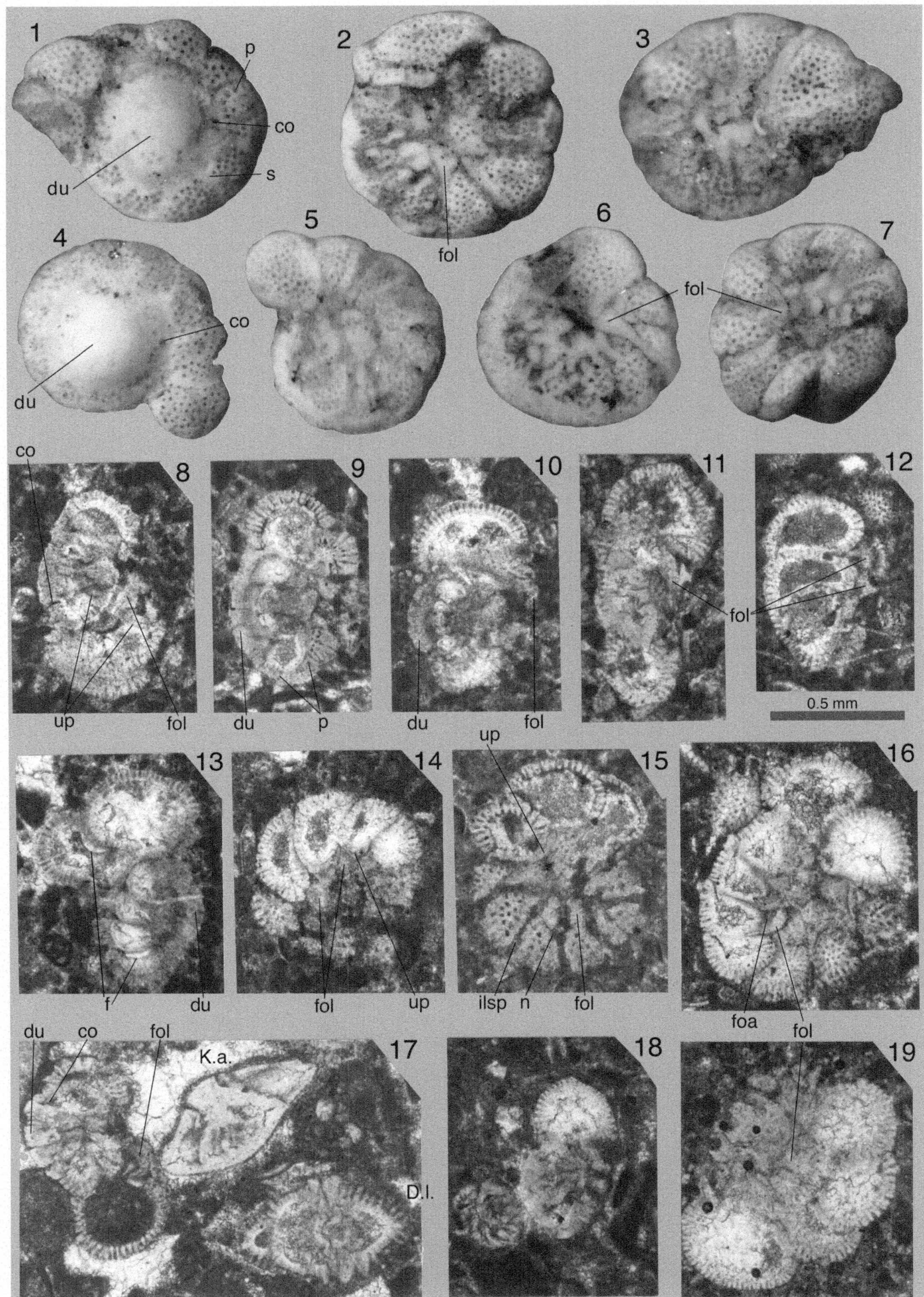

Plate 4.1 *Redmondina henningtoni* Hasson, 1985. (**1–7**) Outer aspect of free specimens from sample Lafarge 9, Western Aquitaine, Southern France, collected by Y. Tambareau; Paleocene (SBZ 3). (**8–10**) Axial sections.

foramina form very low and narrow arches in interiomarginal position (Plate 4.1, Fig. 13). The folia are long and narrow, only a little bit inclined in respect to the axis of the ventral part of the chambers. The narrow umbilicus is empty or filled with a few piles that do not fuse. The piles arise from the imperforate narrow folia along their radial axis (Plate 4.3, Fig. 2) but do not fuse to a solid umbilical filling. In early species, the umbilical plate is small and difficult to see in random sections of the shell. In later species, the umbilical plates are much larger. The spiral interlocular space is obscured by the heavy ornaments produced by the umbilical piles. We have not observed any dimorphic features.

Redmondina henningtoni Hasson, 1985; Plate 4.1, Figs 1–19; Plate 4.2, Figs. 1–13.

1985 *Redmondina henningtoni*—Hasson, p. 352, pl. 3, figs. 4–9.

? 1973 *Epistomaria separans* Le Calvez—Ferrer et al., p. 43, pl. 1, figs. 9–11.

2000 *Redmondina henningtoni* (Hasson)—Peybernès et al., p. 46, pl. 6/7–8.

Remarks: A species with the characteristics of the genus *Redmondina* characterized by a smooth dorsal umbo covering the early whorls of the shell. The umbo has few or no pores forming a kind of white cap that strikes the eye when picking free specimens from washed residues. The adult whorls of the shell have 8–10 chambers. The size of the proloculus is about 0.04 mm. Compare the short, radial folia that do not touch each other with their adaxial tips in Hasson (1985, pl. 3, fig. 4) and in the present paper (Plate 4.1, Fig. 15).

Hasson's (1985) holotype of *R. henningtoni* has a rounded periphery whereas her specimen illustrated pl. 3, fig. 4 shows a strong keel. I consider the latter as an artefact due to the glue used to prepare the specimen for SEM analysis.

***Redmondina garganica* n. sp.**; Fig. 4.1A–J; Plate 4.3, Figs. 1–12.

Syntypes: Specimens figured in Plate 4.3, Fig. 3 (section perpendicular to coiling axis) and Fig. 9 (axial section).

Type locality and type level: Monte Gargano, Italy; Early Eocene, middle Cuisian (SBZ 11).

Derivation of name: Monte Gargano represents the highest point on the Peninsula Garganica, the spur of the Italian boot.

Diagnosis: Shells with the generic characteristics of *Redmondina*, somewhat larger than the type species. 12–14 inflated chambers are counted in the last adult whorl. The septal sutures are sunk without revealing the enlarged intraseptal interlocular space below that reaches from the ventral to the dorsal cameral suture with single orifices on both sides. The equatorial to axial diameter ratio for adult shells varies from 1.5 to 1.7.

Extending over the shell apex, the dorsal umbo is scarcely perforated but distinctly inflated. It may be composed of a certain number of dorsal piles grouped closely around the shell apex. Umbilical filling reduced to few, free umbilical piles standing on the foliar walls that fuse around the shell axis with one another without showing on the shell face any central plug. Proloculus diameter is about 0.04 mm.

Remarks: *Redmondina garganica* n. sp. is distinguished from *R. henningtoni* by the higher number of chambers in the last whorl, by a reduced size of the dorsal umbo and by a widening of the intraseptal interlocular space over the whole septum from ventral to dorsal.

Plate 4.1 (Continued) (**11–13**) Sections parallel to coiling axis. (**14–16**) Sections perpendicular to coiling axis; note imperforate foliar walls and the absence of axial umbilical piles or plugs. (**17**) Oblique section inclined for more than 45° in respect to the coiling axis, associated with *Kathina aquitanica* n. sp. (K. a.) and *Miscellanea juliettae* Leppig, 1988 (M. j.). (**18–19**) Oblique sections with a respective inclination for more or for less than 45°. 8–19 sections in cemented rock from Tena section sample Stop 2 in Robador et al. (1991). Abbreviations: *f* foramen, *up* umbilical plate, *p* pores, *foa* foliar aperture, *fol* folia, *s* septum, *co* canal orifice, *du* dorsal umbo, *ilsp* intraseptal interlocular space, *n* notch

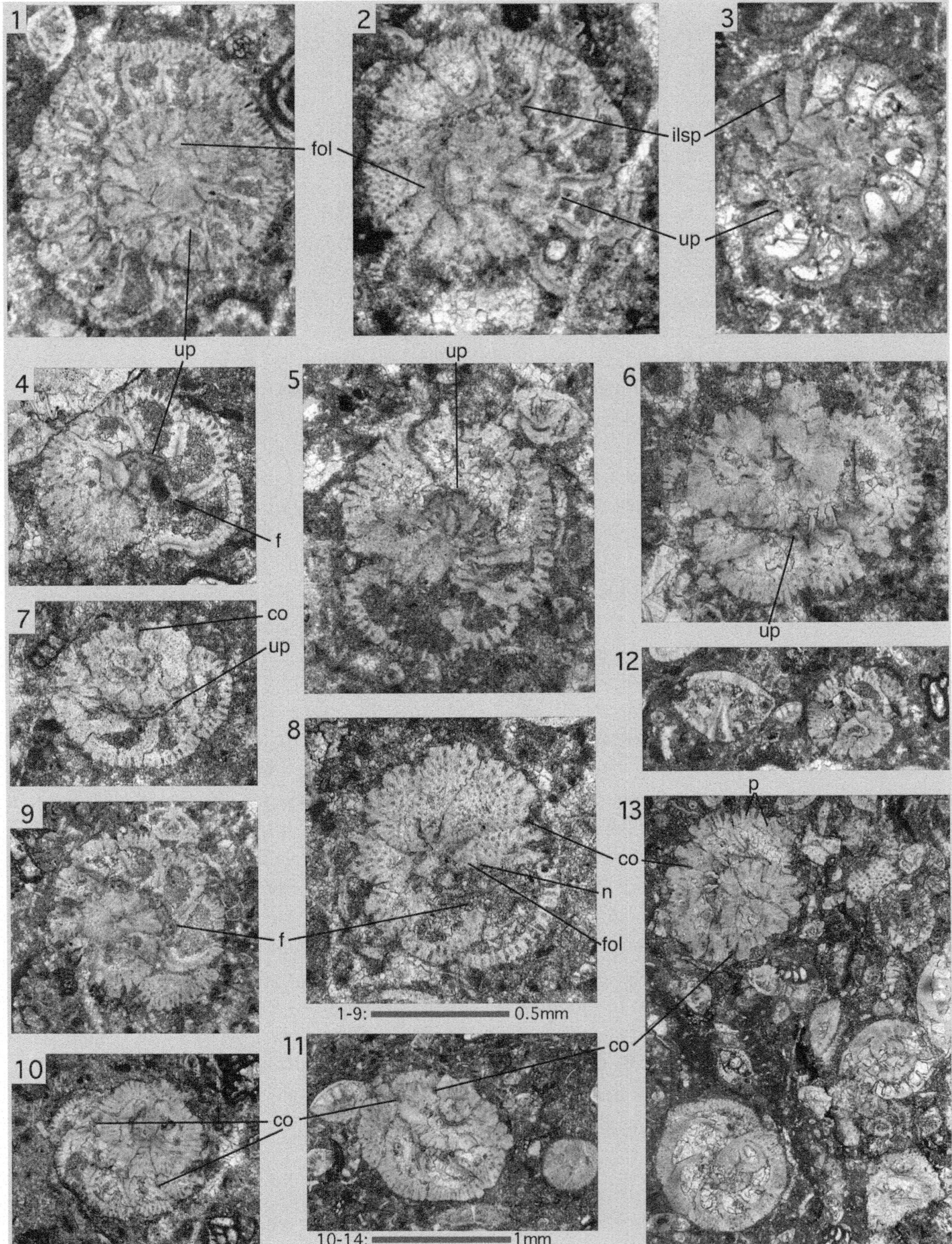

Plate 4.2 *Redmondina henningtoni* Hasson, 1985; sample All77232 collected by F. Allemann, Zhob valley, Baluchistan, western Pakistan; Paleocene (SBZ 4). (**1–3**) Sections perpendicular to the shell's coiling axis. (**4–5**, **7–8**) Oblique sections. (**6**) Oblique section showing the umbilical plate at the base of a chamber in the last whorl. (**9**) Section perpendicular to coiling axis. (**10–11**) Oblique sections respectively inclined for about 60° and 40° against the coiling axis; note the comparatively large canal orifices at the intersection of the spiral whorl sutures

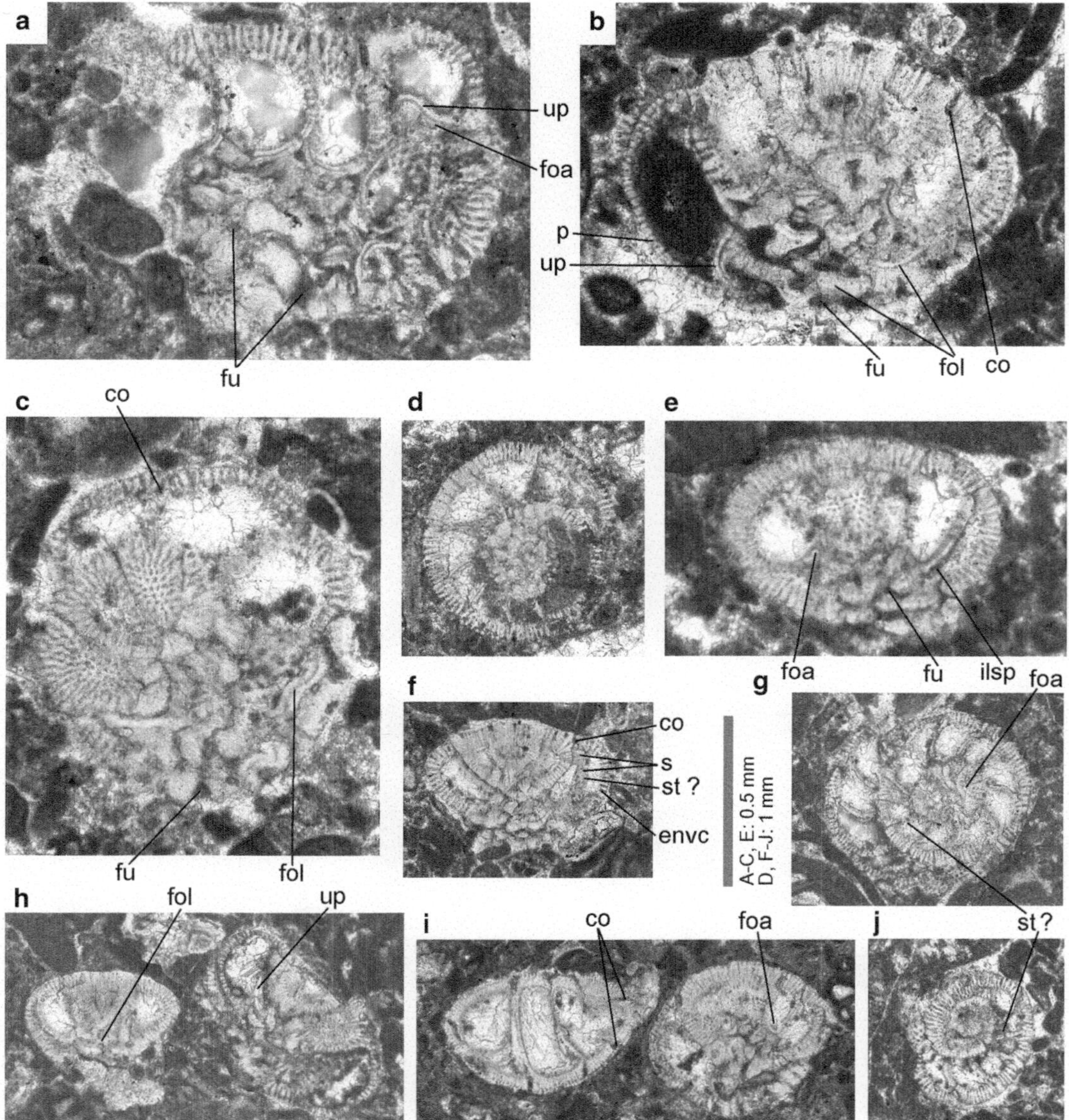

Fig. 4.1 *Redmondina garganica* n. sp.; sample Kar 13, collected by J. Braud, Kuh-e-Kargan, Kermanshah, Zagros, Iran. (**A**, **C**, **E**, **H**, **I**) Oblique sections with different inclination in respect to the shell's coiling axis. (**B**, **F**) Subaxial sections; note the poor development of the apical umbo. (**D**, **G**, **J**) Sections perpendicular to the coiling axis. Abbreviations: *up* umbilical plate, *foa* foliar aperture, *p* pores, *fol* folia, *fu* funnel, *co* canal orifice, *ilsp* intraseptal interlocular space, *s* septum, *st?* suture?, *envc* enveloping canals

Plate 4.2 (Continued) with the more or less radial chamber sutures. (**12**) Oblique section (*left*) associated with a *Kathina delseota* Smout, 1954 (K. s.). (**13**) Oblique section inclined for about 30° in respect of the coiling axis (*top*) and section perpendicular to the coiling axis (*bottom*). Abbreviations: *f* foramen, *up* umbilical plate, *p* pores, *fol* folia, *co* canal orifice, *ilsp* intraseptal interlocular space, *n* notch

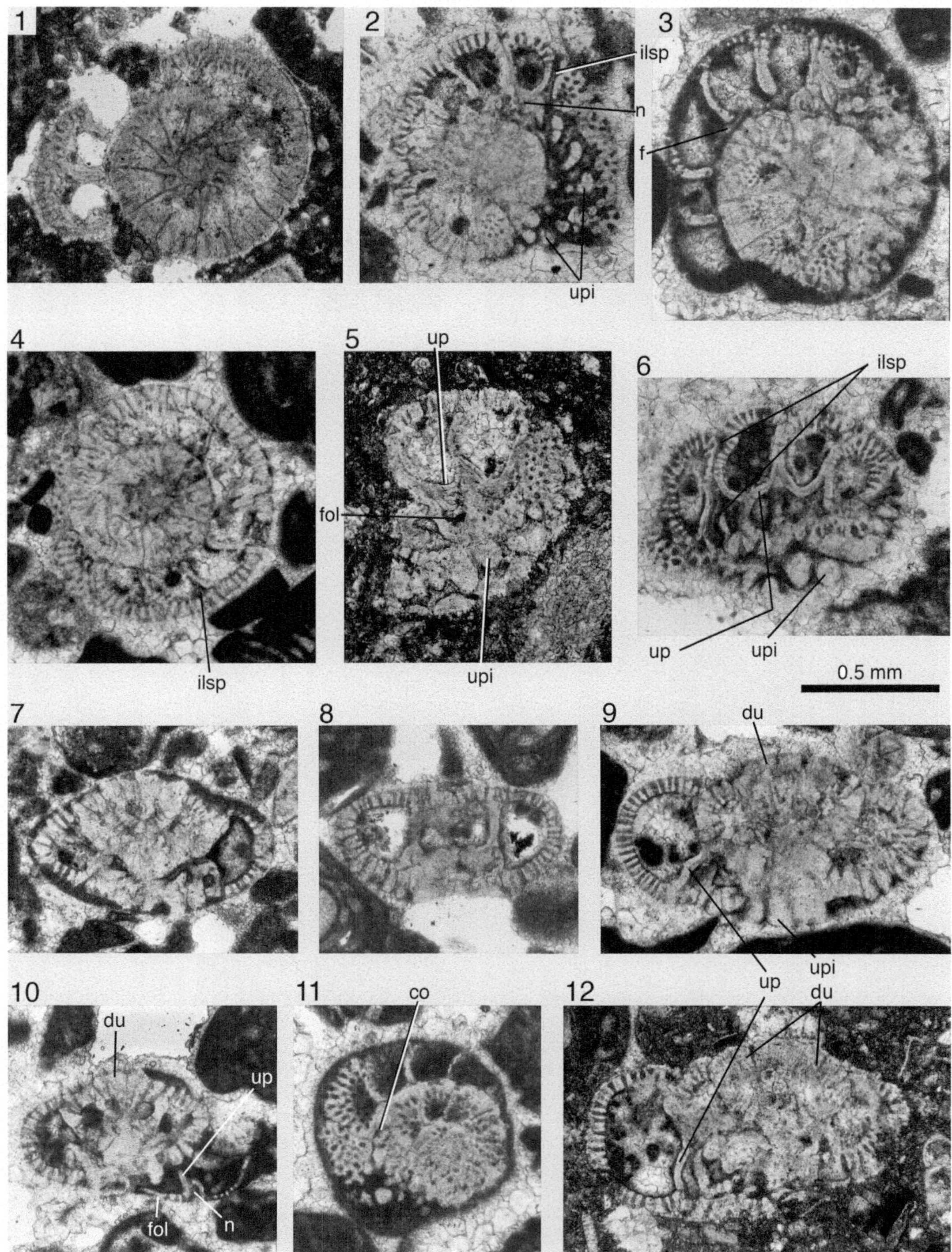

Plate 4.3 *Redmondina garganica* n. sp.; all from *Alveolina* limestone of Monte Gargano (south-eastern Italy), Early Eocene, Middle Cuisian (SBZ 11), collected by Piero De Castro and deposited in the collections of the Dipartimento di Scienze della Terra, University of Naples "Federico II", Italy. (**1–4**) Sections perpendicular to coiling axis. (**5–6**) Oblique sections illustrating the comparatively simple umbilical architecture. (**7**, **9–10**, **12**) Axial sections. (**8**, **11**) Transversal sections parallel and perpendicular to the shell axis. Abbreviations: *f* foramen, *up* umbilical plate, *upi* umbilical piles, *du* dorsal umbo, *fol* folia, *co* canal orifice, *ilsp* intraseptal interlocular space, *n* notch

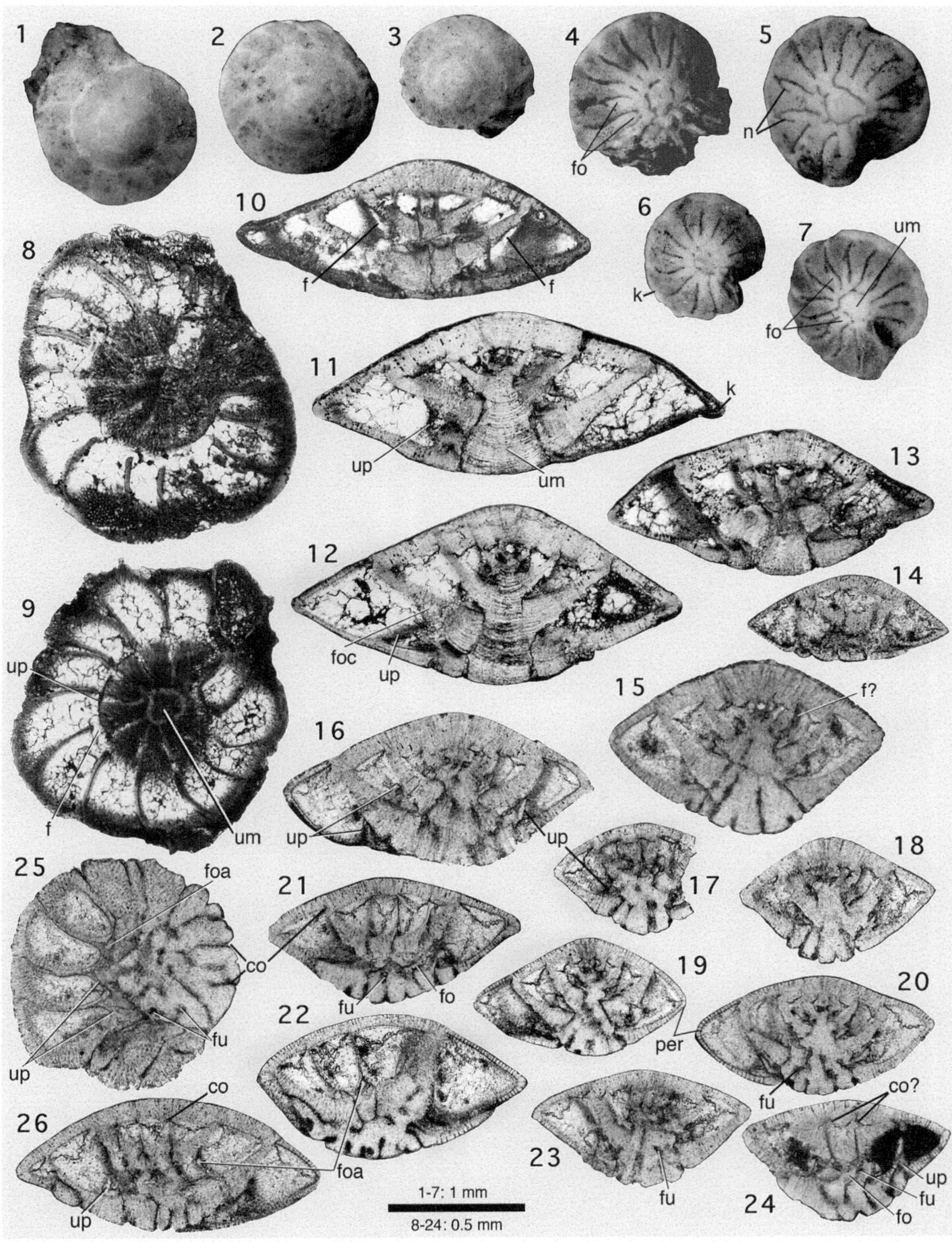

Plate 4.4 (**1–14**) *Kathina aquitanica* n. sp.; isolated specimens from sample 6, Lafarge Quarry, Western Aquitaine, Southern France; Paleocene (SBZ 3). (**1–3**) Dorsal view. (**4–7**) Ventral views; note the simple, slit like openings of the intraseptal interlocular space and the few funnels encircling the undivided, massive umbo. (**8–9**) Sections perpendicular to the coiling axis. (**11–14**) Axial and adaxial sections. (**15–26**) *Slovenites praecursorius* n. sp.; specimens from Cuccuru 'e Flores Conglomerate, Orosei, Sardinia, collected by I. Dieni (see Dieni et al. 1985), associated with *Orduella sphaerica* Sirel, 1998, *Miscellanites globularis* (Rahaghi, 1983) and *Cincoriola* cf. *ovoidea* (Haque, 1958); Paleocene (SBZ 2). (**15**) Centered axial section, syntype. (**16–24**, **26**) Oblique sections with different inclinations in respect to the coiling axis of the shell. (**25**) Section perpendicular to the coiling axis, syntype. Abbreviations: *f* foramen, *up* umbilical plate, *fu* funnel, *foa* foliar aperture, *fo* folium, *per* periphery, *co* canal orifice, *n* notch, *um* umbo, *k* keel, *foc* foliar chamber

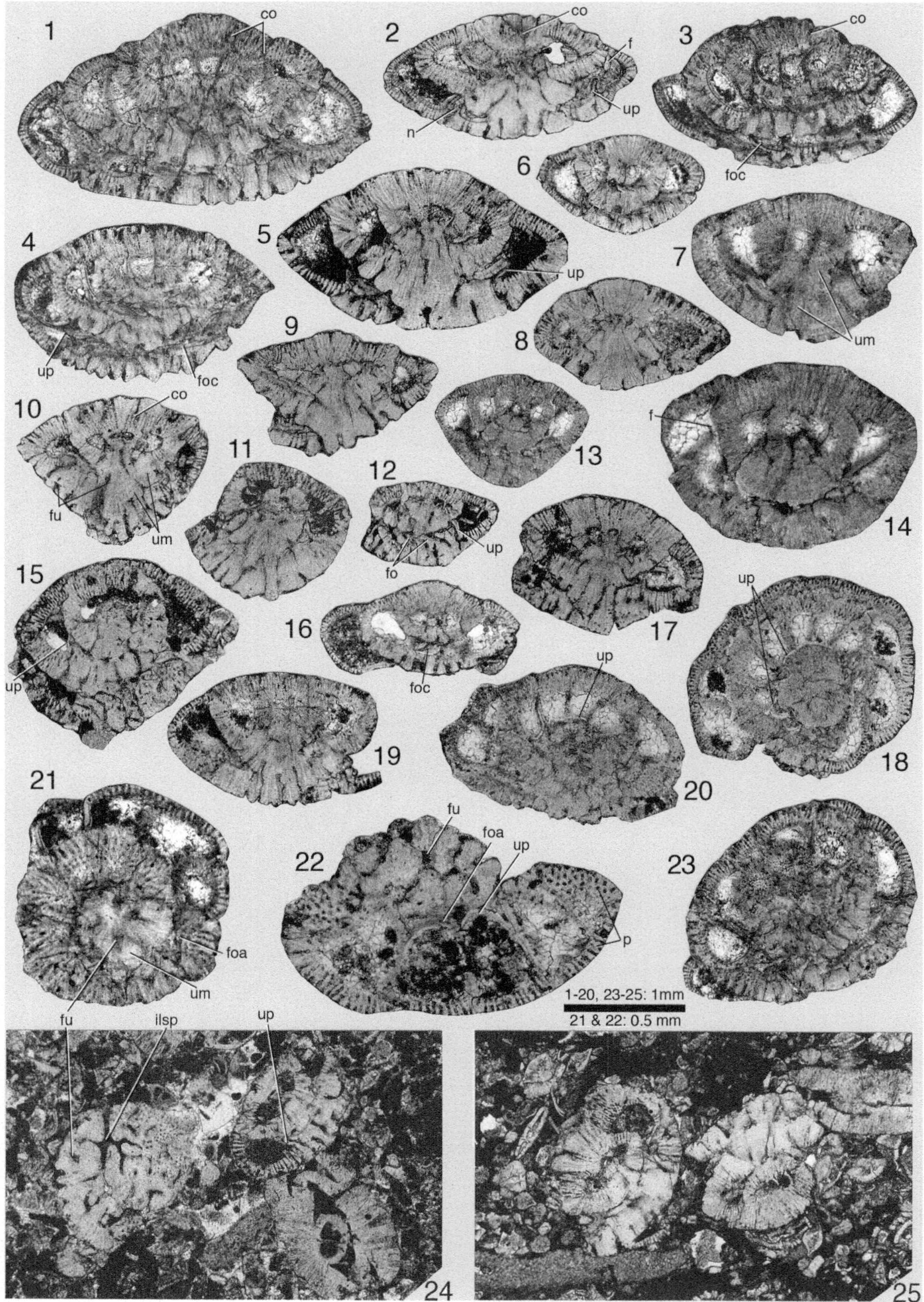

Plate 4.5 (**1–25**) *Slovenites pembaphis* n. sp.; from Sopada, Slovenia, Adriatic platforms, Paleocene (SBZ 3–SBZ 5). (**1–5**) Oblique sections. (**6–10**, **12–13**, **16–17**, **19**, **24–25**) Axial and adaxial sections. (**11**, **20**, **21**, **23**)

4.2 *Slovenites* n. gen.

Type species: *Slovenites pembaphis* n. sp.

Diagnosis: The lenticular shells have a spiral chamber arrangement that is evolute on the dorsal side and involute on the ventral side. The lamellar chamber wall is particularly thick and perforated by very coarse pores. The folia are thick and perforate. They are inclined backwards and fused with their tips to the umbilical filling produced by previous whorls. This umbonal mass is composed of few, thick units separated by furrows in random directions (ilc in Fig. 1. 2C). The foliar lumina are separated from the main chamber lumen by an umbilical plate similar to all Rotaliidae. Comparatively few funnels connect the interlocular space from the junction of the intraseptal with the spiral canals to orifices in the umbilical furrows or at the umbilical surface of the shell. The intraseptal space is reduced to a canal running closely above the rim of the foramen for the entire extension of the interiomargin of the septal face. It opens in an orifice positioned at the proximal junction of whorl and chamber sutures on the dorsal side of the test. The position of this orifice is illustrated in Fig. 1.2C (dsc: dorsal sutural canal orifice).

Remarks: *Slovenites* n. gen. is closest to *Medocia* Parvati, 1971 by its similar, coarse funnels. It differs from *Medocia* by its much coarser perforation, its ornamented dorsal side of the shell with sutural canal orifices, and by its fissures in the umbilical mass where part of the funnels open onto the substrate of the shell. *Rotalia sensu stricto* has no funnels in its umbilical filling, nor has it dorsal sutural orifices of the canal system. However, the foliar characteristics including the foliar apertures in *Slovenites* and *Rotalia* are so similar that we consider this new genus as a member of the Rotaliinae.

Slovenites is a conservative genus with a long range from Middle Paleocene (SBZ 2) to Lutetian (SBZ 13 at least). The earliest species, *S. praecursorius*, from SBZ 2 has been found in Sardinia. *Slovenites pembaphis* is widespread on the Adriatic platform in SBZ 4 and 5. *Slovenites decastroi* is present at several levels of the Monte Gargano Peninsula and of the Adriatic platform with alveolinids from SBZ 11–13.

***Slovenites praecursorius* n. sp.**; Plate 4.4, Figs. 15–26.

Syntypes: Axial section figured in Plate 4.4, Fig. 15 and section almost perpendicular to the coiling axis figured in Plate 4.4, Fig. 25.

Type locality and type level: From sample F 4a, Sardinia, see Dieni et al. (1985); Paleocene (SBZ 2).

Derivation of name: precursor species of a genus discovered first in Slovenia.

Diagnosis: The shells are lenticular, equally convex on both sides, composed of three to four whorls of low-trochospiral chambers. The ratio equatorial to axial diameter oscillates between 1.3 and 2.0. The periphery is angular, lacking any keel and marked by a looser disposition of the pores. The cameral septa are curved backwards on their dorsal side, radial on their ventral side. They are flush on the dorsal side of the shell but marked by a deep interlocular fissure on the ventral side. An adult whorl has about 12 chambers. The folia have thick, perforate walls and fuse their proximal tips with the umbilical mass produced by previous whorls. At the

Plate 4.5 (Continued) Oblique sections inclined for more than 45° in respect to the coiling axis. (**14–15**) Oblique sections inclined for less than 45° in respect to the coiling axis. (**18**, **21**) Section perpendicular to the coiling axis. (**22**, **24**) (*left*) Tangential sections to the ventral surface of the cone exhibiting the intraseptal interlocular space. Abbreviations: *f* foramen, *up* umbilical plate, *um* umbo, *foc* foliar chamber, *foa* foliar aperture, *fu* funnel, *co* canal orifice, *n* notch, *fo* folium, *p* pores, *ilsp* intraseptal interlocular space

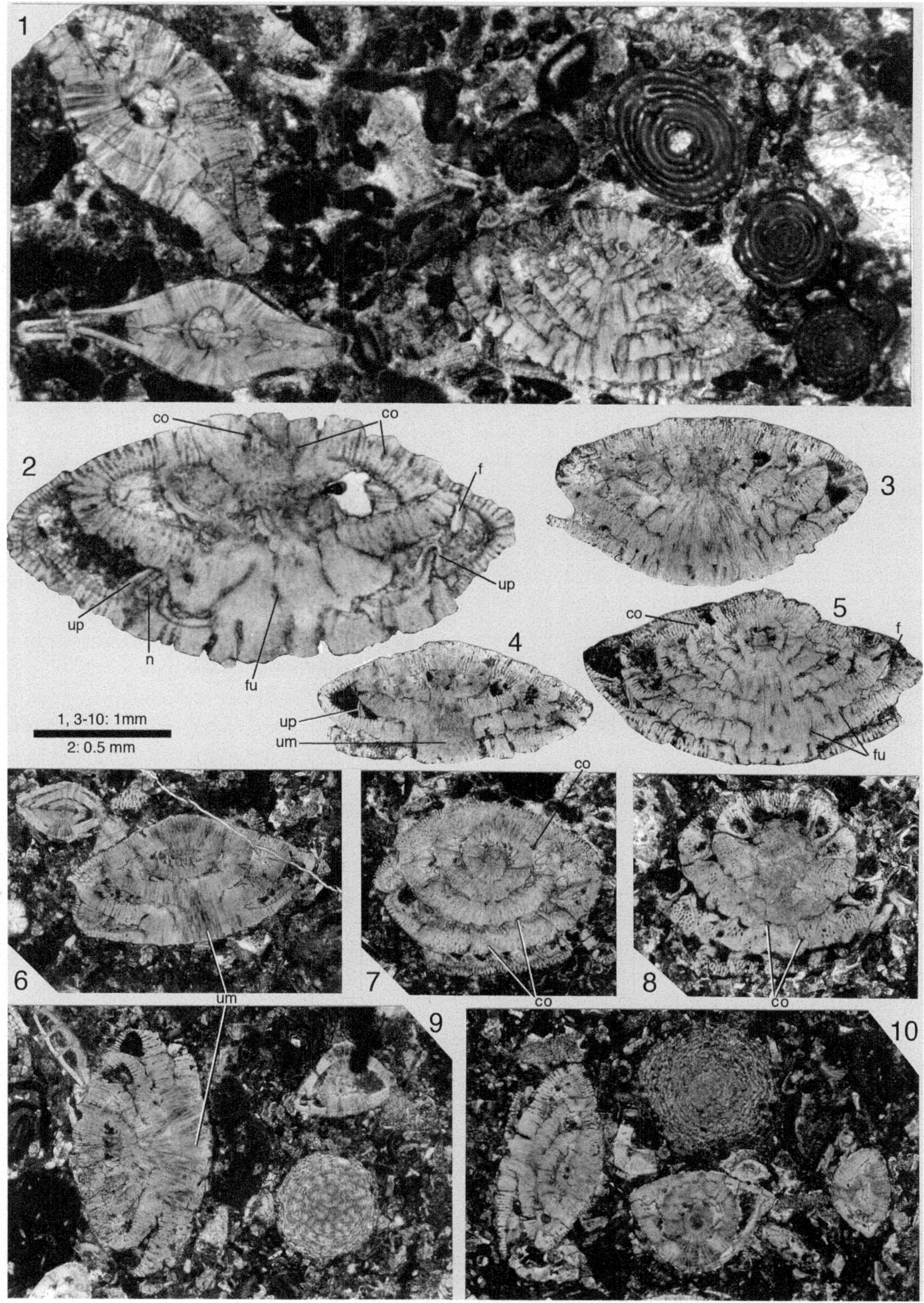

Plate 4.6 (**1–2**) *Slovenites pembaphis* n. sp.; from Sopada, Slovenia, Adriatic platforms, Paleocene (SBZ 3-SBZ 5). (**1**) Holotype, megalospheric specimen axial section. (**2**) Axial uncentered section associated

junction of the intraseptal space with the umbilical interlocular space and in some places in the sutures of neighboring folia, wide open funnels are generated that seem to distribute their orifices at random over the ventral surface of the umbilical fill. The foramen is a very low, slit-like arch in interiomarginal position followed in ventral direction by a small, low umbilical plate.

Remarks: All known specimens of his taxon are from cemented carbonate rock. Their morphology can be defined only in thin sections. The minimum number of sections that permit to support a morphological definition of the taxon is two. These sections must be perpendicular to each other, one of them centered. Therefore, two syntypes are selected from the restricted material at our disposal.

***Slovenites pembaphis* n. sp.**; Plate 4.5, Figs. 1–25; Plate 4.6, Figs. 1–2.

1969 *Lockhartia tipperi* (Davies, 1926)—Butterlin and Monod, p. 586, pl. 1, fig. 7.

1998 ?*Kathina* sp.—Accordi et al., p. 202, pl. 17, fig. 5.

Holotype: Megalospheric specimen figured in Plate 4.6, Fig. 1.

Type locality and type level: Sopada, Slovenia, Adriatic platforms, Paleocene (SBZ 3–SBZ 5).

Derivation of name: *Slovenites*, first discovery in Slovenia; *pembaphis*, Greek allusion to swelling walls and coarse pores like the state of a corpse floating for some time in the water.

Diagnosis: Lenticular shells composed of about four whorls of spiral chambers with thick, coarsely perforate walls. The lens is equally convex on both sides, reaching an equatorial diameter of 2.5 mm. The ratio of equatorial to axial diameter of the lens varies from 1.65 to 1.85. There are 10–12 slightly inflated chambers in an adult whorl. The umbilical architecture is described in the definition of the genus *Slovenites* above. The proloculus is small, reaching 0.08 mm in diameter. No dimorphism has been observed so far.

***Slovenites decastroi* n. sp.**; Plate 4.6, Figs. 3–10; Plate 4.7, Figs. 1–12.

Holotype: Megalospheric specimen figured in Plate 4.6, Fig. 5.

Type locality and type level: Monte Gargano, Italy, samples A 4340, A 4344, collected by P. De Castro; Cuisian (SBZ 11–12) by association with alveolinids.

Derivation of name: in honour of Piero De Castro (Naples) and his work on the Monte Gargano.

Diagnosis: The lenticular shells are composed of five to six whorls of low-trochospiral chambers that are dorsally evolute and ventrally involute, with a wide umbilicus filled by thick foliar walls that are fused to a solid mass. The lens is almost equally convex (index 1.9–2.2); the ventral side however may be flatted in the center of the umbilicus. The lenticular shells reach an equatorial diameter of about 1.8 mm, with 20 dorsally isometric chambers in the adult last whorl. The megalosphere reaches a diameter of 0.12 mm at most. No dimorphism has been observed so far.

Remarks: *Slovenites decastroi* is distinguished from *S. pembaphis* by its larger number of spiral chambers that are arranged in a spiral with more whorls. The differences in the umbilical architecture all are due to the higher

Plate 4.6 (Continued) with *Ranikothalia* sp., *Lacazinella* sp. and alveolinids. (**3–10**) *Slovenites decastroi* n. sp.; Specimens from Bencovac, Croatia, associated with nummulitids and alveolinids, Cuisian (SBZ 11–12). (**3, 5**) Axial uncentered sections. (**4, 6**) Axial sections showing axial umbilical plug with few or without funnels. (**7–8**) Oblique sections inclined for more than 45° in respect to the coiling axis. (**9**) Axial slightly off center. (**10**) Oblique section inclined for about 45° in respect to the coiling axis. Abbreviations: *f* foramen, *up* umbilical plate, *fu* funnel, *co* canal orifice, *n* notch

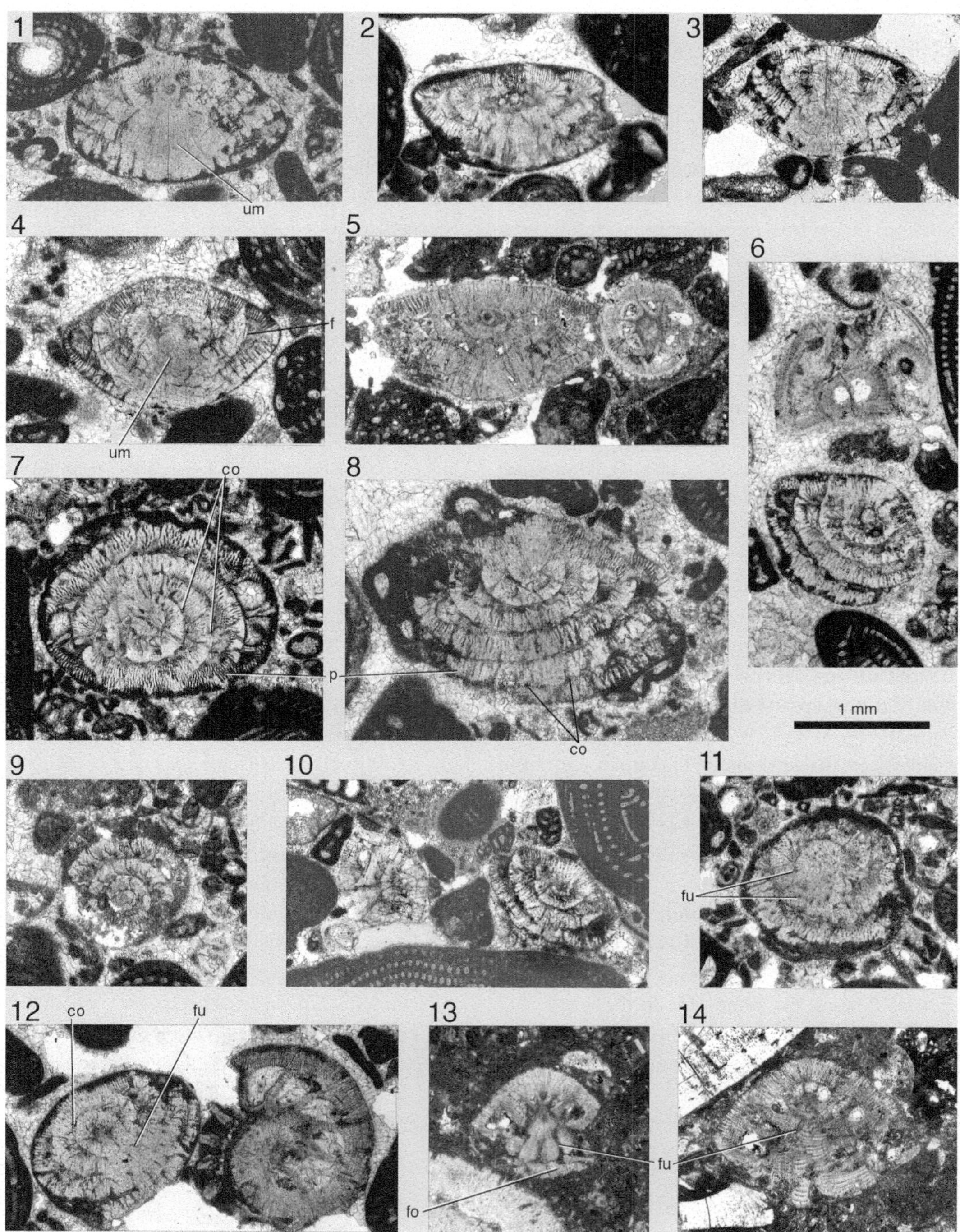

Plate 4.7 (**1–12**) *Slovenites decastroi* n. sp.; all specimens from Monte Gargano (Italy), samples A 4340-A 4344, collected by Piero De Castro; Cuisian (SBZ 11–11) by association with alveolinids. (**1**) Axial section. (**2–5**) Oblique sections with inclinations by far less than 45° in respect to the coiling axis. (**6–8**) Oblique sections inclined for much more than 45°, associated in 6 (*top*) to (G. m.) *Gyroidinella magna* Le Calvez, 1949. (**9**) Centered section perpendicular to the coiling axis. (**10–12**) Oblique sections with various inclinations associated in 10 with Cuisian alveolinids and in 12 with *Gyroidinella* sp. (**13–14**) *Medocia* sp. (**13**) Oblique section inclined at the same angle as the cone mantel defining the umbilical fill with fused foliar tips (compare with Hottinger 2007, pl. 12, fig. 4 and pl. 13, fig. 4). (**14**) Axial section. Abbreviations: *f* foramen, *um* umbo, *fu* funnel, *co* canal orifice, *p* pores, *fo* folium

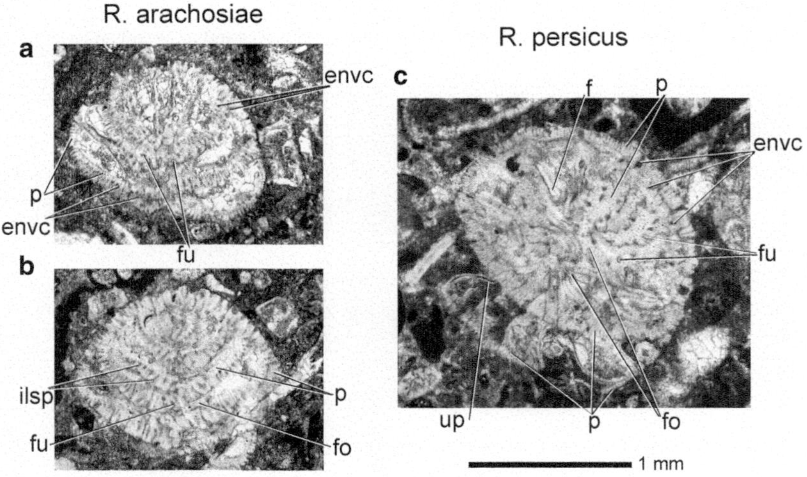

Fig. 4.2 *Rotaliconus arachosiae* n. sp. (**A**) and (**B**) compared to *R. persicus* Hottinger, 2007 (**C**) in oblique sections with similar inclination in respect to the shell's coiling axis. Note the much denser primary subdivision of the intraseptal interlocular space (envc) in *R. arachosiae* that gives access to a ventral enveloping canal system whereas in *R. persicus* the much coarser primary subdivision of the intraseptal interlocular space produces only a corresponding row of funnels. (**A**) and (**B**) from Zhob valley, Baluchistan, Pakistan, Paleocene (SBZ 4); (**C**) from Shiraz, Iran, SBZ 17 or 18, late Bartonian (Hottinger 2007). Abbreviations: *f* foramen, *p* pores, *fu* funnels, *fo* folium, *up* umbilical plate, *ilsp* intraseptal interlocular space, *envc* enveloping canals

number of whorls and spiral chambers: the number of funnels increases with the number of septa in the adult whorls while the very first, small sized nepionic whorls do not develop any funnels and keep therefore the center of the umbilicus free of these structural elements.

4.3 *Rotaliconus* Hottinger, 2007

Type species: *Rotaliconus persicus* Hottinger, 2007

***Rotaliconus arachosiae* n. sp.**; Fig. 4.2A–B; Plate 4.8, Figs. 1–11, 14–15.

Syntypes: Specimen figured in Plate 4.8, Figs. 1, 4. *Rotaliconus arachosiae* is known only from cemented rock of SBZ 4 (Late Paleocene) in Pakistan. Since there are only thin-sections available, syntypes have to be designated to fix the main diagnostic morphological elements.

Type locality and type level: AI177232 and 77233, Zhob valley, Baluchistan, western Pakistan; Late Paleocene (SBZ 4), dated by association with "*Orbitolina*" *daviesi* Hofker jun., 1966.

Derivation of name: Arachosia, classical name of the country West of the Indus in times of Persian dominance (500 a. s.), corresponding today to the western Pakistani border provinces.

Diagnosis: High-conical shells with a subspherical outline produced by a rounded apex and a round periphery. The cone base is slightly flattened. The chamber walls are thick and coarsely perforate. In sections perpendicular to the wall surface, they seem to be ornate by numerous pustules, each carrying many pores. The "pustules" are intersections of the perforate shell between the enveloping canals covering the free chamber walls. The foramen seems to correspond to a very narrow slit (Plate 4.8, Fig. 1) in interiomarginal position placed on the dorsal angle of the septal face. The intraseptal interlocular space is widely open and undivided. The enveloping canals appear only below the surface

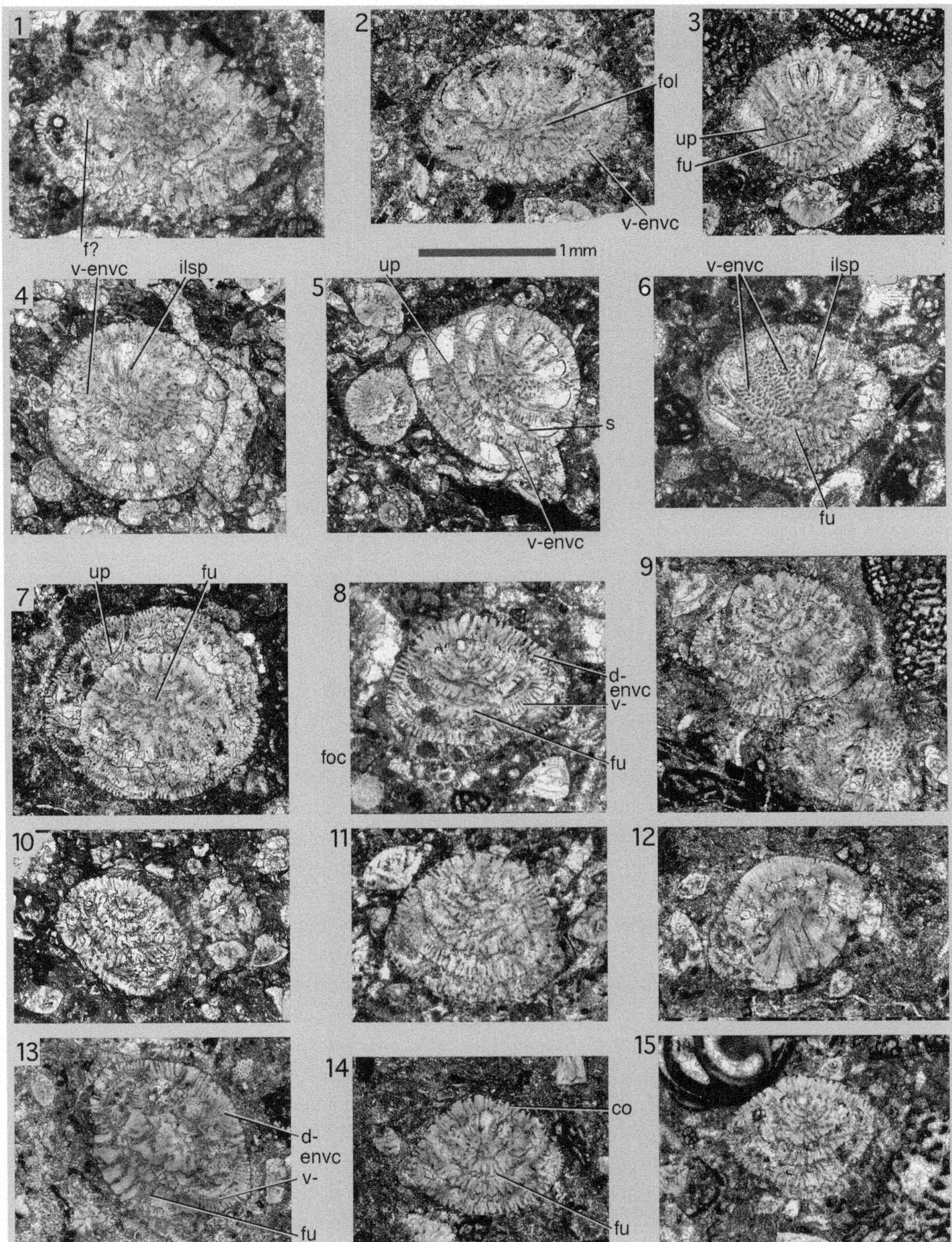

Plate 4.8 (**1–11**, **14–15**) *Rotaliconus arachosiae* n. sp. (**1**) Oblique section inclined for about 20° in respect to the shell's coiling axis, syntype; note the ornamental dorsal pustules with their regular, dense perforation. (**2**) Oblique section inclined for about 35°. Note in the ventral part of the section the minute imperforate folia. (**3**) Oblique section inclined for about 35°; note the umbilical plate and the funnels communicating with each other by a network of fissures. (**4**) Syntype, section perpendicular to the coiling axis showing the ventral intraseptal interlocular space subdivided in a row of canals that give rise to an enveloping canal system invading all the

of the chambers, whereas the interlocular space is bridged by numerous ponticuli. The folia are radial, short and loosely fuse at their tips to an umbilical mass that admits some funnels placed over the sutures of neighboring folia. Over the septal sutures of the main chambers the interlocular space is subdivided and transformed into a ventral enveloping canal system that is denser than on the dorsal side (Plate 4.8, Fig. 6).

The equatorial to axial diameter ratio is 1.0–1.5. There are 16–18 spiral chambers in the last whorl. The nepiont exhibits 9–10 chambers per whorl following a proloculus of 0.08 mm in diameter. No dimorphism of generations has been observed.

Remarks: *Rotaliconus persicus* Hottinger, 2007 from the late Lutetian-Bartonian of Iran has an outline of the shell and a habit of the umbilical architecture that is similar to the ones observed in *Rotaliconus arachosiae* from the Late Paleocene. *R. persicus* (Fig. 4.2C) is distinguished from *R. arachosiae* by the absence of a ventral canal system that envelops the main spiral chamber walls. There is only a single row of orifices for communication of the interlocular space in the septa with the ambient environment in front of the shell's face.

4.4 *Pachyrotalia* n. gen.

Type species: *Pachyrotalia massa* n. sp.

Remarks: The test is coarsely perforated and has very thick walls. The trochospiral arrangement of the chambers throughout does not impede the subglobular shape of the shell.

***Pachyrotalia massa* n. sp.**; Plate 4.8, Figs. 12–13; Plate 4.9, Figs. 1–15.

1998 unidentified rotaliid—Accordi et al., p. 182, pl. 7, fig. c.

Syntypes: Specimens illustrated in Plate 4.9, Figs. 1–2 (top left), 4.

Type locality and type level: Sample All77232 collected by F. Allemann; Zhob valley, Baluchistan, western Pakistan; Paleocene (SBZ 4).

Derivation of name: *Pachyrotalia* for thick (walled) *Rotalia*, massa for heavy, massive umbilical plugs.

Diagnosis: Trochospiral shells with a subglobular outline (equatorial to axial diameter ratio is 1.1–1.3). The walls are very thick and perforate. The outer shell surface exposed to the ambient environment exhibits a loose meshwork of enveloping canals. Their radial parts interrupt the much finer, regular perforation of the walls. The periphery is rounded or slightly angular, always without any keeled marks. The narrow umbilicus is filed with a very massive, composed umbilical plug composed of several piles. Their separate origin is marked by funnels or fissures. The umbilical architecture is limited to a narrow space around the axial umbilical plug showing minute and thin folia covering a foliar chamberlet by small umbilical plates. The intraseptal interlocular space is comparatively wide and undivided, giving rise to the enveloping canal system on the abaxial, ventral as well as on the

Plate 4.8 (Continued) free surfaces of the shell and in particular also the base of the cone, the shell's face. (**5**) Oblique section inclined for about 45°, associated with *Kathina selveri* Smout, 1954 (*left*). (**6**) Syntype, section tangential to the ventral surface of the shell cone exhibiting the intraseptal interlocular space transformed into a ventral enveloping network of canals. (**7**) Section perpendicular to the coiling axis; note the oblique pathway of the umbilical plate and the network of communications between neighboring funnels. (**8**) Nearly axial, centered section showing the dense enveloping canal system on both sides of the shell; note the size of the proloculus. (**9**) Oblique section with an inclination of about 35°, associated with "*Orbitolina*" *douvillei* J. Hofker jun., 1966. (**10**) Oblique section inclined for about 60°. (**11**) Oblique section inclined for about 30°. (**14–15**) Centered oblique sections inclined for about 45°. (**12–13**) *Medocia* n. sp. (**12–13**) adaxial and oblique sections. All specimens on this plate from the samples All77232 and All77233, Zhob Valley, Baluchistan, western Pakistan, collected by F. Allemann; Paleocene (SBZ 4). Abbreviations: *f* foramen, *v-envc* ventral enveloping canals, *d-envc* dorsal enveloping canals, *up* umbilical plate, *ilsp* intraseptal interlocular space, *fu*, funnel, *s* septum, *co* canal orifice, *fol* folia, *foc* foliar chamber

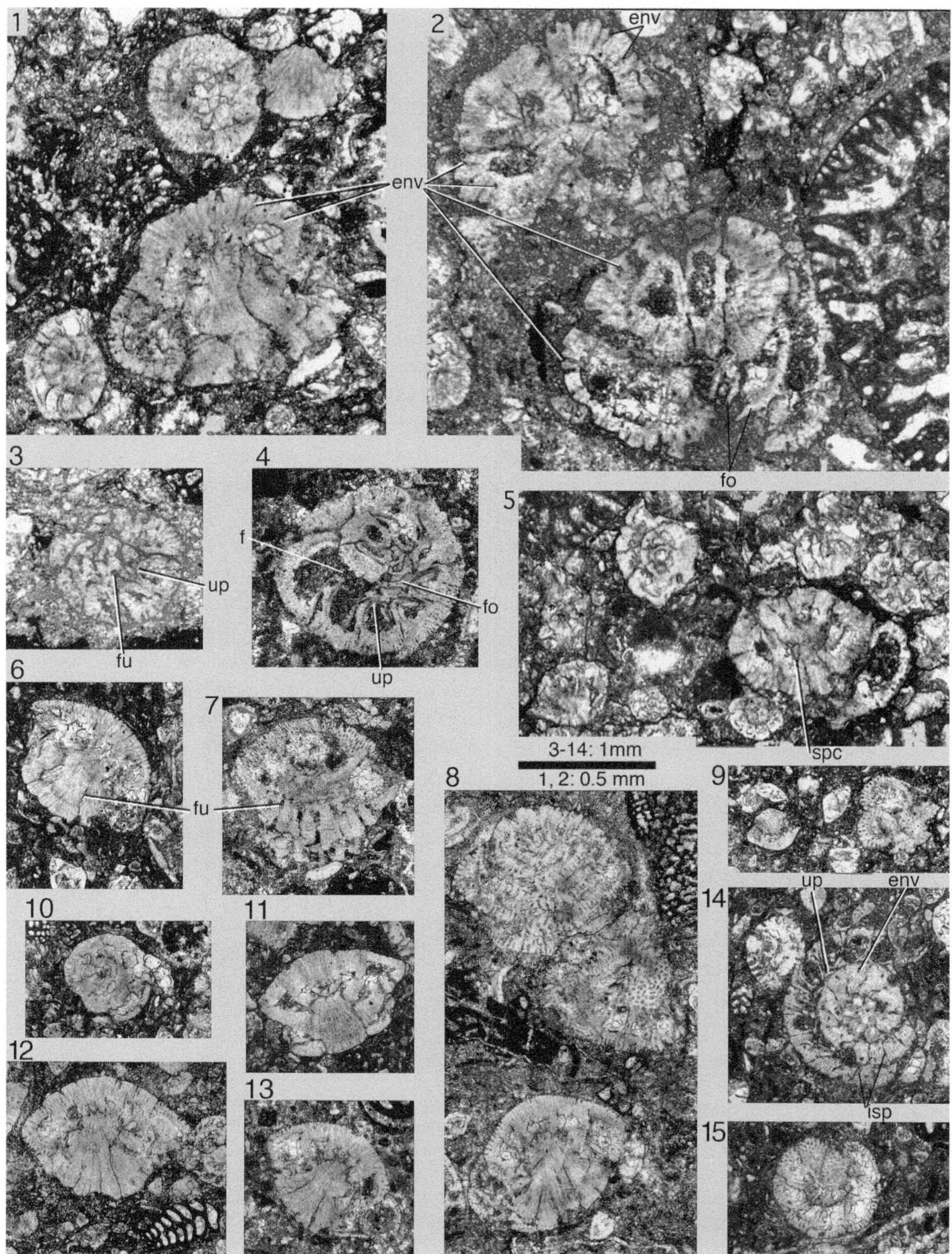

Plate 4.9 *Pachyrotalia massa* n. sp.; sample All77232, collected by F. Allemann; western Pakistan, Paleocene (SBZ 4). (**1**) Syntype, axial section slightly off center showing the massive umbilical plug. (**2**) Section parallel to the shell's coiling axis (*upper left*, syntype) and oblique section (*bottom center*); both sections are out of reach from the adaxial umbilical plug; note the enveloping canal system and the tiny foliar walls; associated with "*Orbitolina*" *daviesi* J. Hofker jun., 1966 (*right*). (**3**, **5**, **9–10**) Oblique sections. (**4**, **14–15**) Sections perpendicular to coiling axis. (**6–8**, **11–13**) Axial sections. Abbreviations: *f* foramen, *env* enveloping canals, *up* umbilical plate, *fu* funnel, *spc* spiral canal, *fo* folium, *isp* intraseptal space

dorsal side of the shell. A single centered section (Plate 4.9, Fig. 4) reveals a megalosphere of 0.08 mm. At an early equatorial diameter of about 0.8 mm, the shell has 8 chambers per whorl.

References

Accordi G, Carbone F, Pignatti J (1998) Depositional history of a Paleogene ramp (Western Cephalonia, Ionian islands, Greece). Geol Romana 34:131–205

Butterlin J, Monod O (1969) Biostratigraphie (Paléocène à Eocène moyen) d'une coupe dans le Taurus de Beysehir (Turquie). Etude des Nummulites cordelées et révision de ce groupe. Eclogae geol Helv 62(2): 583–604

Davies LM (1926) Remarks on Carter's genus *Conulites–Dictyoconoides* Nuttall, with description of some new species from the Eocene of North West India. Rec Geol Surv India Calcutta 59(2):237–257, 16–20 pls

Dieni I, Massari F, Radoicic R (1985) Marine Paleocene pebbles included in the Cuccuru'e Flores Conglomerate (Post-Cuisian) of Orosei (Sardinia). In: 19th European micropaleontological colloquium excursion guide. AGIP Mineraria, San Donato Milanese, Milano, pp 213–214

Ferrer J, Le Calvez Y, Luterbacher H-P, Premoli-Silva I (1973) Contribution à l'étude des foraminifères Ilerdiens de la region de Tremp (Catalogne). Mém Mus Nat Hist Nat Paris, Sér C 29:3–107

Haque A (1958) *Cincoriola*, a new generic name for *Punjabia* Haque, 1956, Contr Cushman Lab Foram Res 9(4):103

Hasson PF (1985) New observations on the biostratigraphy of the Saudi Arabian Umm er Radhuma Formation (Paleogene) and its correlation with neighboring regions. Micropaleontology 31(4): 335–364

Hottinger L (2007) Revision of the foraminiferal genus *Globoreticulina* Rahaghi, 1978 and of its associated fauna of larger foraminifera from the late Middle Eocene of Iran. Carnets Géol Article 2007/06, CG2007_A06

Parvati S (1971) A study of some rotaliid Foraminifera. Proc Kon Ned Akad Wetensch Ser B74(1):1–26, 4 pls

Peybernès B, Fondecave-Wallez M-J, Hottinger L, Eichène P, Segonzac G (2000) Limite Crétacé-Tertiaire et biozonation micropaléontologique du Danien-Sélandien dans le Béarn occidental et la Haute-Soule (Pyrénées Atlantiques). Geobios 33: 35–48

Rahaghi A (1983) Stratigraphy and faunal assemblage of Paleocene-Lower Eocene in Iran. Nat Iranian Oil Comp Geol Lab 10, 73 pp, 49 pls

Robador A, Samsó JM, Serra-Kiel J, Tosquella J (1991) Field Guide. In: Introduction to the Early Paleogene of the south Pyrenean Basin. Early Paleogene Benthos 1st meeting IGCP nº 286, Jaca 1990, 131–159 pp

Sirel E (1998) Foraminiferal description and biostratigraphy of the Paleocene-Eocene shallow water limestones and discussion of the Cretaceous-Tertiary boundary in Turkey. Gen Dir Min Res Expl (MTA), Monograph ser 2, 117 pp, 68 pls

Smout AH (1954) Lower Tertiary foraminifera of the Qatar peninsula. Brit Mus (Nat Hist), 96 pp, 44 figs, 15 pls

5 New Subfamily Lockhartiinae

Abstract
The new subfamily Lockhartiinae presents a peculiar umbilical structure characterized by umbilical cavities that are delimited by successive foliar walls and numerous parallel umbilical piles. The shells are always low-trochospiral, with an equatorial to axial diameter ratio always above 1. The dorsal side of the shell bears heavy ornaments from more or less thickened limbate spiral and cameral sutures to ribbed or cancellate ornaments. Four genera (*Rotalispira* n. gen., *Lockhartia*, *Dictyoconoides*, *Sakesaria*) and 19 species (*R. scarsellai*, *R. pyrenaica* n. sp., *L. praehaimei*, *L. haimei*, *L. retiata*, *L. diversa*, *L. roeae*, *L. conditi*, *L. hunti*, *L. tipperi*, *D. flemingi*, *D. kohaticus*, *D. cooki*, *S. cotteri*, *S. costulata*, *S. cylindrata*, *S. pyrum*, *S. somalica*, *S. trichilata*) are described and illustrated.

Rotalispira n. gen. is a Cretaceous to Paleocene group of smaller benthics with a strong, thickened spiral suture on the dorsal side of the shell. The umbilicus is filled with large, oblique folia that fuse to a few axial piles standing in a narrow group around the shell axis. We tentatively consider this genus as the phylogenetic root of the larger lockhartiines appearing in the Paleocene. These present shells with a bilamellar-perforate primary chamber wall, coarse keels and raised sutures. The generations are dimorphic. There may be multiple spirals in advanced genera that are restricted or not to the agamontic generation. Neither supplemental skeleton nor apertures are observed on the always evolute dorsal side of the shell. The septa have radially short interlocular spaces. The umbilical plate ("umbilical flap" in Müller-Merz 1980) is large, heavy and soldered to the chamber wall of the previous whorl in oblique direction in respect to the axial median plane. The foramina consist of a low arch in interiomarginal position.

There is a folium ("lip" in Müller-Merz 1980) for each chamber that covers almost the entire umbilical area. It bears pustules that grow to piles in successive chambers by blueprinting the pustule axes in subsequent folia. The umbilical piles in *Lockhartia* and consorts produce a pattern that is very similar to the pillared architecture of the agglutinated-conical shells of *Dictyoconus* or *Fallotella* (Vecchio and Hottinger 2007). The folia overlap each other in distal and in proximal direction with their peripheral margins. Between the piles layers of low umbilical spaces remain open and form a kind of umbilical "chamberlets" by subdivision of the umbilical interlocular

L. Hottinger, *Paleogene larger rotaliid foraminifera from the western and central Neotethys*,
DOI 10.1007/978-3-319-02853-8_5,

space. These "chamberlets" communicate between successive layers with vertical funnels and within a layer by horizontal passages. This type of umbilical architecture, produced by agglutinated and lamellar shells in different ways, creates in the umbilicus a single umbilical space. This space is supported by pillars but it is not subdivided into separate compartments with communication restricted to foramina. Morphological features as this, in conical shells of considerable size, might have analogous, autecological significance.

On the surface of the umbilical area of the shell, the pile heads emerge and form a relief. Their number in relation to the diameter of the umbilical area is a useful diagnostic feature on the species level. The folia have large slits as apertures that are positioned almost in the axial median plane of the chamber. They supplement the primary apertures of the shell by a circle of orifices along the circumference of the umbilical area.

5.1 *Rotalispira* n. gen.

Type species: *Rotorbinella scarsellai* Torre, 1966

Remarks: This new genus is proposed here after the revision of *Rotorbinella* by Revets (2001) in order to stress the structural differences on the generic level: *Rotalispira* has an umbilical structure dominated by the extension of the foliar wall over at least half of the shell face. The foliar suture (between main chamber and foliar chamberlet) is thickened by secondary lamellation to form solid projections that are twisted and obliquely superposed like thick propeller blades. The thickened sutures on the umbilical side of the shell seem to match the thickened spiral suture on the dorsal side. The generic name reflects the dorsal aspect of the shell, dominated by the limbate spiral suture. Occasionally, the thickened foliar walls of the previous whorl support the folia of the next whorl. The main chamber lumina are separated from the foliar chamberlet lumina by an umbilical plate. There is a direct communication between subsequent foliar lumina forming a kind of spiral canal. The interlocular space is restricted to the adaxial umbilical part of the septum. There are no canal orifices on the dorsal side of the test.

Rotalispira appears together with true *Rotorbinella* in the Late Cretaceous and might therefore represent a root of the lockhartiines that is separate from the rotaliines.

Rotalispira scarsellai (Torre, 1966); Plate 5.1, Figs. 1–22.

1966 *Rotorbinella scarsellai*—Torre, p. 422, pl. 1, figs. 1–8; pl. 2, fig. 10.

1998 ? *Rotorbinella scarsellai* Torre—Accordi et al., p. 186, pl. 9, figs. 9–10.

Description: Small shells with an equatorial diameter around half a millimeter. The dorsal part of the shell is convex and evolute, bears heavily limbate whorl sutures and slightly limbate or smooth chamber sutures. The latter are curved and bent backwards. The periphery is keeled in accordance with the limbate nature of the whorl suture. The equatorial to axial diameter ratio varies from about 1.5 to 1.75. There are 6 chambers in early, nepionic stages and 8–10 chambers in the last whorl. The only proloculus that is measurable has a diameter of 0.04 mm. The ventral side of the shell is flattened and presents a relatively narrow umbilicus, covered but not closed by long folia. Their thickened adaxial tips fuse to form an annular rim around the axis. There is no umbilical pile in the coiling axis of the shell.

***Rotalispira pyrenaica* n. sp.**; Plate 5.2, Figs. 1–29.

Holotype: Specimen figured on Plate 5.2, Fig. 7.

Type locality and type level: level 9 of the Lassalle Quarry, near Pau, western Aquitaine, southern France; Paleocene (SBZ 3).

Derivation of name: *pyrenaica*, first discovery in north Pyrenean basin.

Diagnosis: Shells with similar size, proportions and dorsal ornamentation by limbate whorl sutures as the generotye *R. scarsellai*. In the axial section, the shell exhibits, however, a higher dorsal and a lower ventral convexity in

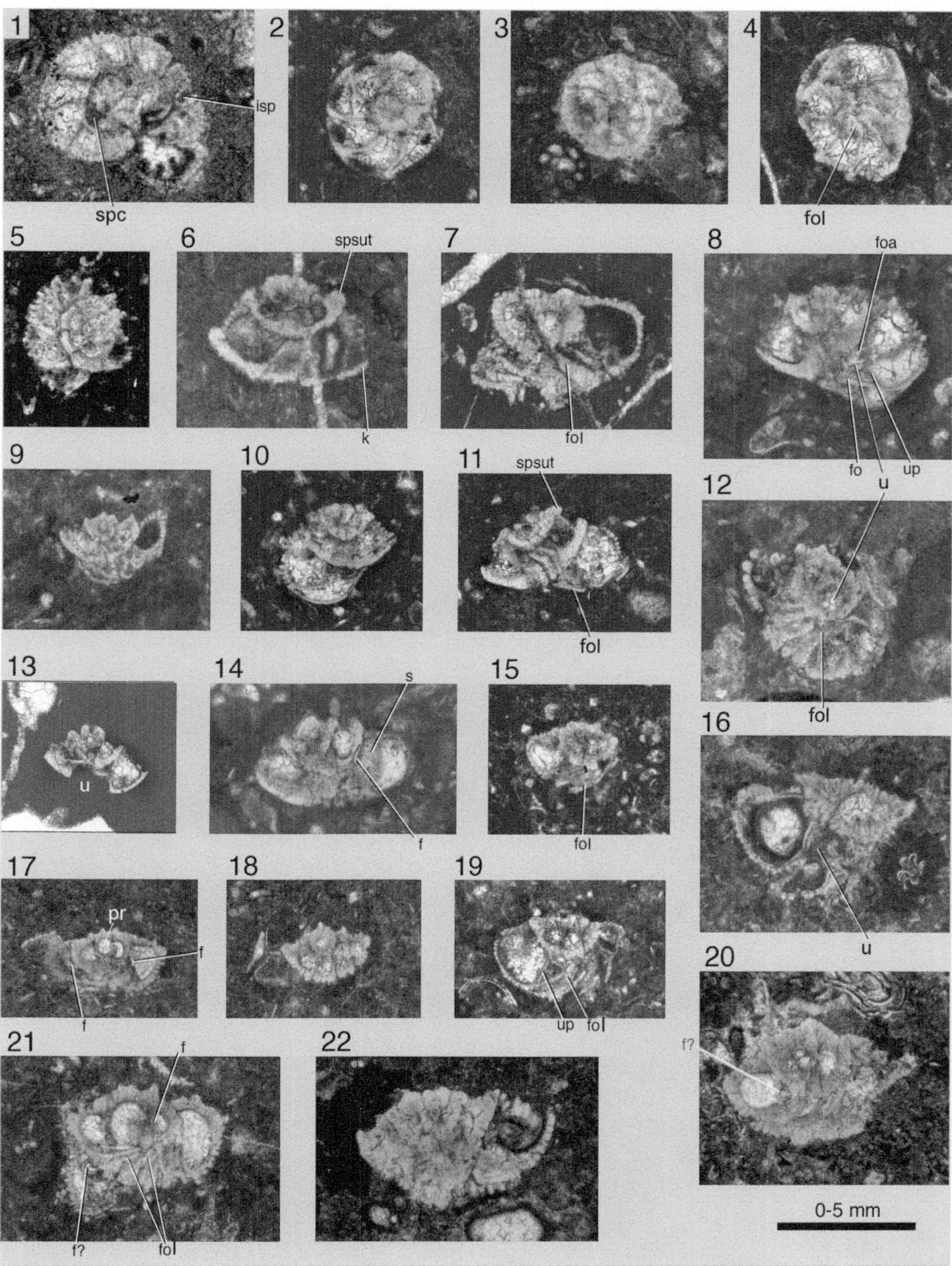

Plate 5.1 *Rotalispira scarsellai* (Torre, 1966); topotypes from the Sorrento peninsula (see Torre 1966), south of Naples, Italy; Late Cretaceous dated by association with *Accordiella conica*. (**1–5**) Sections more or less perpendicular to coiling axis. (**6**) Section tangential to the dorsal surface of the shell showing the backward inclination of the chamber sutures. (**7–22**) Oblique sections with different low and large angles in respect to the coiling axis of the shell. (**17**) Centered section showing the proloculus. (**22**) Section parallel and close to the axial plane. Abbreviations: *f* foramen, *up* umbilical plate, *spc* spiral canal, *fol* folia, *isp* intraseptal space, *u* umbiliculus, *spsut* spiral suture, *pr* proloculus, *s* septum, *k* keel, *fo* folium, *foa* foliar aperture

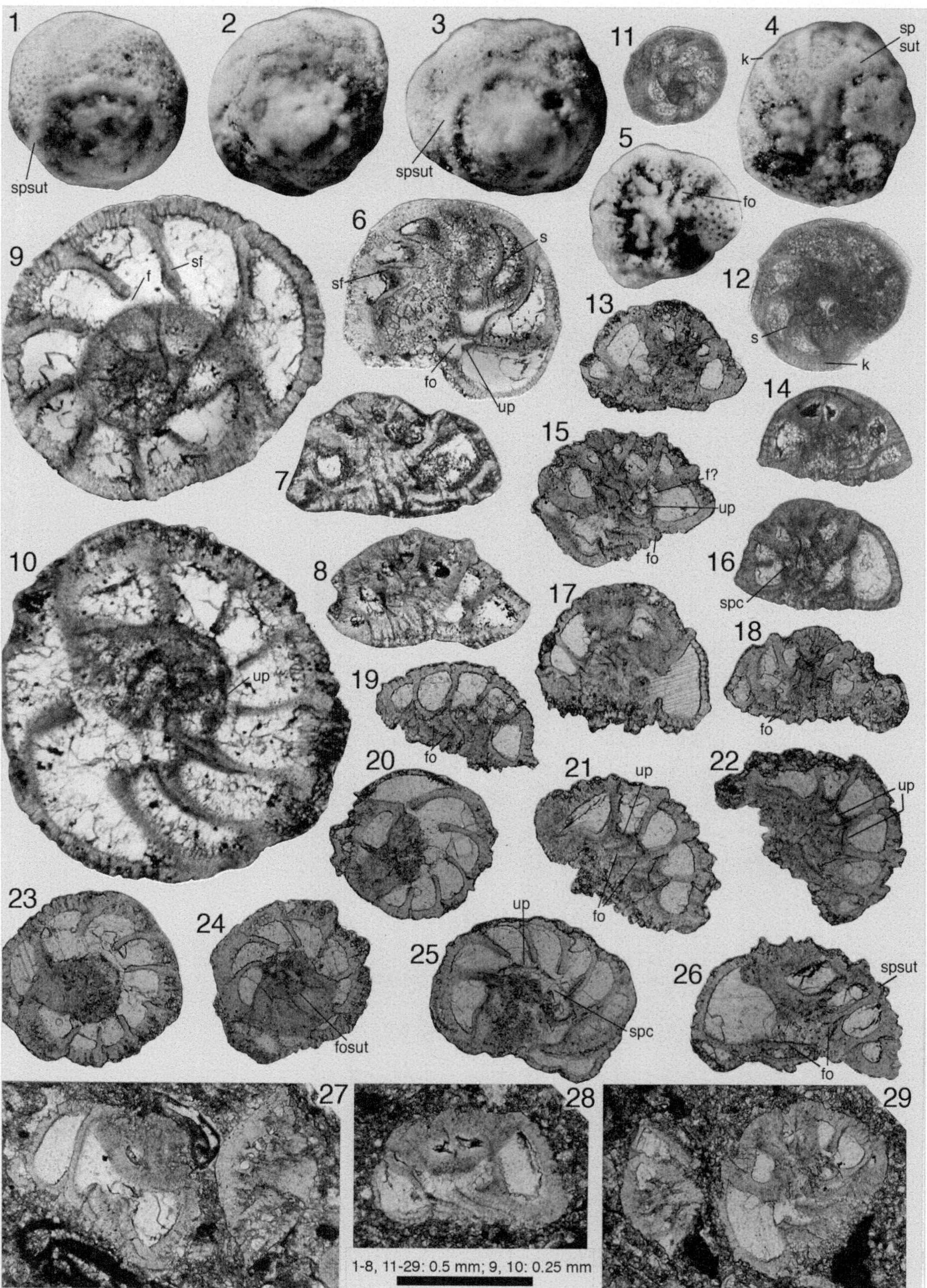

Plate 5.2 *Rotalispira pyrenaica* n. sp.; specimens from the level 9 of the Lassalle Quarry, near Pau, western Aquitaine, southern France; Paleocene (SBZ 3). (**1–3**) External dorsal view. (**4–5**) External ventral view. (**6**, **9–12**, **20**, **23–25**) Sections more or less perpendicular to coiling axis. (**7–8**) Axial sections. (**13–16**, **18**, **27–28**) Oblique sections with different low and large angles in respect to the coiling axis of the shell. (**17**, **19**, **21–22**, **26**, **29**) Section tangential to the dorsal surface of the shell. Abbreviations: *f* foramen, *s* septum, *sf* septal face, *spsut* spiral suture, *up* umbilical plate, *k* keel, *spc* spiral canal, *fo* folium, *fosut* foliar suture

comparison with the Cretaceous species but the ratio of the equatorial to axial diameter remains roughly the same. There are 9 chambers in the last whorl. The only proloculus that is measurable in our material has a diameter of 0.04 mm.

Remarks: The umbilicus is not much wider but the large folia are supported by few piles growing on the foliar adaxial tips and forming by blueprinting few, slender umbilical piles (Plate 5.2, Fig. 7). We see these piles as an argument to consider *Rotalispira* as a genus from which *Lockhartia* may have arisen.

5.2 *Lockhartia* Davies, 1932

Type species: *Dictyoconoides haimei* Davies, 1927

Remarks: A genus with the structural features of the Lockhartiinae. The shells are always low-trochospiral, with an equatorial to axial diameter ratio always above 1. Both generations exhibit a single spiral of chambers during growth. The dorsal side of the shell bears heavy ornaments from more or less thickened limbate spiral and cameral sutures to ribbed or cancellatae ornaments. The umbilicus is filled with a specifically variable number of piles of lamellae supporting the foliar extensions over the shell's face.

Lockhartia praehaimei Smout, 1954; Plate 5.3, Figs. 3–17; Plate 5.4, Figs. 1–17; Plate 5.10, Figs. 16 top–18.

1954 *Lockhartia praehaimei* Smout, p. 51, pl. 2, figs. 21–22; pl. 7, fig. 14.

1993 *Lockhartia* aff. *conditi* (Nuttall)—Weiss, p. 250, pl. 1, fig. 4.

Diagnosis: Topotypes from Smout's collection (Plate 5.4, Figs. 2–5) have lenticular shells with equal convexity of the dorsal and umbilical shell surface. The periphery is sharp and keeled. The ratio of equatorial to axial diameter varies from 1.8 to 2.2. There are 16 chambers in the last whorl of the adult. On the ventral face, an axial section cuts 8 piles per 1 mm radial distance. The ornamentation of the shell's dorsal surface is heavy, dominated by beads marking the spiral sutures. The diameter of the megalosphere does not surpass 0.15 mm.

Remarks: A similar form was found in the marls of the Hangu Formation on the Dahk Pass road crossing the Salt Range; Paleocene (SBZ 3).

Lockhartia haimei (Davies, 1927); Plate 5.3; Figs. 1–2; Plate 5.5, Figs. 1–12; Plate 5.6, Figs. 1–17; Plate 5.7, Figs. 1–12; Plate 5.8, Figs. 1–14.

1932 *Lockhartia haimei* (Davies)—Davies, p. 407, pl. 2, figs. 4–6.

1937 *Lockhartia haimei* (Davies)—Davies and Pinfold, p. 45, pl. 7, figs. 9–13, 15.

1954 *Lockhartia haimei* (Davies)—Smout, p. 49, pl. 2, figs. 1–14.

1956 *Lockhartia haimei* (Davies)—Smout and Haque, p. 57, pl. 11, figs. 3a–b.

1962 *Lockhartia haimei* (Davies)—Sander, p. 19, pl. 5, figs. 1–37.

1976 *Lockhartia haimei* (Davies) pars—Ho et al., p. 50, pl. 27, figs. 1–3, 9–10; non pl. 27, figs. 5–8.

1980 *Lockhartia haimei* (Davies)—Müller-Merz, p. 35, text-figs. 18–19; pl. 1, figs 2–5; pl. 9, figs, 5–6; pl. 10, fig. 7; pl. 15, fig. 2.

1985 *Lockhartia haimei* (Davies)—Hasson, p. 360, pl. 4, fig. 12.

1991 *Lockhartia haimei* (Davies)—Wan, p. 13, pl. 2, figs. 1–2.

1993 *Lockhartia conditi* (Nuttall)—Weiss, p. 250, pl. 5, fig. 3.

1993 *Lockhartia haimei* (Davies) transitional to *Lockhartia altispira* Smout—Weiss, p. 252, pl. 7, fig. 6.

1999a *Lockhartia haimei* (Davies)—Akhtar and Butt, p. 131, pl. 2, figs. 3–5.

1999b *Lockhartia haimei* (Davies)—Akhtar and Butt, p. 188, pl. 2, fig. 2.

2006 *Lockhartia haimei* (Davies)—Hottinger, p. 85, pl. 2, fig. 1.

Description: *Lockhartia haimei* has a conical shell with a more or less pointed apex, an angular periphery with or without keel and a slightly

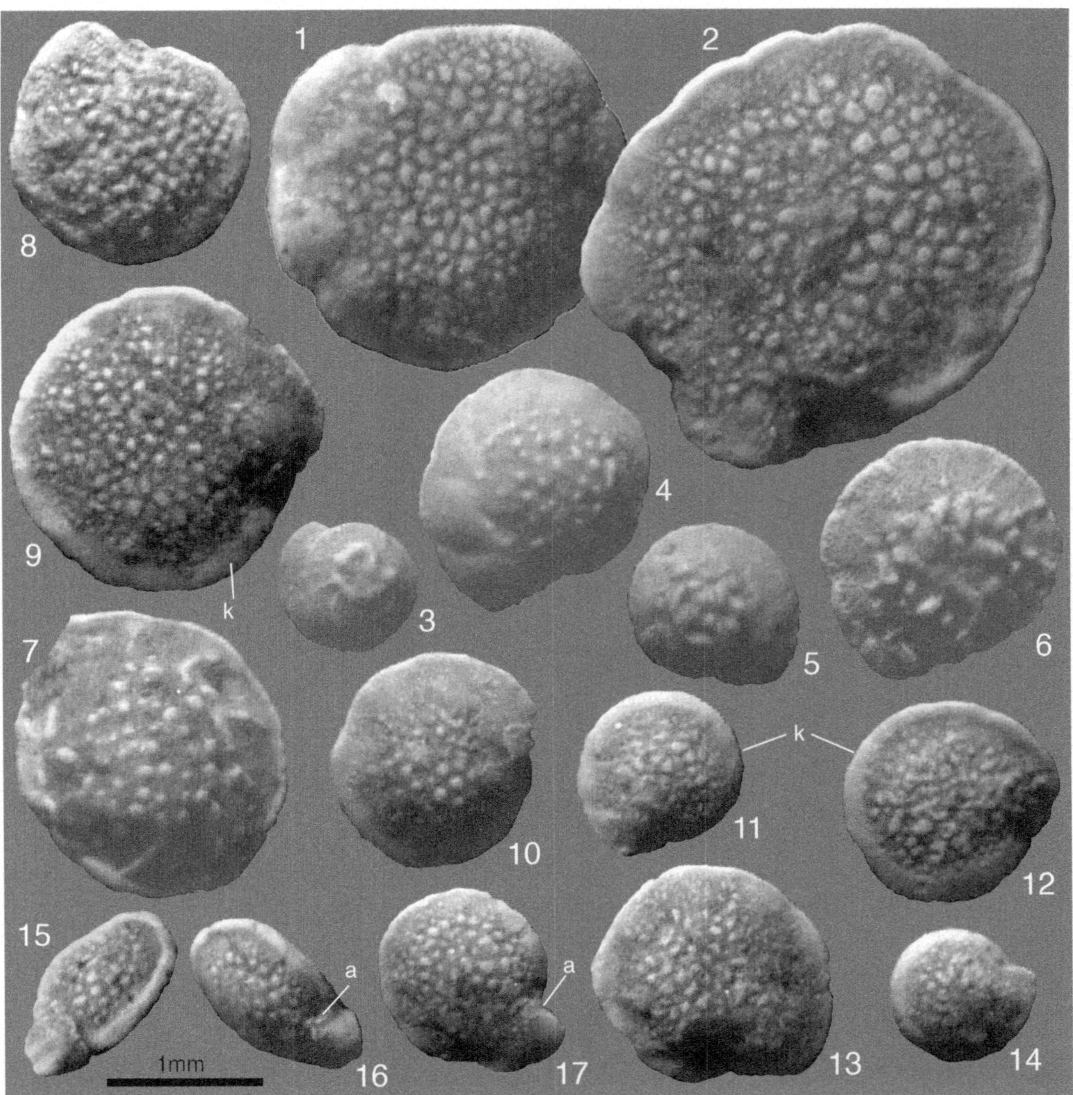

Plate 5.3 *Lockhartia praehaimei* Smout, 1954; sample 93520, Hangu Formation, Dhak Pass, Salt Range, Pakistan; Paleocene (SBZ 3). (**1–2**, **8–14**, **17**) Ventral view, showing the large number of small pile-heads evenly distributed over the flattened face. (**3–6**) Dorsal view. (**15–16**) Two oblique-peripheral views of the same specimen to show the position and shape of the aperture. Abbreviations: *a* aperture, *k* keel

concave, flat or somewhat convex base. The ratio of equatorial to axial diameter varies from 2.0 to 3.0. The umbilicus is wide and shallow, filled with numerous piles. These have equal diameters from the center to the periphery of the umbilical surface of the shell. About 8 piles are cut in an axial section per 1 mm radial distance. Adult shells exhibit 11–13 chambers in the last whorl on the dorsal side of the shell. The septal sutures are much inclined but weakly bent on the dorsal side, radial and straight on the ventral side, as far as visible in the last few chambers where their orientation is not obscured by secondary lamellation.

Remarks: All authors having treated *Lockhartia haimei* in detail agree that this species is extremely variable. Sander (1962) has given names to a certain number of Saudi Arabian varieties that one can recognize also in the Salt Range of Pakistan.

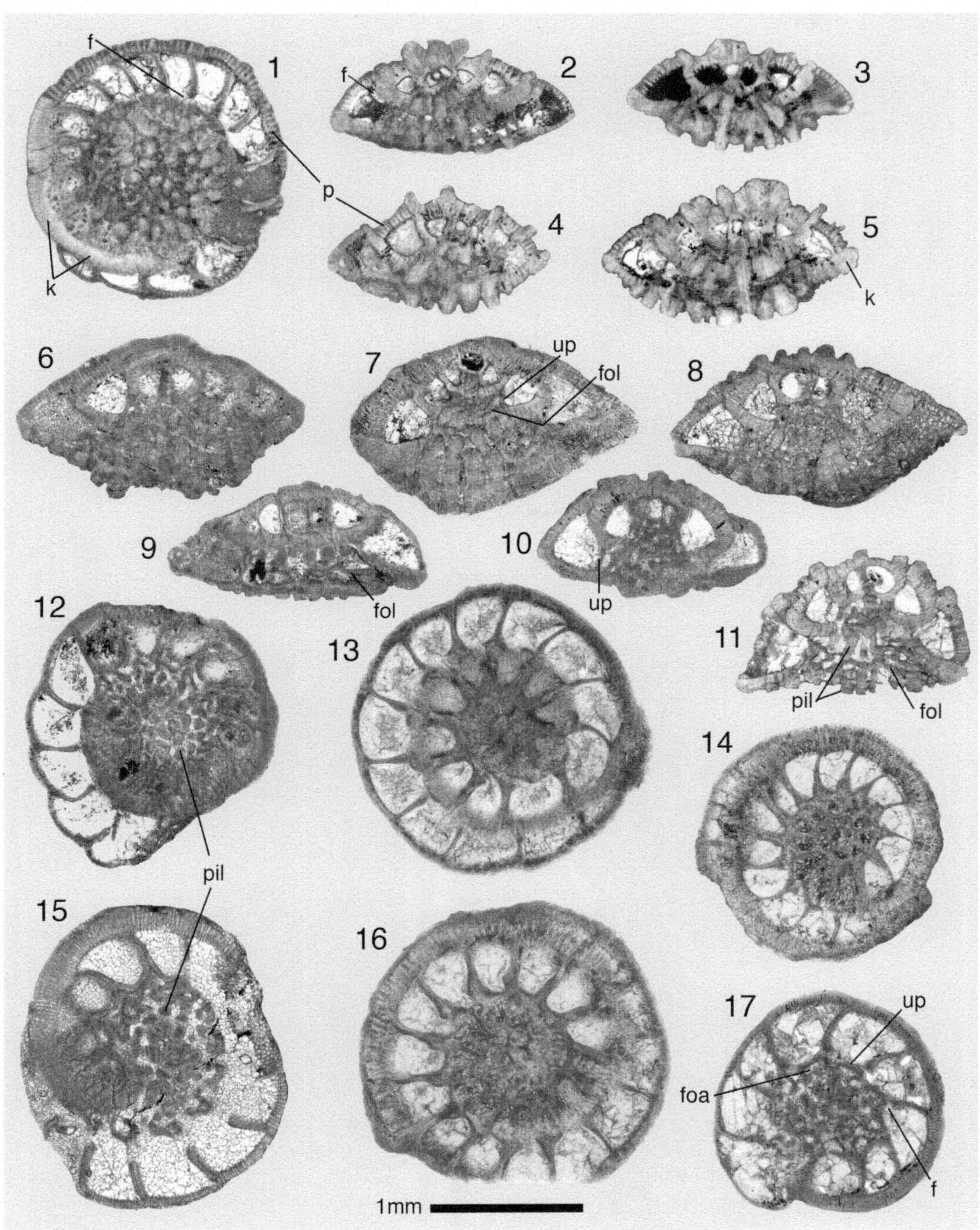

Plate 5.4 *Lockhartia praehaimei* Smout, 1954. (**1**, **12–17**) Sections perpendicular in respect to the shell's coiling axis. (**2–5**, **7**) Centered axial sections. (**6**, **8–11**) Adaxial sections. (**2–5**) topotypes from Smout's collection, the rest of the specimens from the sample 93520, Hangu Formation, Dhak Pass, Salt Range, Pakistan; Paleocene (SBZ 3). Abbreviations: *f* foramen, *up* umbilical plate, *k* keel, *fol* folia, *foa* foliar aperture, *p* pores, *pil* piles

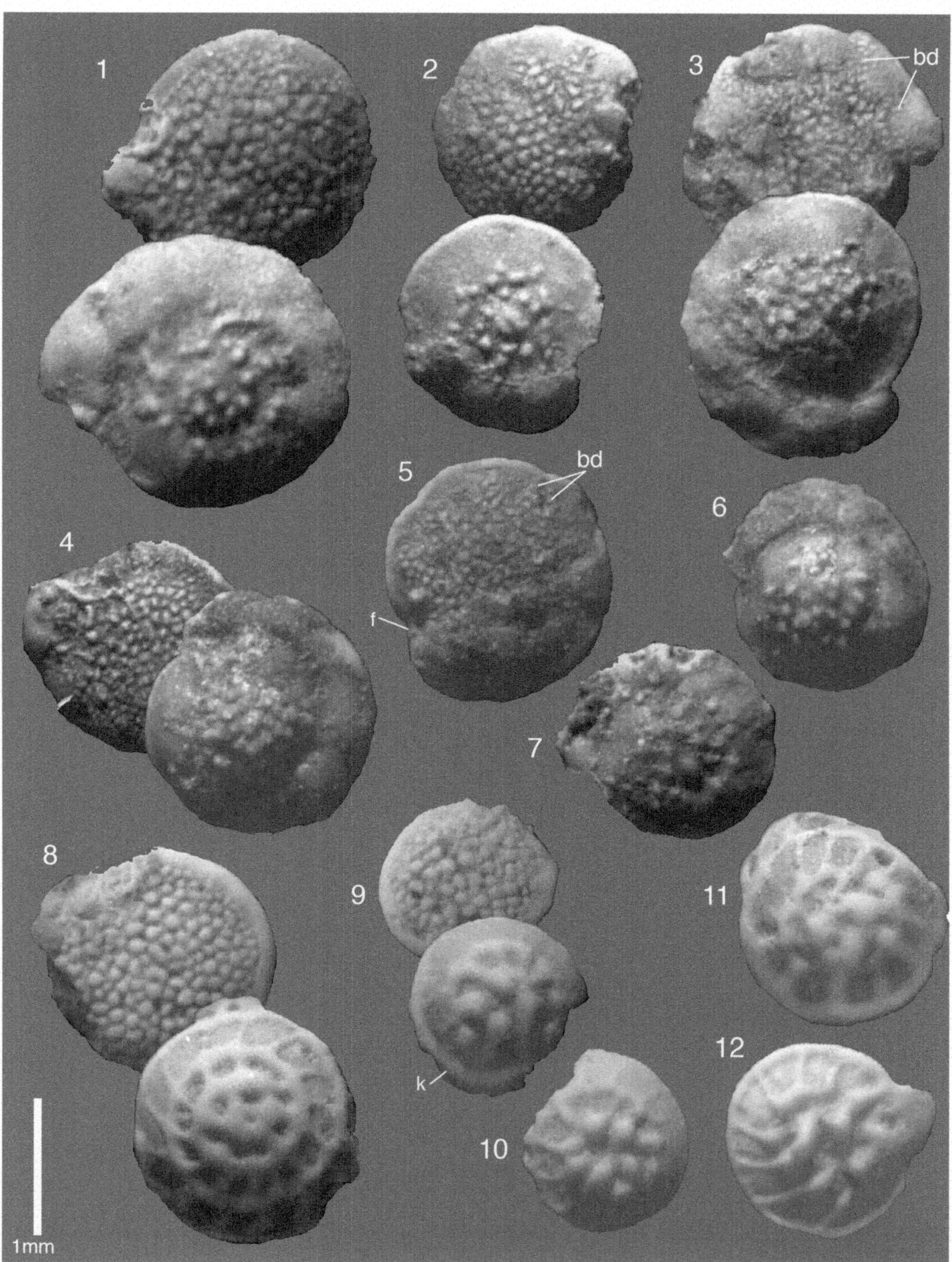

Plate 5.5 *Lockhartia haimei* (Davies, 1927), isolated specimens; variants of dorsal and ventral ornamentation. (**1–7**) *Lockhartia haimei suturadiata* Sander, 1962. (**1–4**) From both sides. (**8**) *L. haimei spirocordata* Sander, 1962, from both sides. (**9–12**) *L. haimei haimei* Sander, 1962. (**9**) From both sides. All specimens from the sample 93563, Patala Formation, Dandot, Salt Range, Pakistan; Paleocene (SBZ 4). Abbreviations: *f* foramen, *bd* beads, *k* keel

Sander's (1962) varieties are based on the dorsal ornamentations of free shells and on their contour in lateral view. This is not sufficient to decide whether Sander's varieties have a biostratigraphic significance on a Tethys-wide scale. A study of the internal morphology including the size of the megalosphere and on the chirality of the shell might help to solve the problem.

Lockhartia haimei with all its varieties is distinguished from all other species groups of the genus by its distribution of piles on the ventral side of the shell: in the ultimate and penultimate ventral chamber sutures, a double row of tiny beads fringe the shoulders of the ventral intraseptal interlocular space almost to the periphery of the shell (bd in Pls. 18, 20). These beads are quickly growing with the subsequent secondary lamellation obscuring the original pattern. This is the reason why in this species the ventral piles reach the periphery of the shell in the last whorl. In all other lockhartia groups, the piles are restricted to the umbilicus as delimited by the foliar sutures in the last whorl.

All varieties of *Lockhartia haimei* have ornaments on the dorsal side of the shell. These may consist of limbate spiral and radial sutures (*L. haimei spirocordata* Sander, 1962; Plate 5.5, Fig. 8), of limbate sutures combined with few, heavy piles in the apical area of the shell (*L. haimei haimei sensu* Sander, 1962; Plate 5.5, Figs. 9–12, compare with sections in Plate 5.6, Figs. 1–8), of a rather irregular group of small beads grouped in the apical area of the shell (*L. haimei nudimarginata* Sander, 1962; Plate 5.7, Figs. 2–3, 5–9) or of low-conical, axially compressed shells with a small number of beads in the apical area and slightly depressed chamber sutures lacking any ornament in the last whorl. Sander's heavily ornate, unnamed variation (1962, pl. 5, figs. 33–34) may correspond to specimens figured here in Plate 5.7, Figs. 7–8, where beads on the spiral suture are combined with limbate chamber sutures.

All these varieties seem to coexist in the Salt Range of Pakistan over a considerable time from the Hangu Formation to the end of the Lockhart Limestone. Within the *suturadiata* variety, there seems to be an evolution in the megalosphere size from about 0.2 to 0.32 mm in diameter. In general, the minimum diameter of the megalospheres is 0.16 mm. Proloculus and deuteroconch form dyads. The larger the megaloshere, the more perfect is the fit of the first two chambers as hemispheres of equal diameter forming a spherical embryo. Microspheres were not observed in spite of the relative abundance of *L. haimei* specimens in the sediments.

Lockhartia retiata Sander, 1962; Figs. 5.1A–G and 5.2A–G; Plate 5.9, Figs. 1–20; Plate 5.10, Figs. 1–16 bottom.

1954 *Lockhartia diversa*—Smout, pars, pl. 3, figs. 16–20.

1960a *Lockhartia haimei* (Davies)—Drooger, p. 458, pl. 2, figs. 4–6.

1960b *Lockhartia haimei* (Davies)—Drooger, p. 302, pl. 3, figs. 1–10.

1962 *Lockhartia retiata*—Sander, p. 22, pl. 4, figs. 1–20.

1972 *Sakesaria* sp.—Sirel, p. 291, pl. 8, fig. 7.

1978 *Lockhartia* sp.—Rahaghi, p. 60, pl. 12, figs. 1–7.

2009 *Lockhartia haimei* Davies—Afzal et al., pl. 1, figs. 7, 9.

Description: The subspherical shells of comparatively small size are produced by chambers arranged in a high spiral. The last whorl of the adult has 8 chambers. The umbilicus is narrow, filled with the usual lockhartiine structure with few piles. The dorsal side of the shell is covered by a heavy ornamental cancellation that provides the axial sections of the shell with a strikingly spinose aspect. The periphery is rounded, without any keel. The megalospheric embryo is a dyad with a proloculus diameter of 0.12–0.24 mm. No microspheric specimens have been discovered.

Remarks: *Lockhartia retiata* clearly is the predecessor of the much better known *L. diversa* that has a similar dorsal ornamentation on the dorsal side of a much larger shell. *L. retiata* is associated

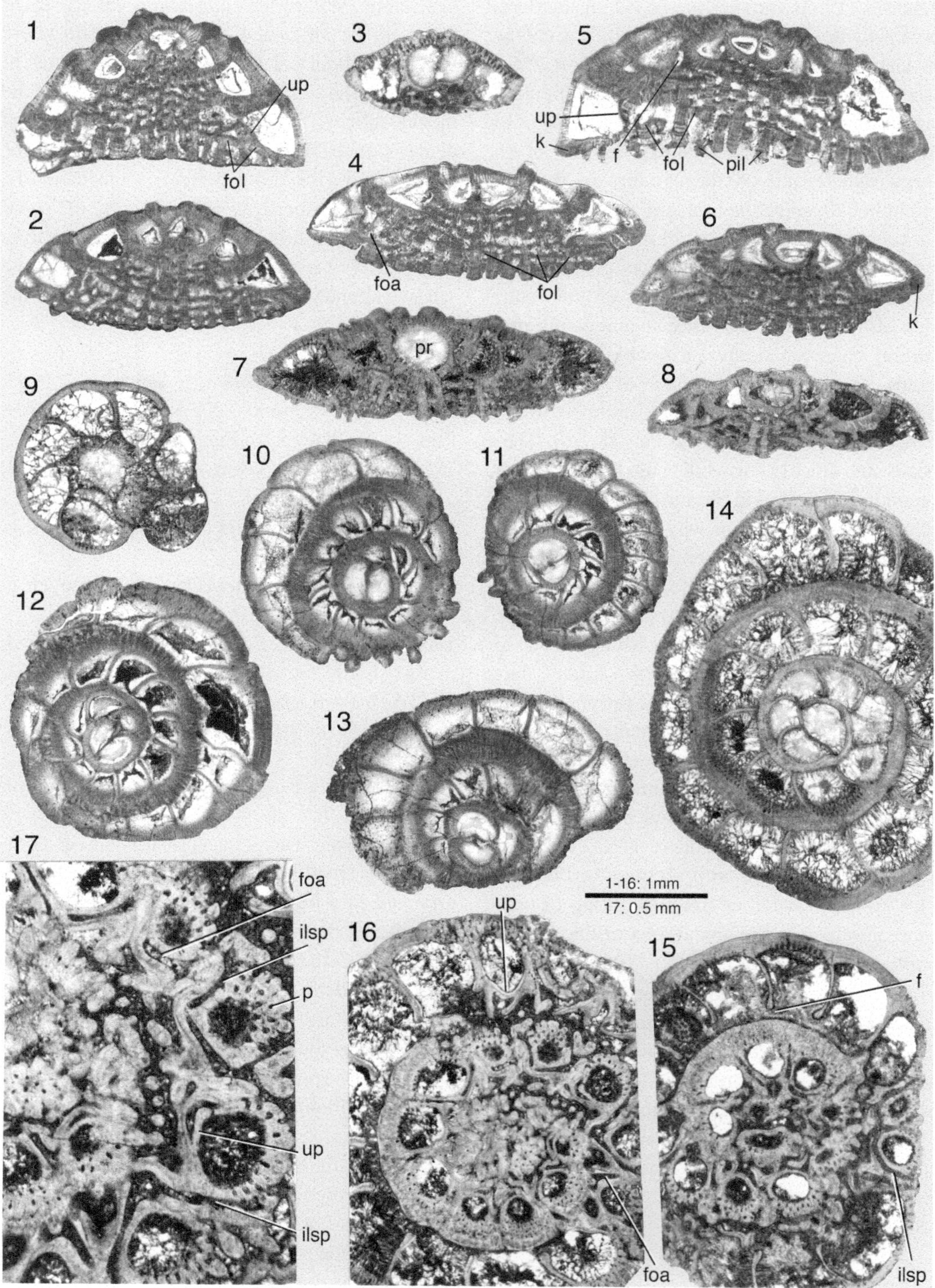

Plate 5.6 *Lockhartia haimei* (Davies 1927), sectioned megalospheric specimens. (**1–8**) Axial sections; (Note small shell size of specimens **2** and **9** representing the nepiont only). (**9–14**) Equatorial sections. (**15–17**) Sections perpendicular in respect to the shell's coiling axis; particularly well preserved specimens showing the details of the

with *Ranikothalia sindensis* (Davies and Major 1927), *Miscellanites globularis* (Rahaghi, 1978), *Lockhartia praehaimei* Smout, 1954 and other index species of SBZ 3. It is reported from French Guyana under the name of *Lockhartia haimei* by Drooger (1960a, b; Fig. 5.1A–G), from the continental shelf on the other side of the Atlantic in the Aquitaine (southern France; Fig. 5.2), in the Zagros Mountains (northern Iran; Pls. 22–23) and on the Arabian Peninsula.

The wide paleogeographic distribution of *Lockhartia retiata* supports the interpretation of the SBZ 3 foraminiferal association as a cosmopolitan community in the early stages of Global Community Maturation.

Lockhartia diversa Smout, 1954; Plate 5.11, Figs. 1–19.

1954 *Lockhartia diversa* Smout, p. 52, pl. 3, figs. 1–13, non figs. 16–20.

1962 *Lockhartia retiata*—Sander, pars, p. 22, pl. 4, fig. 21.

1985 *Lockhartia diversa* Smout—Hasson, p. 360, pl. 4, fig. 11.

1991 *Lockhartia diversa* Smout—Wan, p. 163, pl. 2, figs. 30–32.

Description: The subspherical shell of median size is produced by chambers arranged in a high spire. There are 9 chambers in an adult whorl. The septa are much inclined on the dorsal, radial on the ventral side of the shell. The intercameral foramen is a small slit in interiomarginal position, located at the ventral corner of the chamber lumen. The periphery is rounded but marked by an imperforate band that produces a very weak or no peripheral swelling at all. The ventral chamber wall is very thick, covered with secondary lamellation, similar to the dorsal side of rotaliines without ornamentation.

Remarks: *Lockhartia diversa* exhibits a heavy cancellate ornament all over the dorsal side of the shell. The cancellate depressions are coarser than those of *L. retiata*. Our material is not abundant enough to produce a significant measure of the megalosphere diameter. However, the embryo is a small dyad comparable to *L. retiata* and therefore no diagnostic feature. The single section given in the original description of *L. retiata* by Sander (1962) is interpreted here as *L. diversa* and presents a megaloshere with a diameter of 0.16 mm. Microspheric specimens have not been found.

Lockhartia roeae (Davies, 1930); Plate 5.12, Figs. 1–6; Plate 5.13, Figs. 1–16.

1930 *Dictyoconoides conditi* Nuttall var. *roeae*—Davies, p. 76–77, pl. 10, fig. 9.

1980 *Lockhartia tipperi* Davies—Müller-Merz, pl. 10, figs. 1, 3.

Description: In axial sections the shells of this species have a triangular outline. The ventral side is flat or weakly convex, the dorsal side lacks any ornamentation and delimits a smooth, flat cone with an angular but unkeeled periphery. The ratio of equatorial to axial diameter of the shell varies only from 2.0 to 2.3. The septa are much inclined on the dorsal side, radial on the ventral side and longer in the direction of growth than their radial extension. The last whorl of adult shells has 16–18 chambers. The umbilicus is filled with piles, usually with a central one that is not much thicker than the rest. There are 5–6 piles, cut in an axial section of the shell. The piles initiate on the foliar wall, a single one per chamber, from where they grow in ventral direction (Plate 5.13, Figs. 9–11). The ventral wall of the spiral main chamber is smooth except the sutural depressions. As the shoulders of these

Plate 5.6 (continued) umbilical architecture. (**1**, **3–6**, **10–13**) Sample 93501, base Lockhart Limestone, Nammal Gorge, Salt Range, Pakistan; Paleocene (SBZ 3). (**2**, **8–9**) Sample 93503, on top of Lockhart Limestone, Nammal Gorge, Salt Range, Pakistan; Paleocene (SBZ 4). (**14–17**) Sample 93566, *Alveolina*-bearing marls below Sakesar Limestone, Dandot village, Salt Range, Pakistan; Ilerdian (SBZ 5 or 6). Abbreviations: *f* foramen, *up* umbilical plate, *k* keel, *fol* folia, *foa* foliar aperture, *p* pores, *pil* pile, *pr* proloculus, *ilsp* intraseptal interlocular space

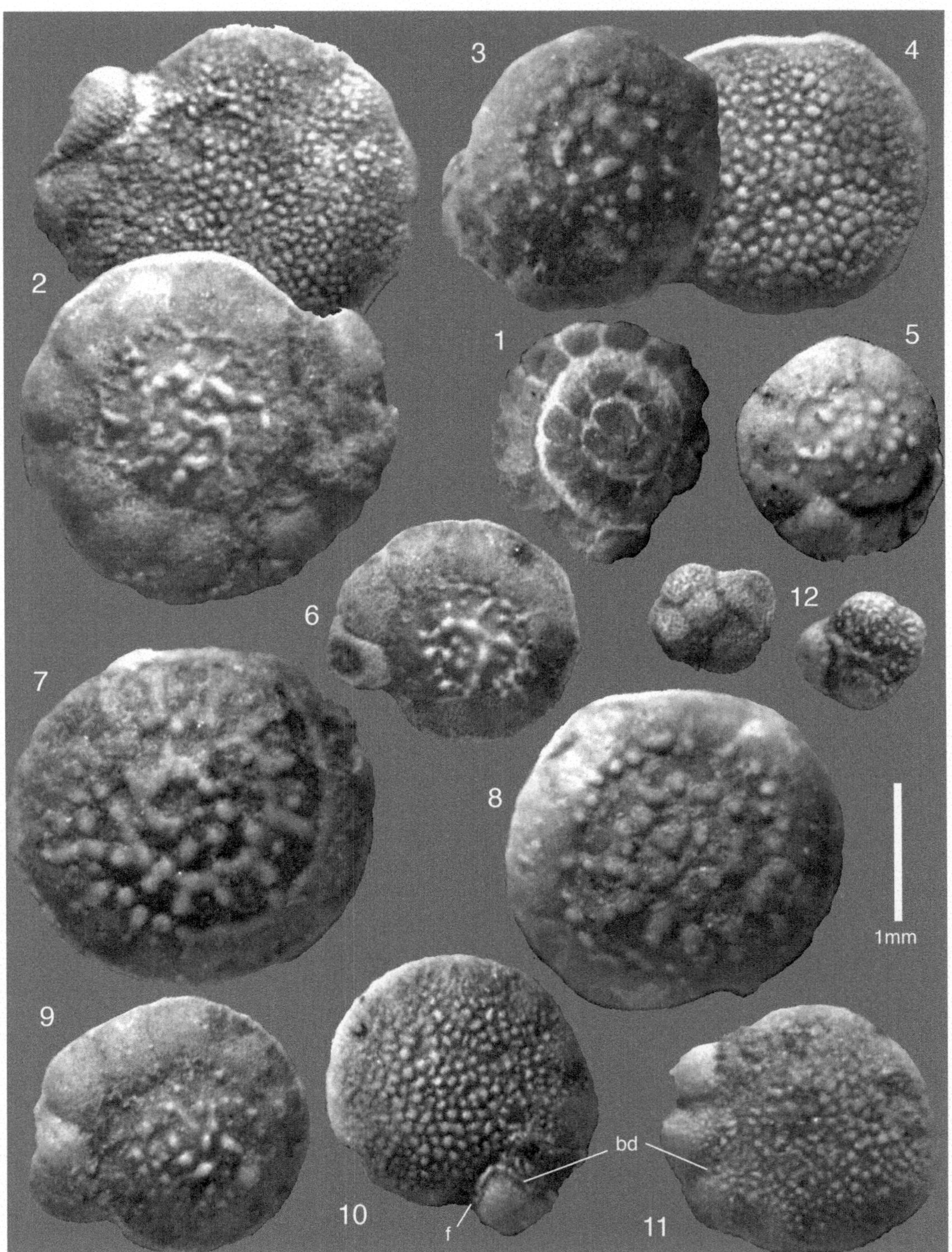

Plate 5.7 Large-sized variants of *Lockhartia haimei* (Davies, 1927). (**1**) Decorticated specimen showing the chamber arrangement on the dorsal side in a megalospheric shell. (**2–6**, **9–11**) Dorsal and ventral views of *Lockhartia haimei suturadiata* Sander, 1962. (**7–8**) Sander's (1962) unnamed variety of *Lockhartia haimei* (Davies, 1927). (**12**) Dorsal and ventral view of a nepiont of *Lockhartia haimei* (Davies, 1927). (**1**) Sample 92014, Nammal Formation, Nammal Gorge, Salt Range, Pakistan. (**2–12**) Sample 93563, Patala Formation, immediately below coal seam, Dandot village, Salt Range, Pakistan; associated with *Daviesina garumnensis* Tambareau, 1972; Paleocene (SBZ 4). Abbreviations: *f* foramen, *bd* beads

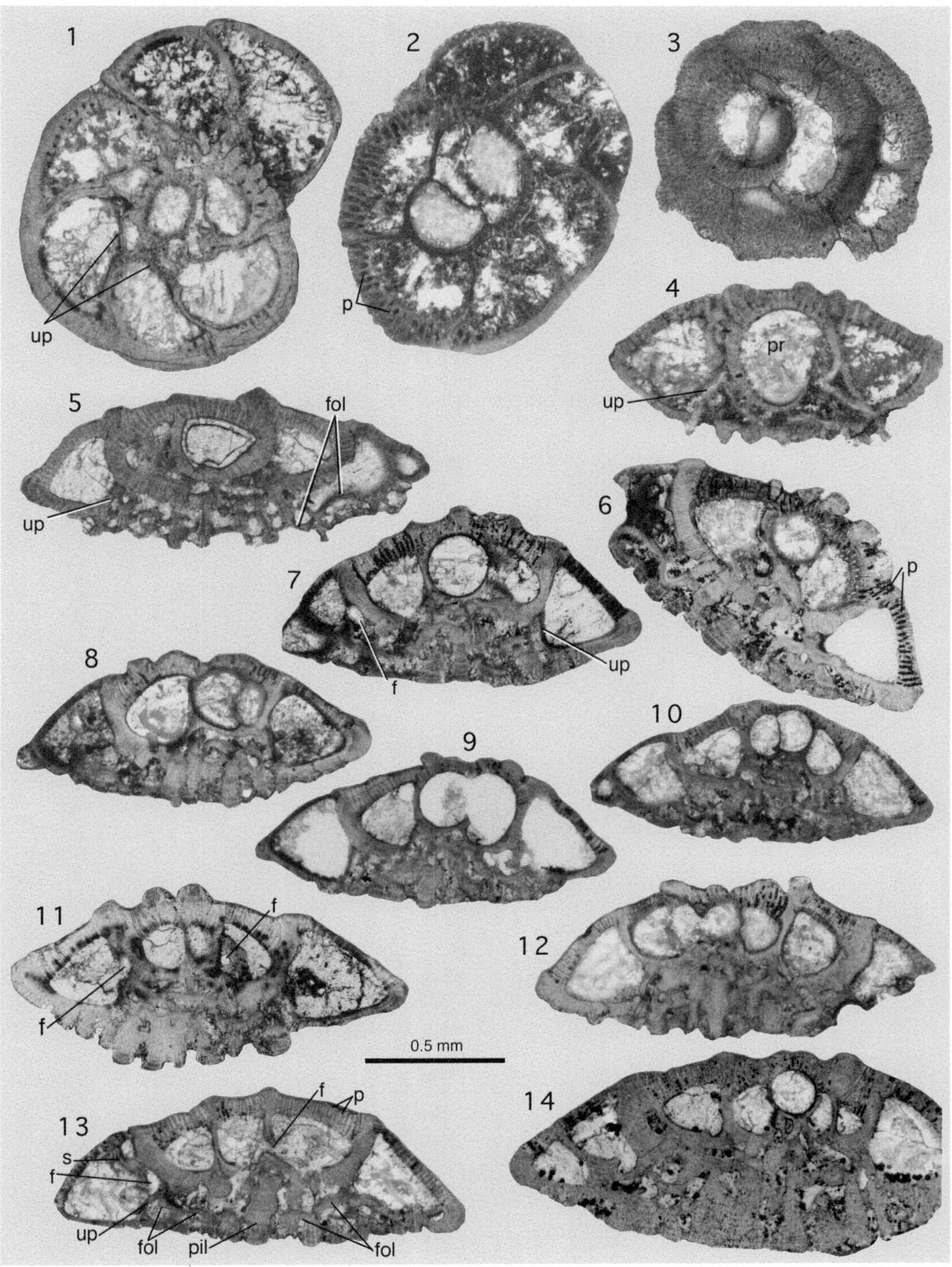

Plate 5.8 *Lockhartia haimei* (Davies, 1927); large megalospheric specimens sectioned. (**1–3**) Equatorial sections, slightly off center. (**5–14**) Axial sections; note the almost lenticular outline of most specimens and the size of their proloculus, corresponding perhaps to Sander's variety *suturadiata*. (**1–2**, **4**, **6–8**) Sample 93563, Patala Formation below coal seam, Dandot village, Salt Range, associated with *Daviesina garumnensis* Tambareau, 1972; Paleocene (SBZ 4). (**3**, **5**, **9–14**) Sample 93503, top Lockhart Limestone, Nammal Gorge, Salt Range; Paleocene (SBZ 4). Abbreviations: *f* foramen, *s* septum, *up* umbilical plate, *fol* folia, *p* pores, *pil* pile, *pr* proloculus

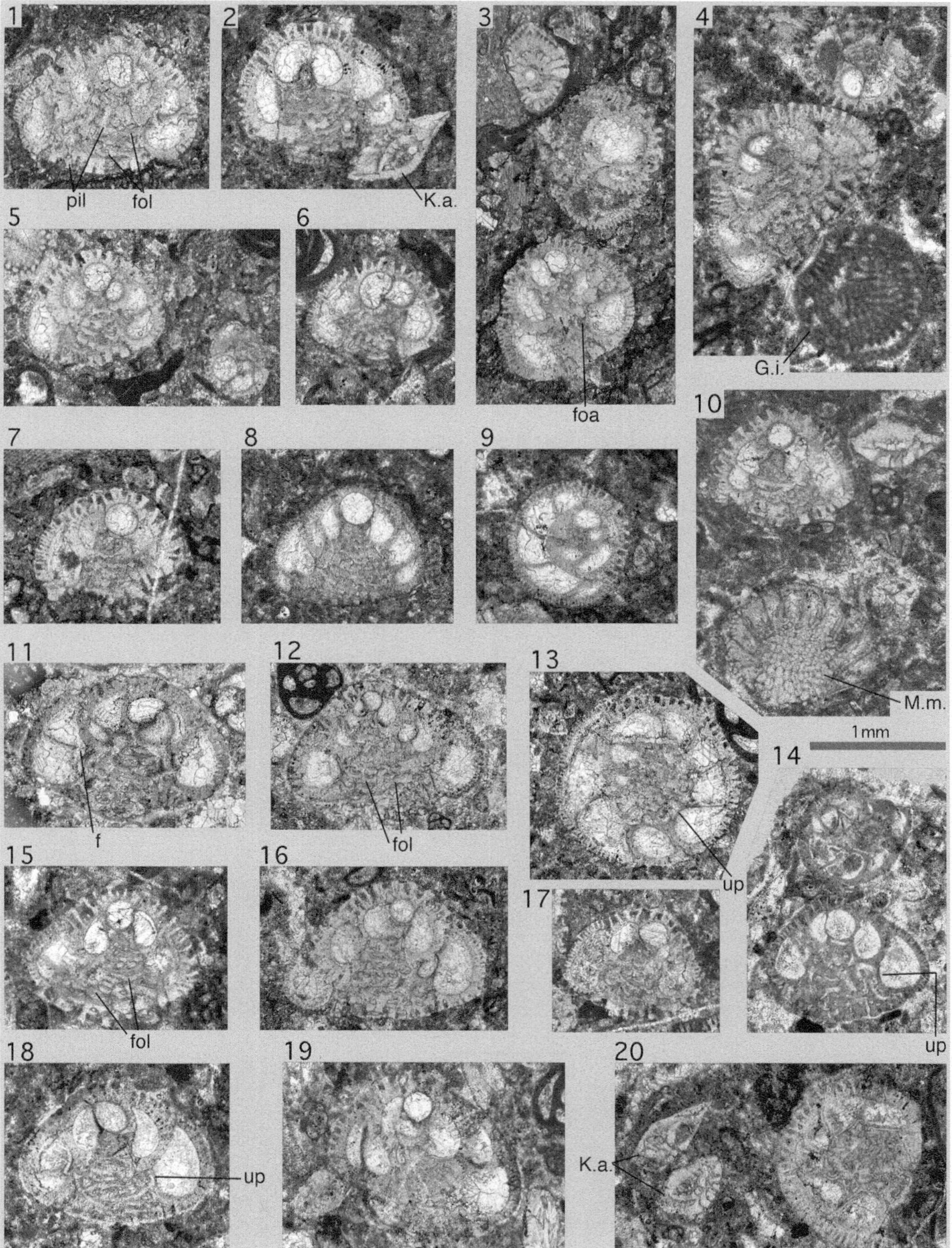

Plate 5.9 *Lockhartia retiata* Sander, 1962. (**2**, **20**) Associated with *Kathina aquitanica* n. sp. (K.a.). (**4**) Associated with "*Globoreticulina*" *paleocaenica* Rahaghi, 1983 (G. p.). (**10**) Associated with *Miscellanites minutus* (Rahaghi, 1983) (M. m.). (**1–3**) Oblique sections. (**9**, **13**, **20**) Sections perpendicular to coiling axis; all others: sections more or less in the coiling axis. All specimens from beds (**11** and **12**) at Kuh-e-Kargan (Kermanshah), Zagros, Iran; Paleocene (SBZ 3); material collected by J. Braud around 1968; see also Hottinger (2009, pls. 12–13, 18–19), where the Miscellaneidae from the same localities are illustrated. Abbreviations: *f* foramen, *up* umbilical plate, *fol* folia, *pil* pile, *foa* foliar aperture

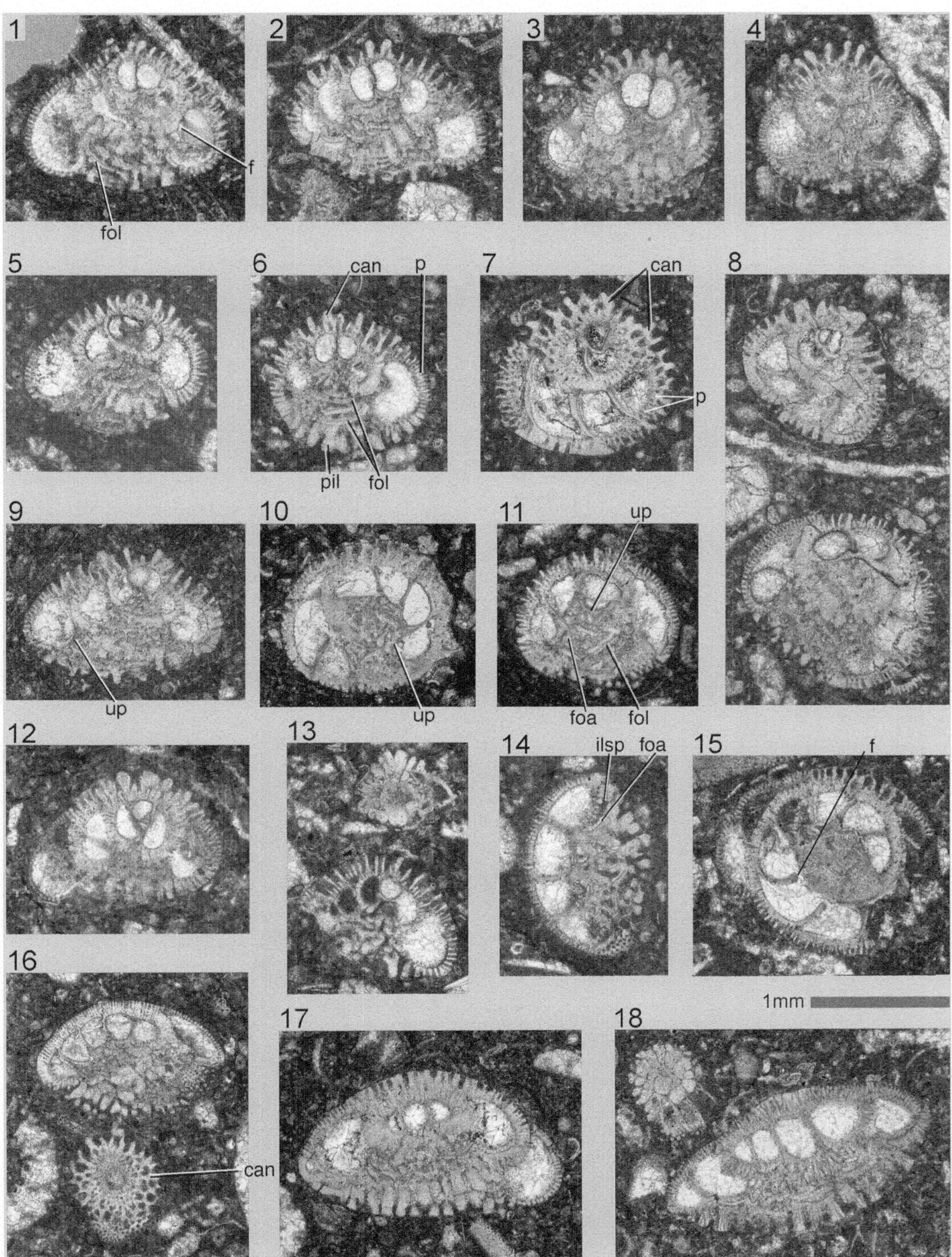

Plate 5.10 (**1–15**) *Lockhartia retiata* Sander, 1962. (**4**, **7**, **8**) (*lower part*), (**13–14**) Oblique sections. (**8**) (*top*), (**10–11**, **15**) Sections perpendicular to coiling axis. (**16**) Lower part, perpendicular to axis and tangential to apex of the shell showing the reticular aspect of the cancellate dorsal ornament. All other (**1–3**, **5–6**, **9**, **12–13**) sections more or less in the axis of coiling. (**16–18**) *Lockhartia praehaimei* Smout, 1954. (**16**) Upper part. (**17**) Section more or less in the axis of coiling. (**18**) Oblique section. All figures on this plate are from bed (**9**) of the sedimentary succession collected by J. Braud from the Kuh-e-Kargan (Kermanshah), Zagros, Iran; Paleocene (SBZ 3). Abbreviations: *f* foramen, *up* umbilical plate, *fol* folia, *pil* pile, *foa* foliar aperture, *p* pores, *can* cancellate, *ilsp* intraseptal interlocular space

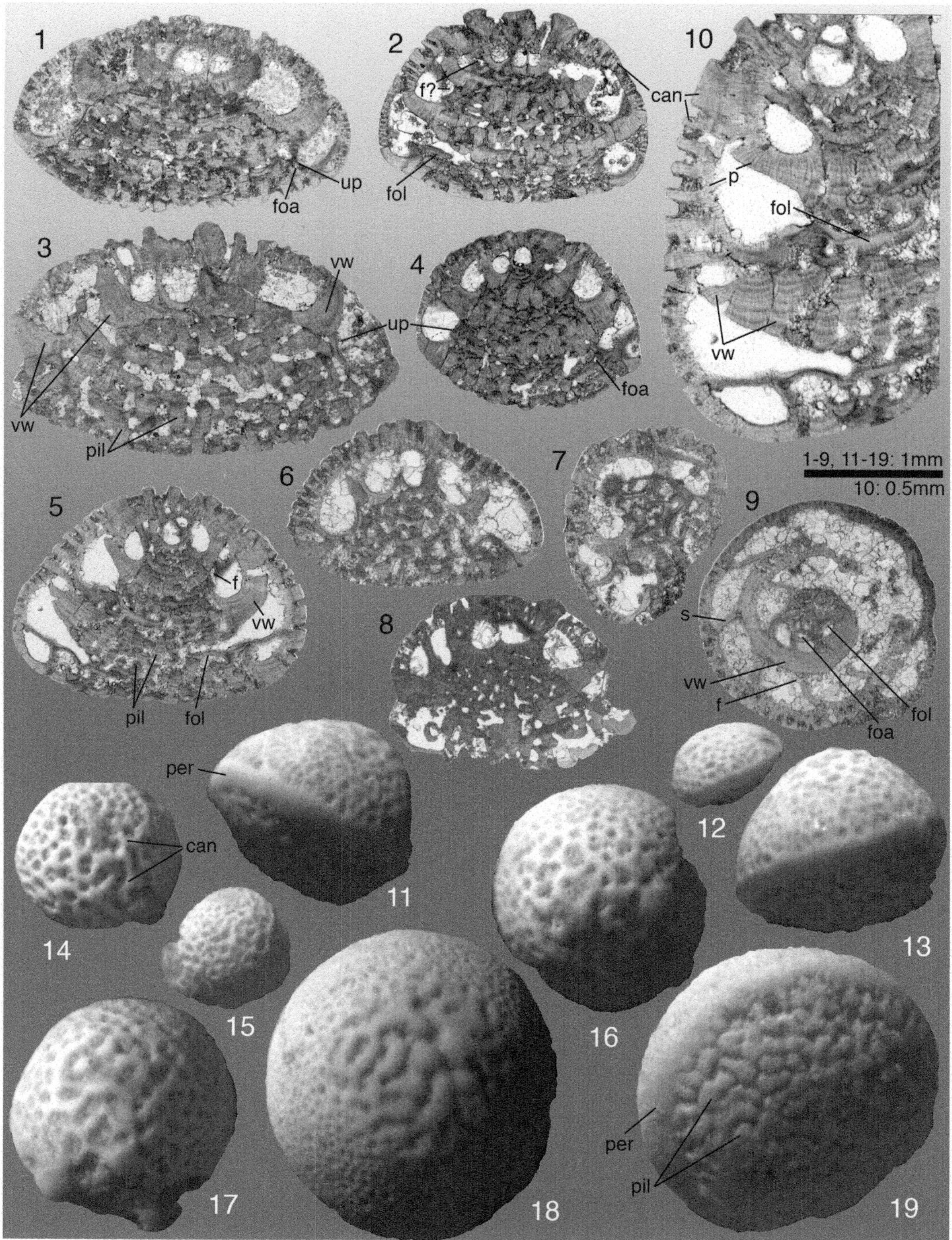

Plate 5.11 *Lockhartia diversa* Smout, 1954; sections of isolated specimens; megalospheric specimens from Qatar, Arabian Peninsula, borehole AQ 6, 1400′–1405′, collected by Max Chatton. (**1–6**) Approximately axial sections. (**7–8**) Oblique sections. (**9**) Section perpendicular to coiling axis. (**10**) Detail of specimen 5 showing thicken ventral chamber walls (vw); external views of isolated specimens. (**11–13**) Peripheral views. (**14–18**) Dorsal view showing cancellate ornament; as can be seen in the sectioned specimens (**7**) the meshes of the cancellate ornament may be filled with sediment that obscures the true depth of the meshes. (**19**) Oblique ventral view showing the density of the ventral piles of lamellae. Abbreviations: *f* foramen, *up* umbilical plate, *fol* folia, *pil* pile, *foa* foliar aperture, *p* pores, *can* cancellate, *per* periphery, *vw* ventral walls, *s* septum

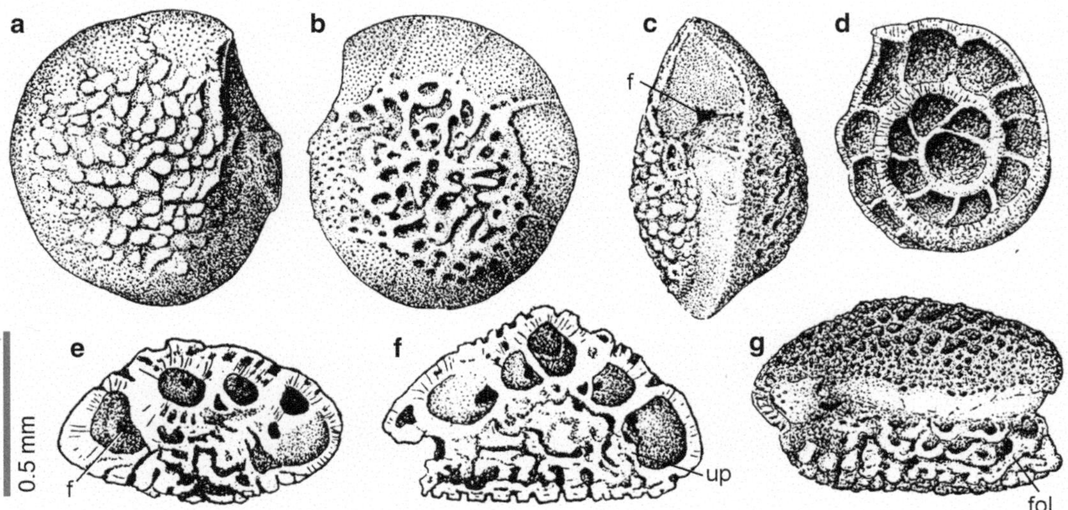

Fig. 5.1 *Lockhartia retiata* Sander, 1962 reported from French Guyana under the name of *Lockhartia haimei* by Drooger (1960b). (**A**) Ventral view. (**B**) Dorsal view. (**C**) Lateral view. (**D**) Equatorial section. (**E–F**) Axial sections. (**G**) Peripheral view

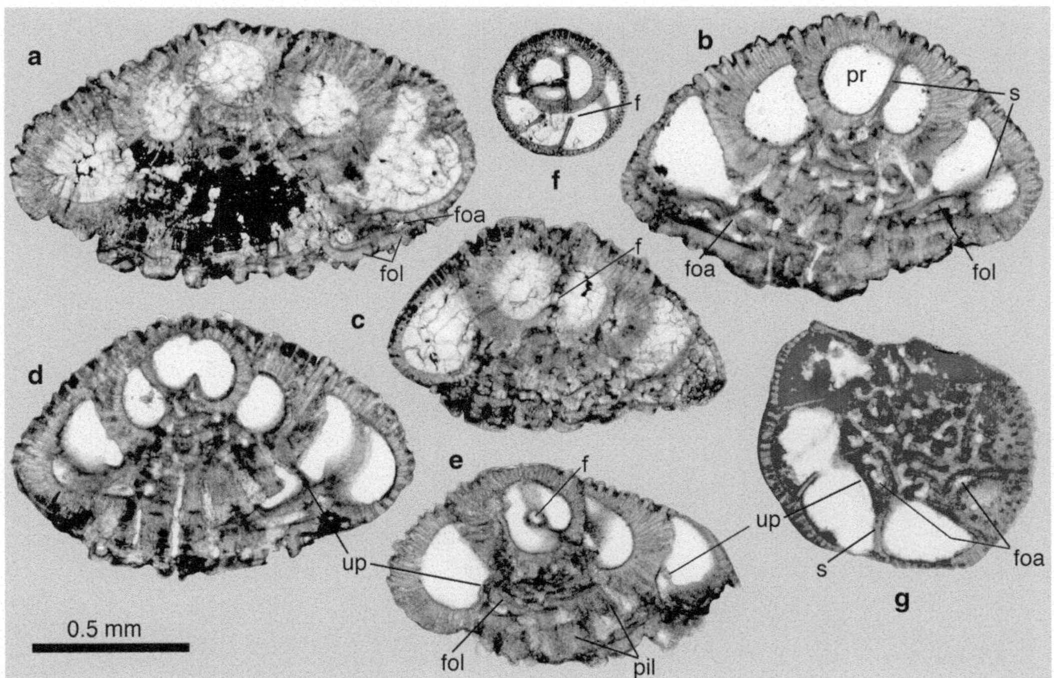

Fig. 5.2 *Lockhartia retiata* Sander, 1962, megalospheric generation; sample Volp 6, near Pau, western Aquitaine, southern France, collected by Y. Tambareau in association with a rich fauna of index species for SBZ 3. (**A–E**) Approximately axial sections; note in (**C**) and (**E**) the central position of the foramen between megalosphere and deuteroloculus. (**F**) Section perpendicular to shell coiling axis near apex, showing megalospheric nepiont. (**G**) Section perpendicular to coiling axis at base of shell cone showing the umbilical elements of the *Lockhartia* architecture. Abbreviations: *f* foramen, *fol* folia, *up* umbilical plate, *foa* foliar aperture, *pr* proloculus, *s* septum, *pil* pile, *fol* folia

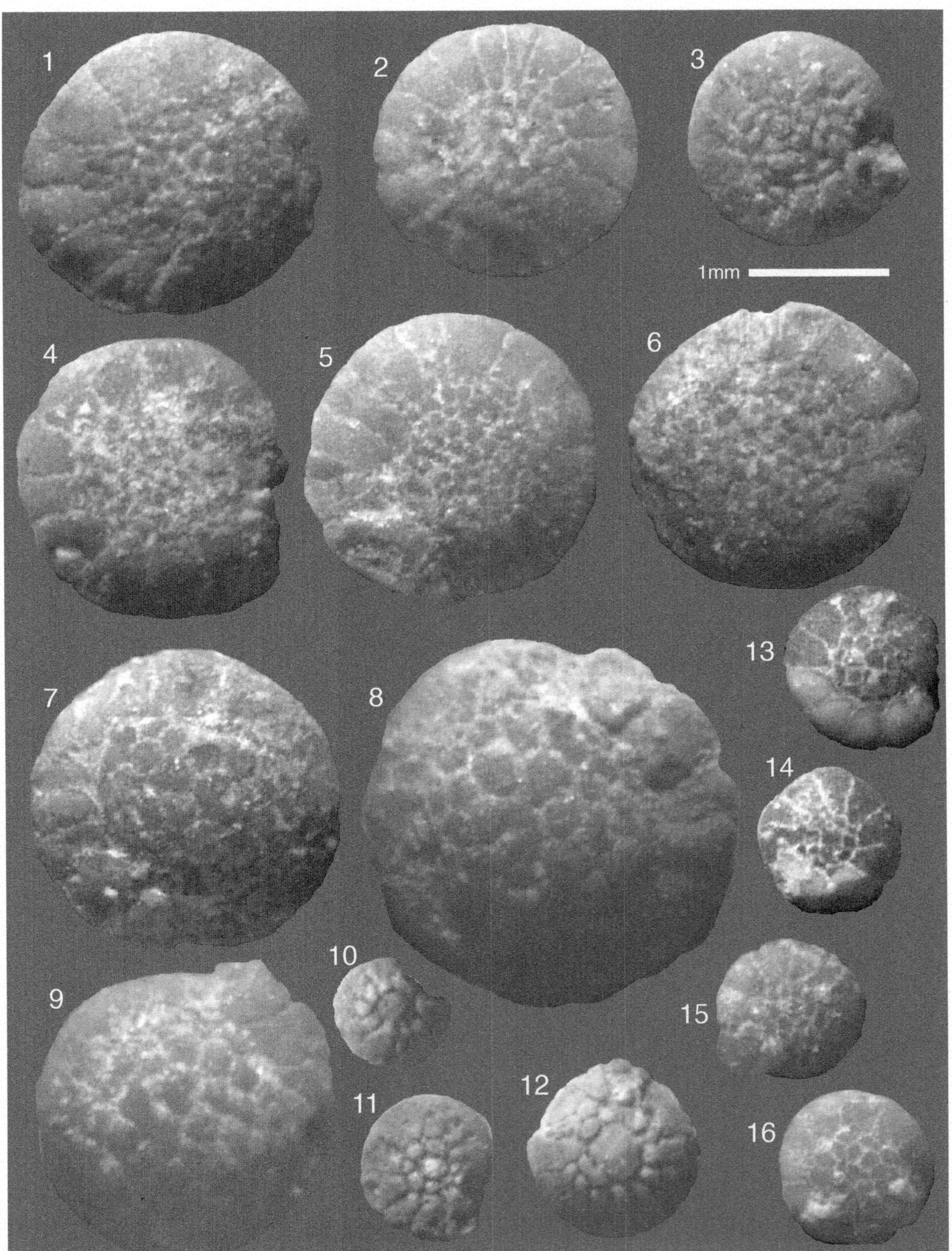

Plate 5.12 (**1–6**) *Lockhartia roeae* (Davies, 1930); external views of the shell face; sample 93520, Hangu Fm below Lockhart Limestone, Dahk Pass, Salt Range, Pakistan; Paleocene (SBZ 3); compared with (**7–9**) *Lockhartia conditi* (Nuttall, 1926) in the same bed. (**10–12**) *Lockhartia* aff. *conditi* but it cannot be excluded that they represent the insufficiently described "*Rotalia*" *dukhani* Smout, 1954; specimens (**10–12**) from sample 95115, Hangu Formation, Dahk Pass, Salt Range, Pakistan; Paleocene (SBZ 13). (**13–16**) *Lockhartia* aff. *conditi* (Nuttall, 1926) with uniformly sized pile heads on their face are considered as true *L. conditi* in their early stage of growth; their late appearance in sample 93545 from the Nammal Formation in the Nilawahan Gorge, Nurpur, Salt Range, supports this interpretation

Plate 5.13 *Lockhartia roeae* (Davies, 1930); sectioned specimens. (**1–5**, **7–16**) Axial sections. (**6**) Section tangential to the dorsal surface of the shell cone, showing elongate spiral chambers with inclined septa and tiny dyads as embryos. (**3**) Microspheric specimen. All specimens from sample 93520 in the Hangu Formation underlying the Lockhart Limestone, Dhak Pass, Salt Range, Pakistan; Paleocene (SBZ 3). Abbreviations: *f* foramen, *up* umbilical plate, *fol* folia, *pil* pile, *foa* foliar aperture, *dy* dyad, *s* septum

depressions do not show any feathering, I suppose that there is not much circulation of protoplasm through the intraseptal interlocular space between the umbilical canal system and the ambient environment. The tiny embryo is a biconch. The megalosphere has a diameter of 0.08–0.16 mm. A single specimen (Plate 5.13, Fig. 3), one and a half times larger but otherwise similar to the numerous megalospheric specimens, might represent the microspheric generation.

Remarks: *Lockhartia roeae* is considered to represent the predecessor of *L. tipperi*. These species have in common a chamber shape that is longer in the direction of growth as compared to their radial extension (Plate 5.13, Fig. 6). The comparatively high number of umbilical piles and the tiny biconch embryos complement the common features.

Lockhartia conditi (Nuttall, 1926); Plate 5.12, Figs. 7–9; Plate 5.14, Figs. 1–24; Plate 5.15, Figs. 1–12.

1926 *Dictyoconoides conditi*. Nuttall, p. 119, plate 4.7, figs. 7–8.

1927 *Dictyoconoides conditi* Nuttall—Davies, p. 279, pl. 21, figs. 10–12; plate 5.9, fig. 5.

1932 *Lockhartia conditi* (Nuttall)—Davies, p. 408, pl. 2, fig. 7; pl. 4, fig. 7.

1937 *Lockhartia conditi* (Nuttall)—Davies and Pinfold, p. 47, pl. 5, fig. 24.

1947 *Lockhartia conditi* (Nuttall)—Ovey, p. 573, pl. 10, figs. 7–8.

1954 *Lockhartia conditi* (Nuttall)—Smout, p. 55, pl. 5, figs. 16–19.

1956 *Lockhartia conditi* (Nuttall)—Smout and Haque, p. 57, pl. 11, figs 5a–b.

1970 *Lockhartia conditi* (Nuttall)—Kaever, p. 85, pl. 8, figs. 7–11.

1970 *Lockhartia haimei* (Davies) pars—Kaever, p. 83, pl. 8, figs. 2–4.

1976 *Lockhartia haimei* (Davies) pars—Ho et al., p. 15, pl. 27, figs. 5–8.

1977 *Lockhartia* sp.—Carbonnel and Blondeau, p. 110, pl. 53, fig. 9.

1980 *Lockhartia conditi* (Nuttall)—Müller-Merz, p. 35, pl. 10, figs. 4, 6.

1987 *Lockhartia* sp. and unidentified rotaliid—Nicora et al., pl. 34, fig. 1, 8.

1991 *Lockhartia conditi* (Nuttall)—Wan, p. 12, pl. 1, figs. 26–27.

1991 *Lockhartia hunti* Ovey—Wan, p. 13, pl. 2, figs. 3–4.

1993 *Lockhartia* aff. *haimei* Davies—Weiss, p. 252, pl. 3, fig. 4.

1999a *Lockhartia conditi* (Nuttall)—Akhtar and Butt, p. 132, pl. 3, figs. 1–2.

1999b *Lockhartia conditi* (Nuttall)—Akhtar and Butt, p. 188, pl. 3, fig. 8.

2008 *Lockhartia conditi* (Nuttall)—BouDagher-Fadel, p. 399, pl. 6.28, fig. 14.

2008 *Lockhartia conditi* Smout (erroneously for (Nuttall))—Ismail and Boukhary, p. 93, pl. 2, figs. 4–5.

2009 *Lockhartia conditi* (Nuttall)—Afzal et al., p. 17, pl. 1, figs. 2, 4.

Description: The bilamellar-perforate shells reach an equatorial diameter of 1.5 mm. The axial section has an oval contour with an equatorial to axial diameter ratio of 1.5–1.8. The convexity of the shell is about equal on both sides. The dorsal side, always evolute, is decorated by few heavy piles forming a low or no relief at all. The periphery is rounded and lacks any keel. There are only three thick umbilical piles or less appearing in axial sections. Sections perpendicular to the coiling axis of the test show a last whorl composed of 10–15 adult spiral chambers. In the equator of the shell the approximately isometric chambers are separated from each other by septa that are slightly inclined backwards in dorsal part and radial in the ventral part of the chamber. The chambers are sometimes slightly inflated (Plate 5.14, Fig. 3). The intercameral foramen is a narrow slit in interiomarginal position bordered by a minute lip.

There are two kinds of proloculi with similar frequency, a larger one with a diameter of 0.04–0.07 mm and a smaller one with a diameter of about 0.02 mm. In the adult stages, no

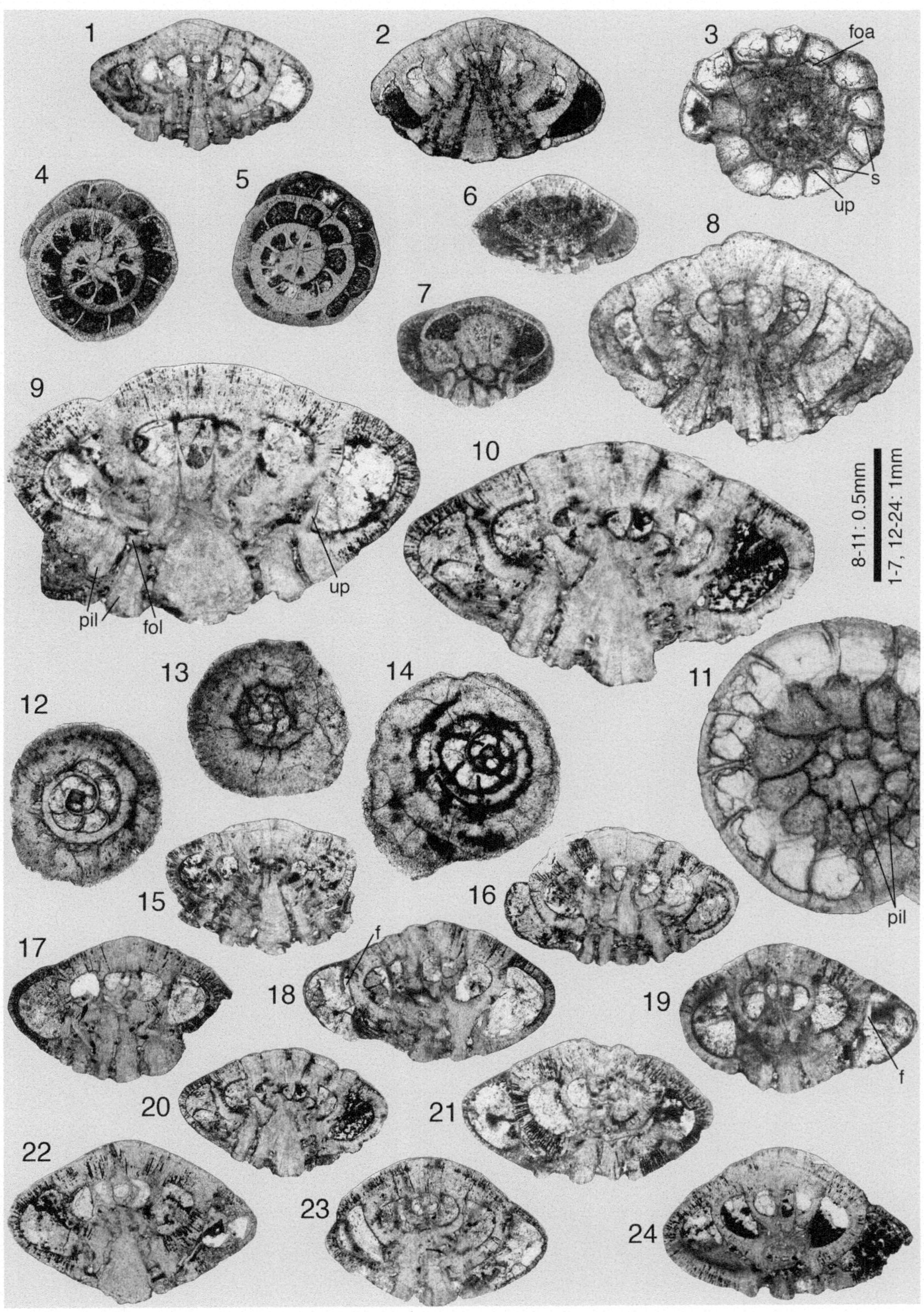

Plate 5.14 *Lockhartia conditi* (Nuttall, 1926); sample 93520 in the Hangu Formation underlying the Lockhart Limestone, Dhak Pass, Salt Range, Pakistan; Paleocene (SBZ 3). (**1–2**, **6**, **8**, **15–16**, **20**) Axial sections. (**3–5**, **11–14**) Sections perpendicular in respect to the shell's coiling axis. (**7**) Tangential section. (**9**, **17**) Adaxial sections. (**18–19**, **21–24**) Oblique sections. Abbreviations: *f* foramen, *up* umbilical plate, *fol* folia, *pil* pile, *foa* foliar aperture, *s* septum

Plate 5.15 *Lockhartia conditi* (Nuttall, 1926); sample 93520 in the Hangu Formation underlying the Lockhart Limestone, Dhak Pass, Salt Range, Pakistan; Paleocene (SBZ 3). (**1**, **5–7**) Sections perpendicular to coiling axis, uncentered. (**2–4**) Sections perpendicular to coiling axis, centered. (**8–13**) Axial sections. Abbreviations: *f* foramen, *up* umbilical plate, *fol* folia, *pil* pile, *foa* foliar aperture, *ilsp* intraseptal interlocular space

Plate 5.16 *Lockhartia hunti* Ovey, 1947; Early Eocene-Middle Eocene (SBZ 11- SBZ 13). (**1**, **3**, **5**) External views from both sides. (**2**, **14**) Eroded specimens showing chamber shape and their spiral arrangement below the dorsal shell surface. (**4**, **6–8**, **15**) Dorsal side of the shells. (**9–11**, **13–14**) Dorsal aspect of tightly coiled specimens with prominent beads at each junction of the spiral with the radial septal sutures. (**12**) Ventral side of a tightly coiled shell with the shell's face

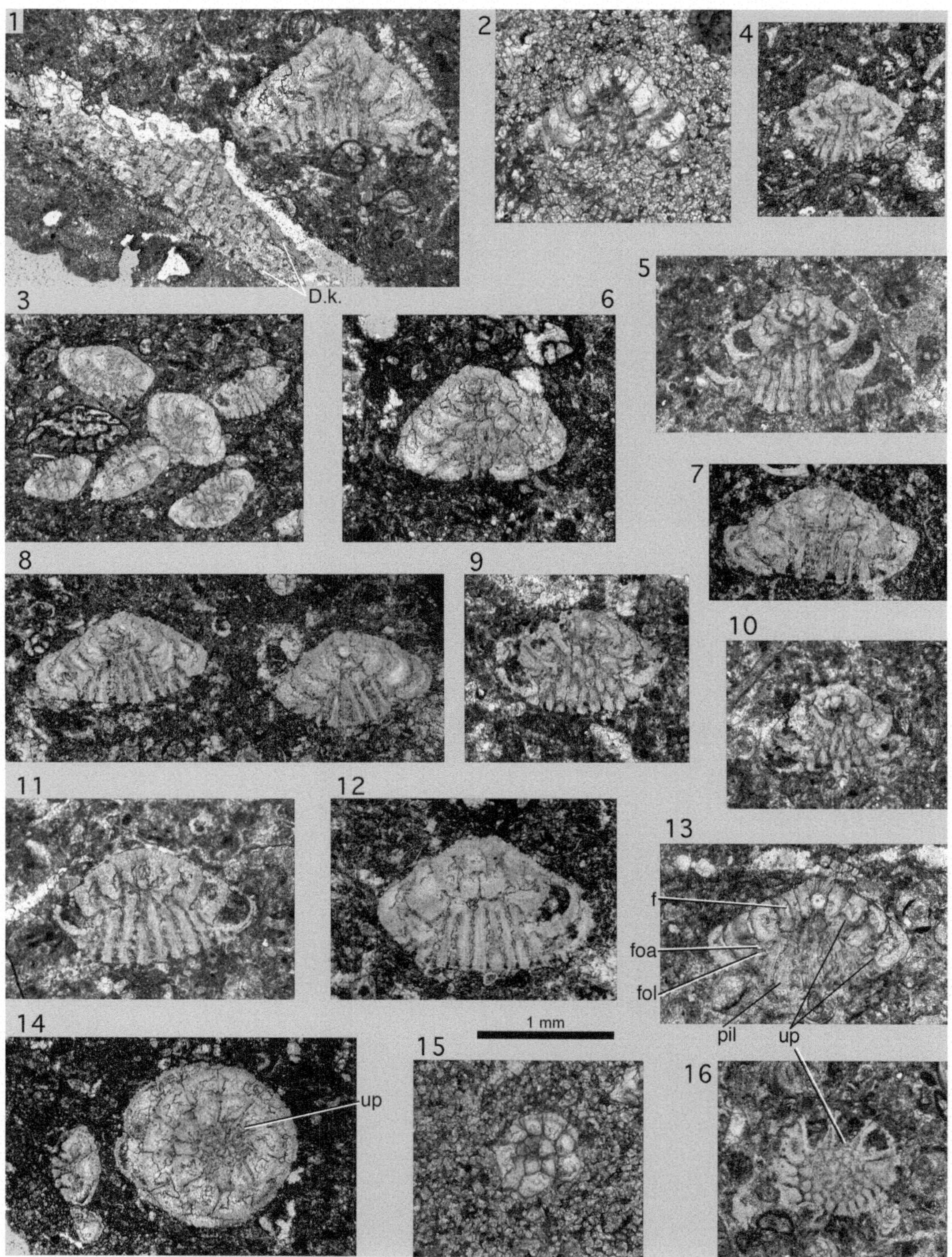

Plate 5.17 *Lockhartia hunti* Ovey, 1947; all specimens from sample All77268, Pakistan, collected by F. Allemann; sections of specimens caught in cemented rock. (**1**) Axial section of conical specimen (*top right*) and damaged fragment (*bottom left*) of a low-conical shell that possibly belongs to (D. k.) *Dictyoconoides kohaticus* (Davies, 1926). (**2**) Axial section. (**3**) Several random sections documenting the local abundance of the specimens of this species. (**4**) Oblique section inclined less than 45° in respect to the shell's coiling axis. (**5**)

differences were observed in the morphology of the shell. We interpret the two kinds of proloculi as a reflection of the two generations of a life cycle in a mesotrophic environment.

Lockhartia hunti Ovey, 1947; Plate 5.16, Figs. 1–15; Plate 5.17, Figs. 1–16; Plate 5.18, Figs. 16–27.

1947 *Lockhartia hunti*—Ovey, p. 573, pl. 10, figs. 1–6, pl. 11, fig. 1.

1954 *Lockhartia hunti* Ovey—Smout, p. 54, pl. 4, fig. 7.

1954 *Lockhartia hunti pustulosa*—Smout, p. 54, pl. 4, figs. 8–10.

1970 *Lockhartia hunti* Ovey var. *pustulosa* Smout—Singh, p. 38, pl. 5, 3a–f.

1979 *Lockhartia conditi* (Nuttall)—Allemann, p. 230, 261, pl. 7, figs. 1, 6.

Description: The bilamellar-perforate shells form a cone with a flattened-convex cone base on the umbilical side and a rounded apex on the dorsal side. The periphery is also rounded and lacks any keel. The dorsal side is decorated by a series of projecting beads located at the intersection of the spiral with the chamber sutures. Thus, they form a spiral of their own. The amount of projection of the beads is increasing towards the shell apex because the height of relief depends not only on the cumulated thickness of the secondary lamellae but also on their number. On the evolute dorsal side of the shell the number of lamellae corresponds to the total number of chambers that have grown after the placement of the bead.

The ventral side of the cone exhibits an umbilical area covered by numerous piles; there are at least five of them that are cut in an axial section. The chambers are longer in the direction of growth than their radial extension and separated by radial, straight septa on both dorsal and ventral sides of the coil. Adult last whorls count 15–20 chambers. The megalosphere reaches 0.13 mm. No microspheric specimens were found.

Remarks: Smout (1954) distinguished *Lockhartia hunti* Ovey, 1947 from a *Lockhartia hunti pustulosa* Smout, 1954 on the basis of heavier beads on the dorsal side of the shell (Plate 5.16, Figs. 9–15). I consider these differences as insufficient to distinguish a separate taxon. Similar differences are observed also in *Lockhartia tipperi*, a species reaching an adult diameter twice as large as in *L. hunti*.

Dictyoconoides kohaticus is not the microspheric generation of *Lockhartia hunti* in spite of their cohabitation. *D. kohaticus* has its own megalospheric generation with multiple spirals.

Lockhartia tipperi (Davies, 1926); Plate 5.19, Figs. 1–14; Plate 5.20, Figs. 7–13.

1926 *Conulites tipperi*—Davies, p. 247, Fig. 8.

1931 *Dictyoconoides tipperi* (Davies)—Nuttall and Brighton, p. 56, pl. 3, figs. 14–17.

1937 *Lockhartia tipperi* (Davies)—Davies and Pinfold, p. 48, pl. 6, fig. 14–16; pl. 7, fig, 17.

1954 *Lockhartia tipperi* (Davies)—Smout, p. 55, pl. 4, figs. 11–13.

1970 *Lockhartia hunti* Ovey—Singh, p. 38, pl. 5, figs. 2a–d.

1976 *Lockhartia tipperi* (Davies)—Ho et al., p. 51, pl. 27, figs. 11–12.

1994 *Lockhartia tipperi* (Davies)—Samanta and Bandopadhyay, p. 182, pl. 3, figs. 1–3.

1998 "*Dictyoconoides*" cf. *tipperi* Davies—Pignatti et al., p. 622, pl. 4, fig. 1.

1999a *Lockhartia tipperi* (Davies)—Akhtar and Butt, p. 132, pl. 4, fig. 4.

1999b *Lockhartia tipperi* (Davies)—Akhtar and Butt, p. 199, pl. 3, fig. 7.

Plate 5.17 (continued) Axial section. (**6**) Oblique section inclined a little less than the cone mantel in respect to the coiling axis of the shell; note size of the proloculus, possibly a megalosphere. (**7–13**) Axial and subaxial sections; note in (**13**) the foramen. (**14–16**) Sections perpendicular to the coiling axis; note in (**16**) the umbilical plate. Abbreviations: *f* foramen, *up* umbilical plate, *fol* folia, *pil* pile, *foa* foliar aperture

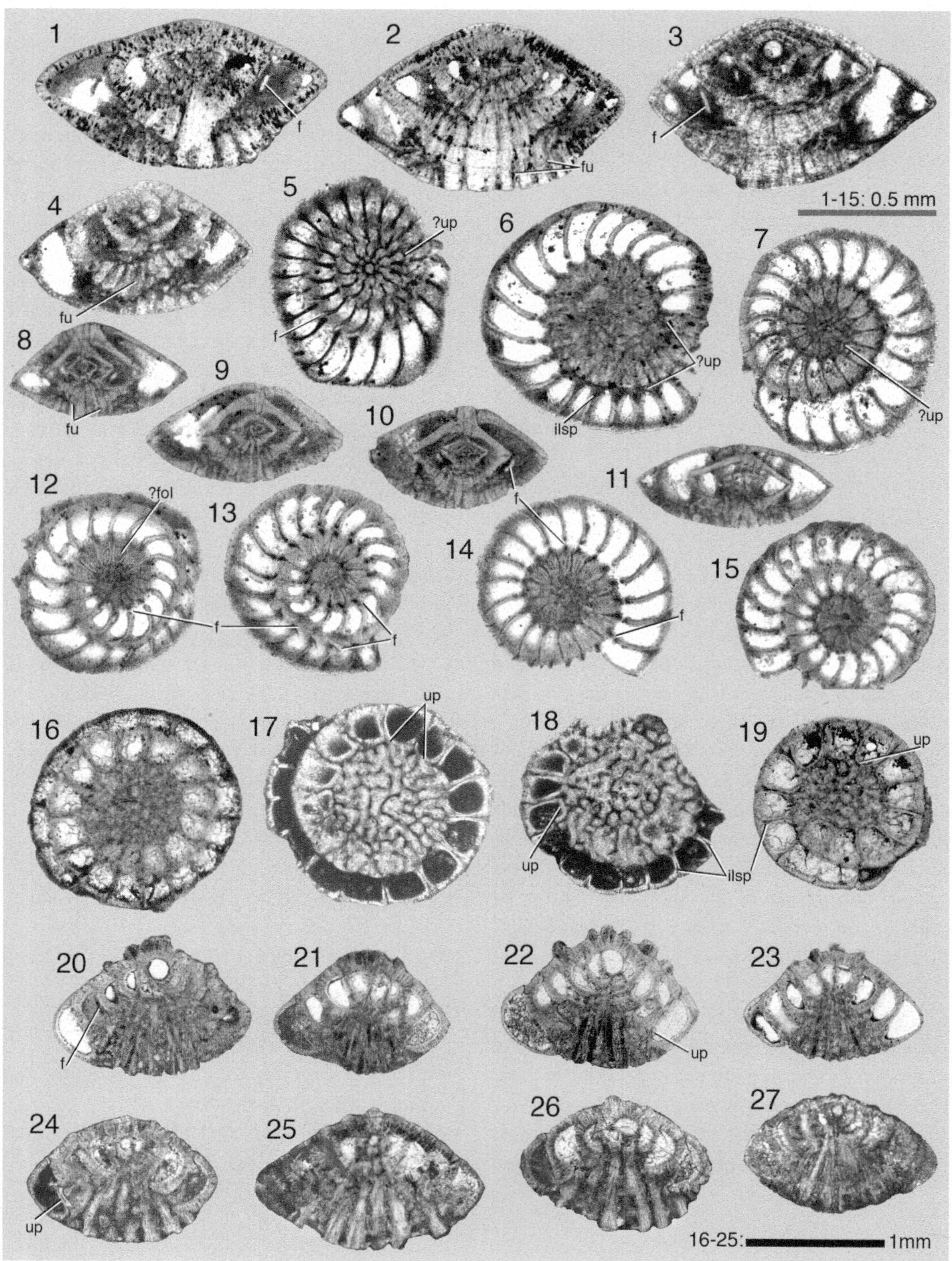

Plate 5.18 (**1–7**) *Smoutina cruysi* Drooger, 1960; topotypes. (**1–4**) Axial sections. (**5–7**) Sections perpendicular to the shell's coiling axis. (**8–15**) *Storrsella haastersi* (Van den Bold, 1946); topotypes. (**8–11**) Axial sections. (**12–15**) Sections perpendicular to the shell's coiling axis. (**16–27**) *Lockhartia hunti* Ovey, 1947; all from the well AQ 6, Qatar, Arabian Peninsula, collected by M. Chatton. (**16–19**) Sections perpendicular to the shell's coiling axis; note the polygonal connections between neighboring funnels. (**20–27**) Axial sections; note the dorsal ornamentation of the shell by comparatively few, heavy pustules with pores that deviate from the radial direction in the regular shell wall to form V-shaped pattern in the sections. Abbreviations: *f* foramen, *up* umbilical plate, *fol* folia, *fu* funnel, *ilsp* intraseptal interlocular space

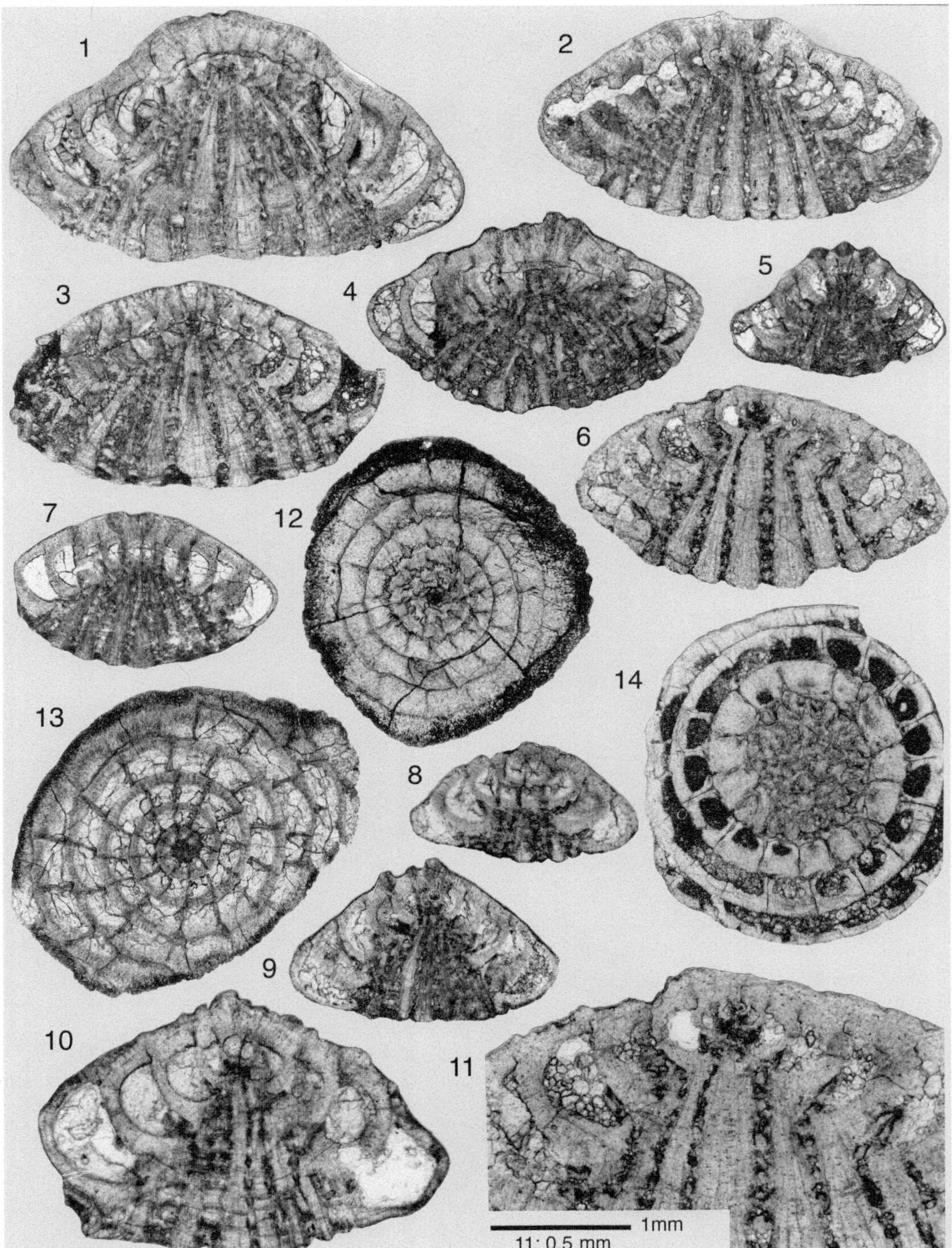

Plate 5.19 *Lockhartia tipperi* (Davies, 1926); isolated specimens; all specimens from samples 93544 and 93545, Nammal Formation, Nilawahan Gorge, Salt Range, Pakistan; sample 93545 contains also *Daviesina ruida* (Schwager, 1863); Early Eocene, Ilerdian (SBZ 6–8). (**1–11**) Axial sections. (**12–14**) Sections perpendicular to the shell's coiling axis; note the chamber shape, longer than high, with almost radial septa below the dorsal shell surface. (**1**, **6**) and detail of 6 in 11 represent the microspheric generation; all other axial sections represent the megalospheric generation. (**5**, **8–9**) are young specimens of *L. tipperi* or specimens to be identified as *Lockhartia* cf. *hunti* Ovey, 1947

Plate 5.20 (**1–6**) *Lockhartia* cf. *hunti* Ovey, 1947; samples 95166 and 95168, both from the Panoba shales, Chechan village, Kohat Basin, northwestern Pakistan; Cuisian (SBZ 11–12?). (**1–6**) Axial sections of *Lockhartia* cf. *hunti* or young specimens of *L. tipperi* (Davies, 1926). (**7–13**) *Lockhartia tipperi* (Davies, 1926). (**7**) Section perpendicular to coiling axis showing the shape of the chambers in outer whorls: much longer than high and with almost radial septa. (**8–13**) Axial sections; note the high number of umbilical piles. (**10**) Possible microspheric specimen. (**9**) Sample

2008 *Lockhartia conditi* (Nuttall)—BouDhagher-Fadel, p. 362, pl. 6.28, fig. 14.

Description: The bilamellar-perforate shells form cones of large size that reach an equatorial diameter of 3.3 mm and an axial height of 1.8 mm. The dorsal shell surface is decorated by heavy pustules of low projection. They may form a kind of cobblestone pavement. The dorsal contour of the chambers is longer in the direction of growth in respect to their radial extension. The last whorl of an adult specimen counts 15–17 chambers. The ventral side may be less convex as compared to the dorsal side and is dominated by a wide open umbilicus filled with umbilical files. An axial section cuts six or more piles in their axis. The megalosphere is small, 0.12 mm wide whereas the microsphere is 0.04 in diameter.

Remarks: The microspheric generation is relatively frequent, as in *Lockhartia conditi* and the *Dictyoconoides* species, indicating mesotrophic strategies of life. *Lockhartia tipperi* is not the megalospheric generation of *Dictyoconoides flemingi* (see below) because there are two generations in *L. tipperi* and their stratigraphic distribution, although overlapping, is not identical.

5.3 *Dictyoconoides* Nuttall, 1926

Type species: *Conulites cooki* Carter, 1861

Remarks: The bilamellar-perforate shell produces a low cone by a chamber arrangement in two or more simultaneous spirals in the adult stage of growth. The dorsal side of the cone has neither supplemental skeleton nor orifices of the intraseptal canal system or sutural apertures. The dorsal ornamentation consists of isolated beads located at the junction of cameral and whorl sutures. The ventral (umbilical) side of the shell has a flattened cone base covered by numerous pile heads. The cone base is slightly convex or may be slightly concave in advanced very large species. Subsequent chambers of a whorl communicate by a foramen formed by an interiomarginal arch located immediately below the periphery. The first chamber of an additional spiral is fed by the intraseptal interlocular space between two adjacent chambers of the previous whorl. The umbilical space is subdivided by the numerous folia in horizontal direction and the piles in vertical direction into stacks of "chamberlets" characteristic for the lockhartiines. There is a curious inversion of the usual relative number of gamonts (A-forms in high numbers) and agamonts (B-forms in low numbers) in the populations of *Dictyoconoides* that must have some biological meaning beyond possible artefacts caused by preferential collection of the much larger microspheric shells.

Dictyoconoides flemingi Davies and Pinfold, 1937; Plate 5.21, Figs. 1–9.

1937 *Dictyoconoides flemingi*—Davies and Pinfold, p. 51, pl. 6, figs. 11–13.

2006 *Dictyoconoides flemingi* Davies and Pinfold—Hottinger, p. 85, pl. 2, figs. 2–3.

2007 *Dictyoconoides flemingi* Davies and Pinfold—Singh, pl. 1, fig. 2.

Description: The bilamellar-perforate low cone reaches an equatorial diameter of 4.8 mm. The chambers are arranged in several spirals. Their start is marked by arrows in Plate 5.21, Figs. 1–2. They accelerate the growth of the shell by the simultaneous addition of new chambers at several locations on the periphery. The dorsal side of the shell is marked by sutural

Plate 5.20 (continued) 95166; (**7–8**, **10–11**) sample 95169; both samples from Panoba shales, Chechan village, Kohat Basin, northwestern Pakistan. (**12**) Sample 93549, Badrar beds, Nilawahan Gorge, Nurpur, Salt Range; Cuisian (SBZ 11–12). (**13**) Sample 93557, *Assilina* beds, Badrar beds, Ara village, Salt Range, Pakistan. Abbreviations: *f* foramen, *up* umbilical plate, *fol* folia, *pil* pile

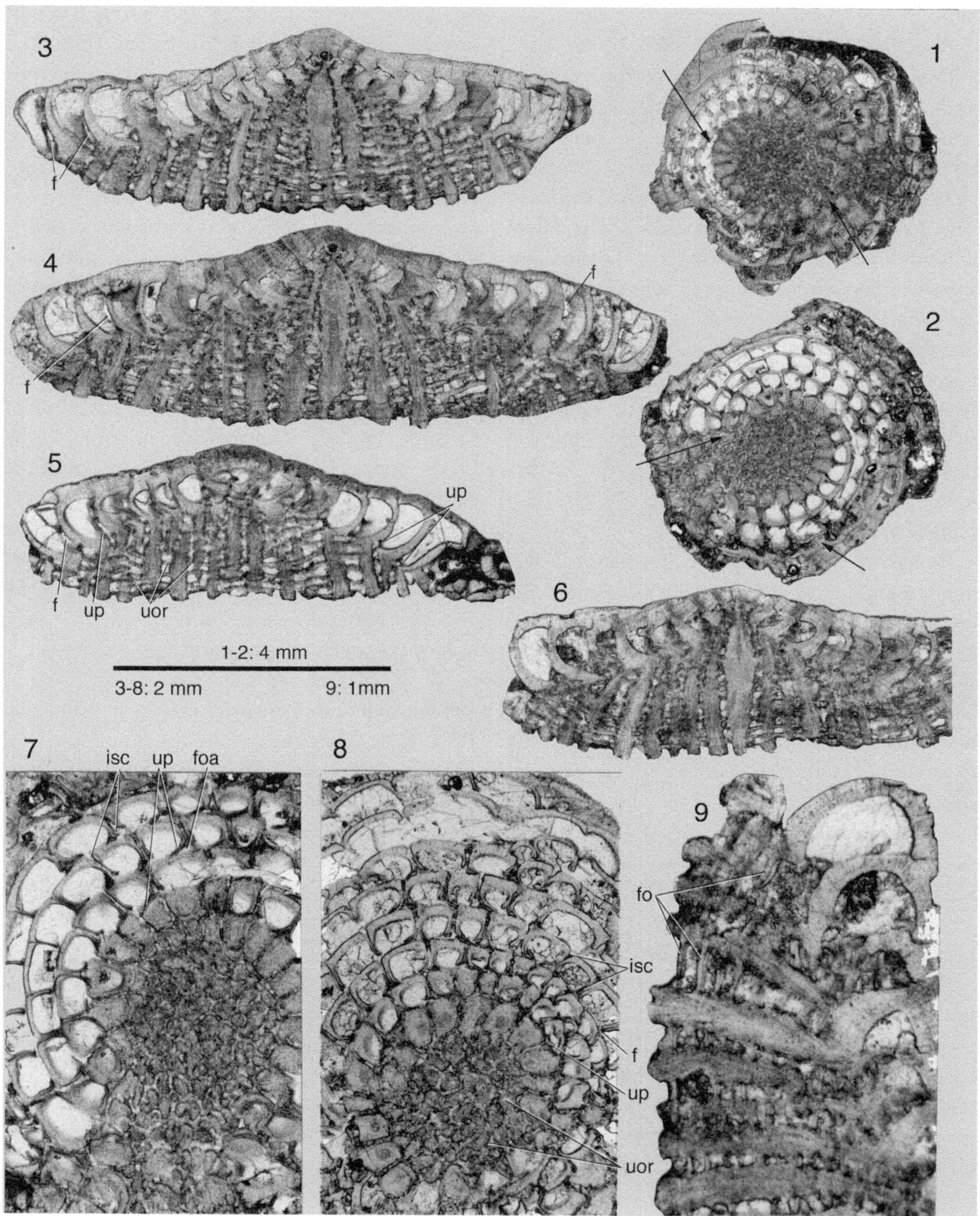

Plate 5.21 *Dictyoconoides flemingi* Davies and Pinfold, 1937; all specimens are from the sample 92010a, base Patala Formation, Dhak Pass, Salt Range, Pakistan, associated with *Ranikothalia nuttalli* (Davies, 1927); Paleocene (SBZ 4). (**1–2**) Sections perpendicular to the coiling axis showing the start of supplemental spirals (*arrows*). (**3–6**) Axial sections. (**7–8**) Details of (**1–2**). (**9**) Detail of **6**. Abbreviations: *f* foramen, *up* umbilical plate, *uor* umbilical orifices, *isc* intraseptal canal system, *foa* foliar aperture, *fo* folium

depressions and a rising apex. The ventral side, covered by the heads of the umbilical piles over the large umbilical extension, is more convex than the dorsal side. The ratio of the equatorial to the axial diameter varies from 3.4 to 3.6. The periphery without keels is rounded. The chambers are isometric or somewhat longer in the direction of growth than their radial extension. The septa are straight, radial on the ventral side, slightly inclined backwards on the dorsal side of the spiral. In an adult whorl, there are about 24 chambers. In axial sections of adult shells, 4–7 umbilical piles are sectioned per 1 mm radius.

The foramen, seen in axial sections of the shell, is forming a low arch over the ventral interiomargin of the chamber. The folia are apparently bilamellar (Plate 5.21, Fig. 9) and can therefore bear regular pores. There is a foliar aperture providing a direct umbilical passage from the foliar chamberlet to the ambient environment. It is complemented by additional orifices along the umbilical piles that connect the umbilical lumina with each other and with the substrate of the organism. Compared to the adult size of the shell, the proloculus is small, never exceeding 0.08 mm. Keeping in mind the dimorphism of the proloculi in *Lockhartia tipperi* described above, a proloculus of 0.08 mm size is interpreted here also as microspheric.

Dictyoconoides kohaticus (Davies, 1926); Plate 5.17, Fig. 1; Plate 5.22, Figs. 1–9; Plate 5.23, Figs. 1–12.

1926 *Conulites kohaticus*—Davies, p. 240, Figs. 1–5.

1930 *Rotalia kohaticus* (Davies)—Rijsinge van, p. 116, text-figs. 1–12, pl. 5, fig. 1; pl. 6, figs. 1–5.

1931 *Dictyoconoides kohaticus* (Davies)—Nuttall and Brighton, p. 54, text-fig. 2, pl. 3, figs. 9–13.

1932 *Dictyoconoides kohaticus* (Davies)—Davies, p. 405, pl. 1, figs. 2, 6, 8, 10–11, 13–14; pl. 2, fig. 1.

1948 *Dictyoconoides cooki* (Carter)—Gill, pars?, p. 174, pl. 9, figs. 1–11; pl. 10, figs. 1–12.

1962 *Dictyoconoides kohaticus* (Davies)—Sander, p. 25, pl. 4, figs. 22–24.

1970 *Dictyoconoides cooki* (Carter)—Kaever, p. 90, pl. 9, figs. 1–7.

2008 *Dictyoconoides kohaticus* (Davies)—BouDagher-Fadel, p. 399, pl. 6.28, figs. 6–7.

Remarks: Van Rijsinge (1930), a student of J. Hofker sen., correctly identified the basic architecture of *Dictyoconoides* as similar to the one of *Rotalia* and advocated therefore its transfer to the genus *Rotalia*. But he did not carry out this transfer himself and described the fossil under the heading *Dictyoconoides*. Rijsinge claimed the species to be trimorphic. There is no convincing evidence in his material and the much more numerous specimens at my disposal also do not show any trimorphic features.

In his material from Pakistan that seems to have the same early Lutetian age, Gill (1948) did not find significant morphological differences to separate *Dictyoconoides cooki* from *D. kohaticus*. As long as the detailed biostratigraphic relationships of the various occurrences of the Middle Eocene *Dictyoconoides* are not cleared by their nummulitic associates, I prefer to keep the two species names in use. The lower and smaller cones of *D. kohaticus* have a dimorphism that is documented by a comparatively large number of megalospheric specimens. Possibly the smaller and flatter specimens described by Davies (1926) as *Conulites vredenburgi* represent the megalospheric generation of *D. kohaticus*. In *D. cooki* only microspheric specimens are known so far. They are associated to *Lockhartia hunti* that is not much smaller than megalospheric *D. kohaticus* but lacks intercalary spirals.

Dictyoconoides cooki (Carter, 1861); Plate 5.24, Figs. 1–4.

1861 *Conulites cooki*—Carter, p. 457, pl. 15, fig. 7.

1862 *Patellina cooki* (Carter)—Carpenter et al., p. 233, text-fig. 38.

1932 *Dictyoconoides cooki* (Carter)—Davies, p. 405, pl. 1, figs. 9, 12c, 12e.

Plate 5.22 *Dictyoconoides kohaticus* (Davies, 1926); sample 00001, Oman. (**1–3**) External view of dorsal side of the shell. (**1**) Microspheric specimen. (**2–3**) Megalospheric specimen. (**4**) Both sides of microspheric specimen. (**5**) Both sides of megalospheric specimen; note on the shell's ventral side the four separate apertural faces corresponding to four spirals growing simultaneously. (**6–9**) Sectioned megalospheric specimens in axial (**6**) and in directions perpendicular to the coiling axis (**7–9**). Abbreviations: *af* apertural face, *pil* pile

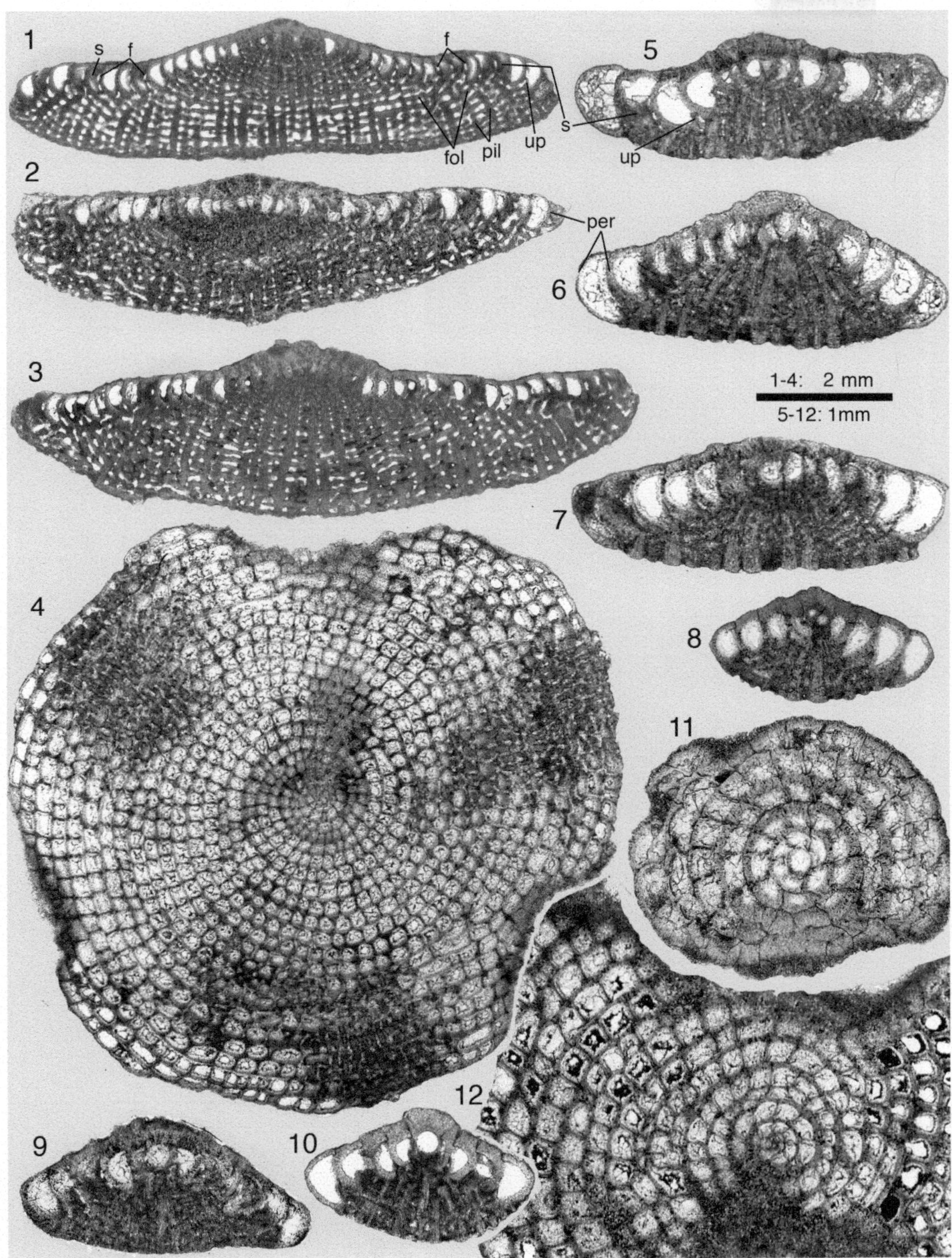

Plate 5.23 *Dictyoconoides kohaticus* (Davies, 1926); sample 00001, Oman. (**1–4**, **12**) Microspheric generation. (**5–11**) Megalospheric generation. (**1–3**, **5–10**) Axial sections. (**4**, **11–12**) Sections perpendicular to coiling axis. Abbreviations: *f* foramen, *up* umbilical plate, *per* periphery, *s* septum, *pil* pile, *fol* folia

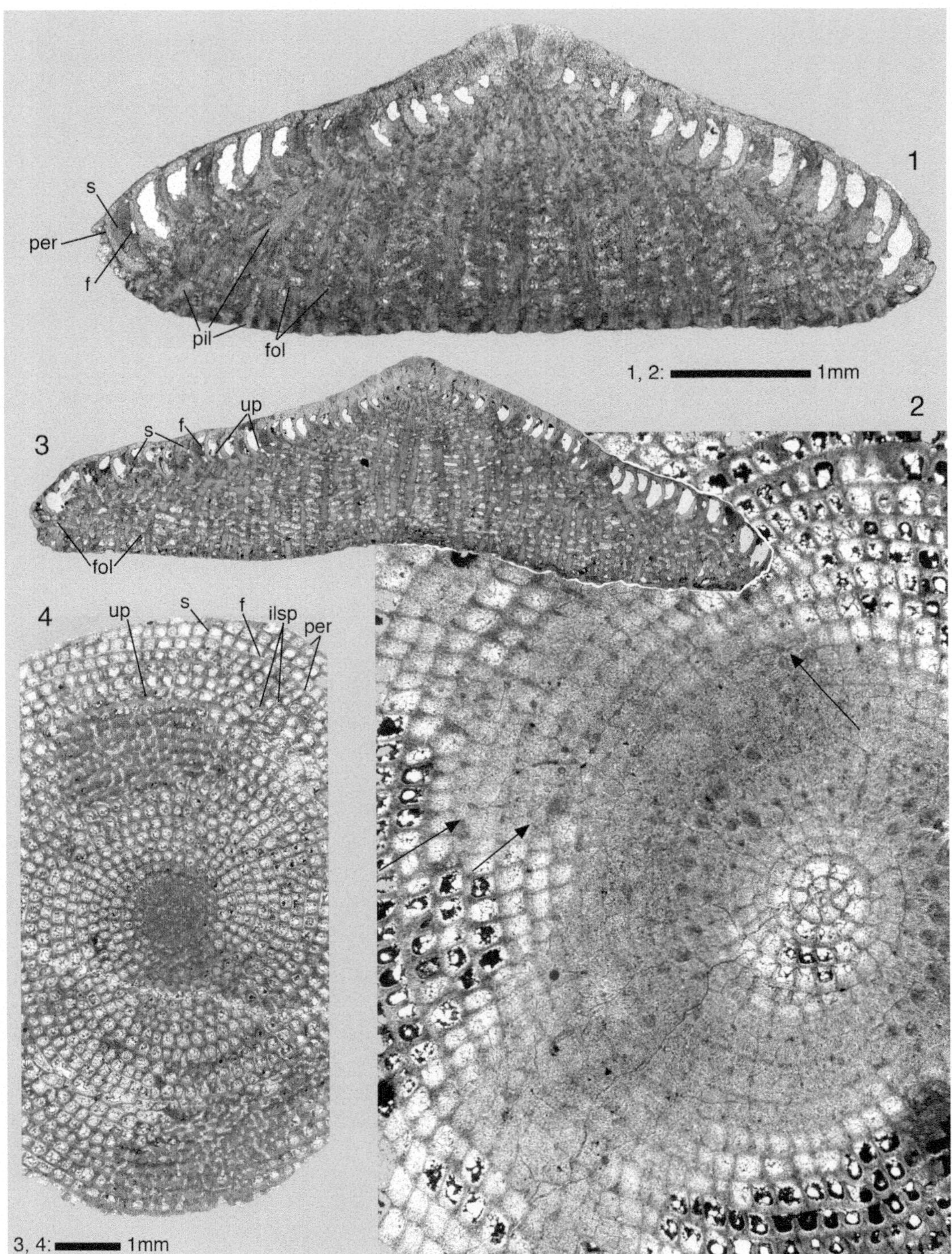

Plate 5.24 *Dictyoconoides cooki* (Carter, 1861); sample with exclusively microspheric specimens, associated with *Alveolina elliptica nuttalli* Davies, 1940; Qatar, collected by Ch. Pomerol; Lutetian (SBZ 13–14). Microspheric, free specimens sectioned in axial (**1**, **3**) direction and perpendicular to the coiling axis (**2**, **4**). *Arrows*: initiation of supplemental spirals. Abbreviations: *f* foramen, *up* umbilical plate, *per* periphery, *s* septum, *pil* pile, *fol* folia, *ilsp* intraseptal interlocular space

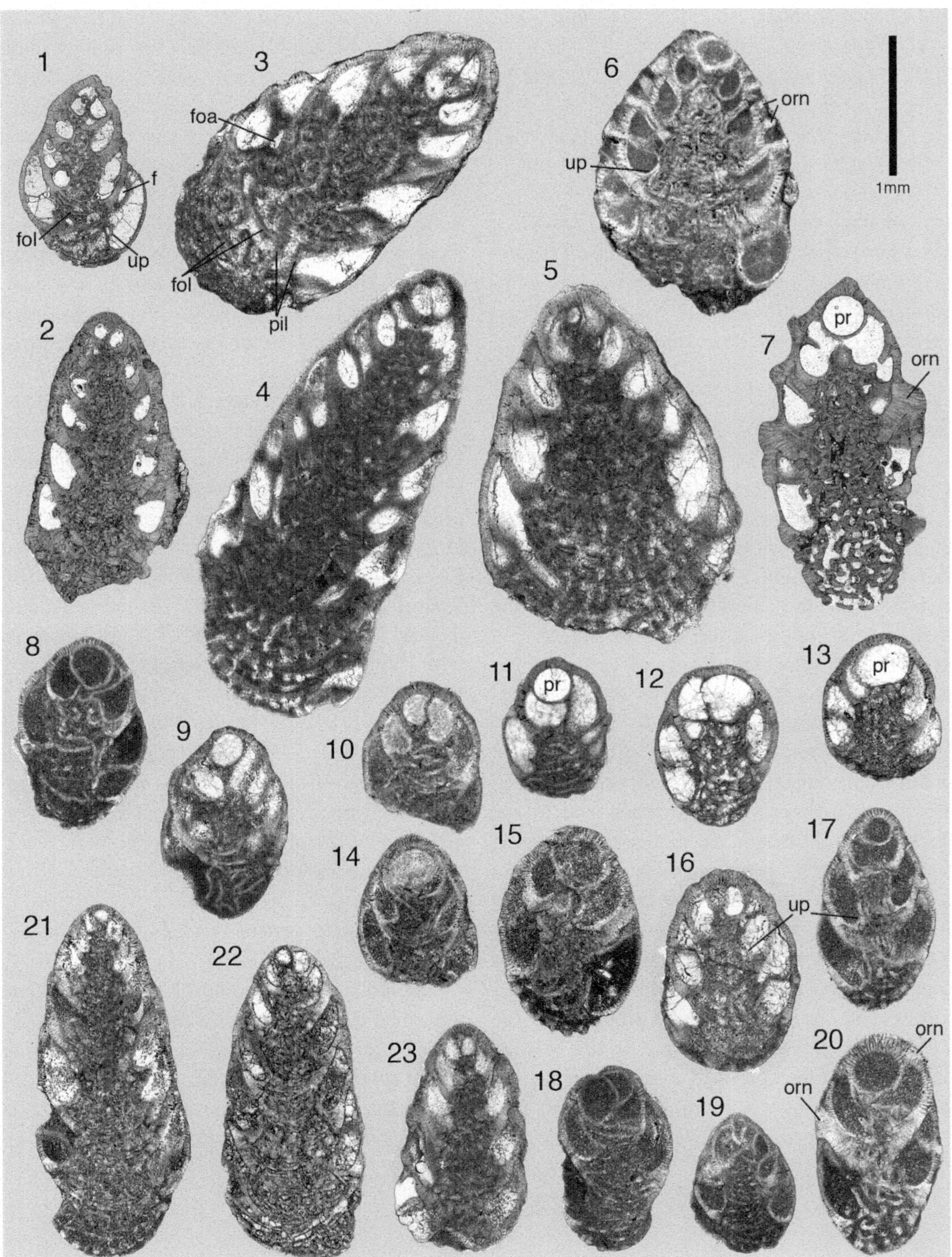

Plate 5.25 (**1**) *Sakesaria pyrum* Ruggieri, 1950, axial section. (**2**, **7**, **20**) *Sakesaria costulata* Ruggieri, 1950, axial sections; note the thickened walls produced by limbate spiral sutures crossing longitudinal ribs; specimens **2** and **20** have eroded surfaces where the ornamentation has been worn off. (**6**) *Sakesaria trichilata* Sander, 1962, axial section; note ornamentation. (**3–5**) *Sakesaria cotteri* Davies and Pinfold, 1937, axial sections. (**8–19**) *Sakesaria somalica* Ruggieri, 1950; (**8–18**) axial sections; (**19**) section parallel to coiling axis, missing the proloculus. (**21–23**) *Sakesaria cylindrata* Ruggieri, 1950, axial sections. (**1**) From well

1948 *Dictyoconoides cooki* (Carter)—Gill, pars? p. 174, pl. 8, figs. 1–11.
2007 *Dictyoconoides cooki* (Carter)—Singh, pl. 9, figs. 1–8.
non: 2008 *Dictyoconoides cooki* (Carter)—BouDagher-Fadel, pl. 6.28, figs. 5–6 (= *Dictyoconus* cf. *indicus* Davies).

Remarks: Whereas *Dictyoconoides kohaticus* has a horizontal diameter to axial thickness ratio of 4 or more, *D. cooki* is larger and thicker, with a corresponding ratio 2.7–3.8. No megalospheric specimens have been discovered so far.

5.4 *Sakesaria* Davies, 1937

Type species: *Sakesaria cotteri* Davies, 1937

Remarks: The bilamellar-perforate shells are high-trochospiral with a ratio of horizontal diameter to axial height always lower than 1. The convexity of the umbilical area provides the shell with an upright-oval, egg- to pear-shaped axial outline. The dorsal side of the shell has neither supplemental skeleton nor sutural supplementary apertures but frequently the chamber and whorl sutures are heavily limbate. Most ornamental elements, beads or ribs, are perforate. The pores diverge from the inner side of the thickened wall towards the shell surface keeping a more or less perpendicular orientation in respect of the lamellar layering of the wall (Plate 5.25, Fig. 7). The umbilical area is comparatively narrow, with only few pile heads that support a convex umbilical filling. The latter follows the principles of construction in the lockhartiine subfamily adapting to the relative narrowness of the umbilical space by a reduction of the number of umbilical piles. No microspheric specimens have been found so far.

The question arises whether a difference in disposition of the chambers in a low or high spiral warrants the use of separate generic names. In the Neogene–Recent genus *Pseudorotalia*, an analogous question was negatively answered by Billman et al. (1980). The corresponding placement of the prioritarian *Asanoina* Finlay, 1939 in synonymy with *Pseudorotalia* Reiss and Merling, 1958 was incorrect as pointed out by Loeblich and Tappan (1987). However, they kept the two genera apart. In this paper, the analogous situation between *Lockhartia* and *Sakesaria* is treated in the same way.

In the Paleogene, there is a particular species with a similar habit of chamber disposition in a high trochospiral. It is characterized by limbate spiral and chamber sutures that produce a heavy ornamentation of the spiral side of the shell. Originally described as *Asanoina eocaenica* Sacal and Debourle, 1957, it was attributed to *Sakesaria* by Sztràkos (2000). The internal architecture, however, reveals its relationships with *Gyroidinella* (Victoriellidae, see below, Pls. 9.10–9.11).

Sakesaria cotteri Davies and Pinfold, 1937; Plate 5.25, Figs. 3–5; Plate 5.26, Figs. 1–7, 9–12.
1937 *Sakesaria cotteri*—Davies and Pinfold, p. 49, text-fig. 2, pl. 7, figs. 18, 21–24.
1950 *Sakesaria cotteri* Davies—Ruggieri, p. 95.
1954 *Sakesaria cotteri* Davies—Smout, p. 57, pl. 5, figs. 1–3.
1962 *Sakesaria cotteri* Davies—Sander, p. 32, pl. 4, fig. 27–32.

Remarks: *Sakesaria cotteri*, for the first time described from Pakistan's Salt Range by Davies in 1937, represents the genus all alone during early Lower Eocene (Ilerdian) all over the central and western Neotethys from Pakistan to Somalia and Egypt. These shells have the shape of an

Plate 5.25 (continued) AQ 6, 780′, Qatar, collected by M. Chatton. (**2, 7**) From well AQ 6, 790′ Qatar, collected by M. Chatton. (**6**) From sample G.123–3, 50′ from base of Khairabad Limestone, South of Dalwal, Salt Range, Pakistan, collected by W.D. Gill, 1948. (**3, 5**) From sample 93545, Nammal Formation, Nilawahan Gorge, Salt Range, Pakistan; Ilerdian. (**4**) From sample 93567, top Nammal Formation, Dandot village, Salt Range, Pakistan. (**8–23**) From Well AQ 6, 875′, Qatar, collected by M. Chatton. Abbreviations: *pr* proloculus, *orn* ornamentation, *pil* pile, *fol* folia, *foa* foliar aperture, *f* foramen, *up* umbilical plate

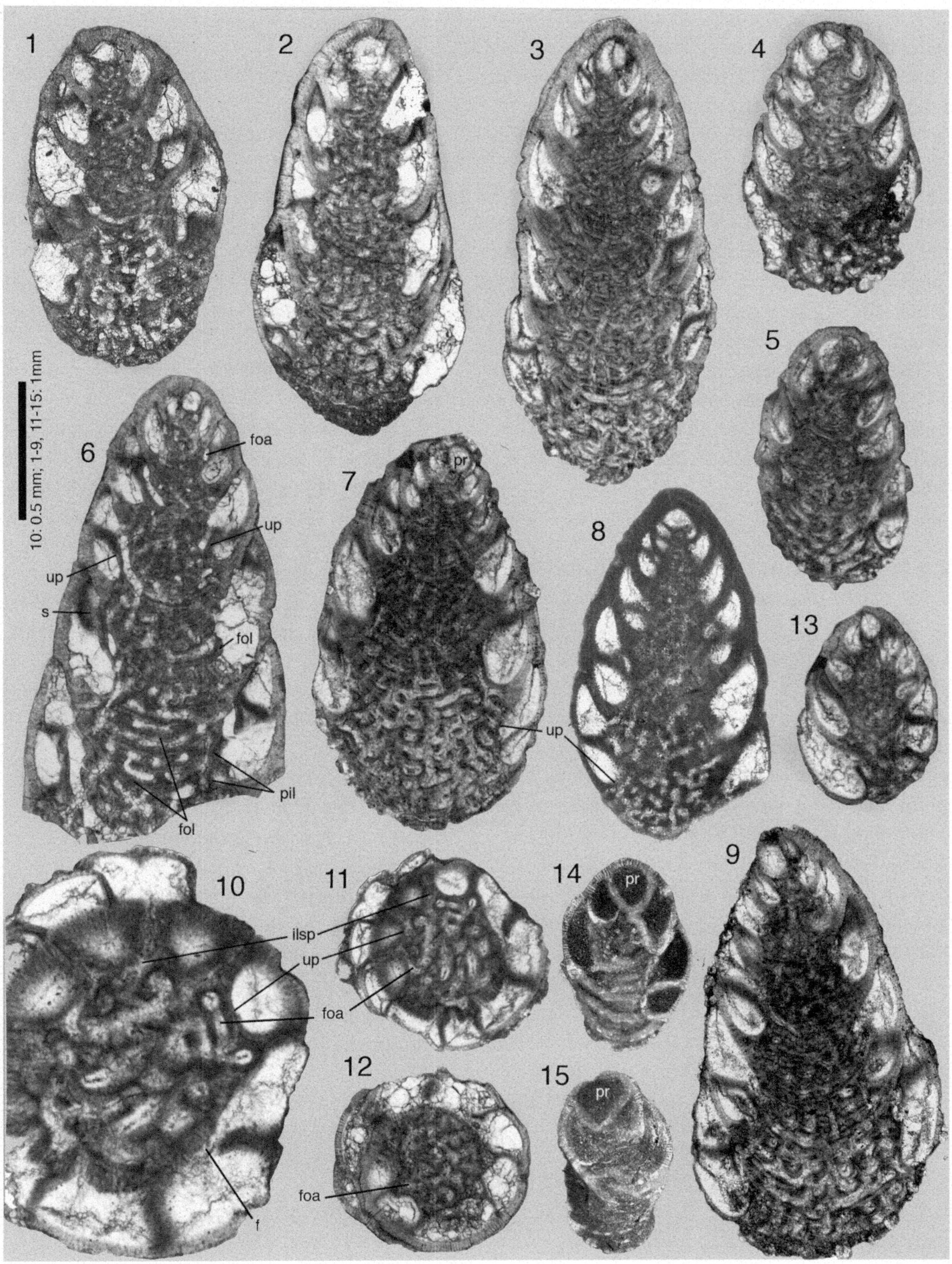

Plate 5.26 (**1–7**, **9–12**) *Sakesaria cotteri* Davies and Pinfold, 1937. (**1–7**, **9**) Axial sections. (**10–12**) Sections perpendicular to the shell's coiling axis. (**1–3**, **7**, **9–12**) From samples 93544 and 93545, Nammal Formation, Nilawahan Gorge, Salt Range, Pakistan. (**2–6**) From sample 93567, Nammal Formation, Dandot village, Salt Range, Pakistan. (**8**, **13**) *Sakesaria pyrum* Ruggieri, 1950; axial sections, from sample 93545, Nammal Formation, Nilawahan Gorge, Salt Range, Pakistan; Ilerdian. Abbreviations: *f* foramen, *up* umbilical plate, *pr* proloculus, *s* septum, *pil* pile, *fol* folia, *foa* foliar aperture, *ilsp* intraseptal interlocular space

inverted carrot with a ratio of equatorial to axial diameter of 0.42–0.66. The carrot reaches 3.8 mm in length. The base of the high cone is much convex and corresponds to a hemisphere. Accordingly, the ventral walls of the spiral chambers are much inclined in respect to the coiling axis. There are about 8–10 spiral chambers in the last whorl of shells with an equatorial diameter of 1.4 mm. The umbilicus is narrow and admits only few piles. Their number is difficult to measure in plane sections of the shell because their direction is oblique to the coiling axis.

The megalosphere reaches a diameter of 0.17 mm, takes a hemispheric shape and forms with an identical deuteroconch a dyad as embryo. Microspheric specimens have never been found.

5.4.1 Additional Species or Varieties of *Sakesaria*

Three authors (Ruggieri 1950; Smout 1954; Sander 1962) have introduced additional specific names for *Sakesaria*. Their papers are not clear about the taxonomic rank to be given to these names, species, subspecies or varieties. The junior authors apparently were not aware of the senior author's paper. The authors used morphological features visible on the exterior of free specimens. The material at my disposal is insufficient and often too poorly preserved to permit a thorough revision of the *Sakesaria* species. However, we feel obliged to present, in alphabetic order, an illustration of specimens that suggest at least the existence of several additional species in Paleocene strata. Their taxonomic treatment including their synonymies must be considered as tentative.

Sakesaria costulata Ruggieri, 1950; Fig. 5.3A; Plate 5.25, Figs. 2, 7, 20.

1950 *Sakesaria cylindrica costulata*—Ruggieri, p. 96, text-fig. 3; reproduced here at standard enlargements in Fig. 5.3A.

1962 *Sakesaria nodulifera*—Sander, p. 29, pl. 5, figs. 46–50.

1985 *Sakesaria cotteri* Davies—Hasson, p. 360, pl. 6, fig. 11.

Remarks: This species is characterised by heavy costae forming limbate sutures of the spiral chambers. Their oblique appearance in axial sections of the shell shows the pattern of divergence in the perforation of the thickened shell material that forms the ribs. Note also the large diameter of the megalosphere.

Sakesaria cylindrata Ruggieri, 1950; Fig. 5.3B–C; Plate 5.25, Figs. 21–23.

1950 *Sakesaria cylindrata*—Ruggieri, p. 96, text-figs. 1–2, 4.

1962 *Sakesaria nodulifera*—Sander, p. 29, pl. 5, figs. 46–50.

Description: Adult specimens have subcylindrical shapes and limbate chamber and whorl sutures of variable thickness. Axial sections of specimens that are identified accordingly as *Sakesaria cylindrica* in our material have strikingly small megalospheres reaching only 0.1 mm in diameter.

Sakesaria pyrum Ruggieri, 1950; Fig. 5.3D; Plate 5.25, Fig. 1; Plate 5.26, Figs. 8, 13.

1950 *Sakesaria pyrum*—Ruggieri, p. 98, text-fig. 7.

Description: Pear-shaped shells with slightly inflated chambers. The umbilicus is particularly narrow. The narrowness of the umbilicus increases where the shell has its widest circumference. The intercameral foramen, visible in a single specimen, is a comparatively high arch in interiomarginal position (Plate 5.25, Fig. 1).

Sakesaria somalica Ruggieri, 1950; Fig. 5.3E–F; Plate 5.25, Figs. 8–19.

1950 *Sakesaria somalica*—Ruggieri, p. 97, text-figs. 5–6.

1954 *Sakesaria dukhani*—Smout, p. 57, pl. 5, figs. 9–12.

1962 *Sakesaria teretra*—Sander, p. 28, pl. 5, figs. 42–45.

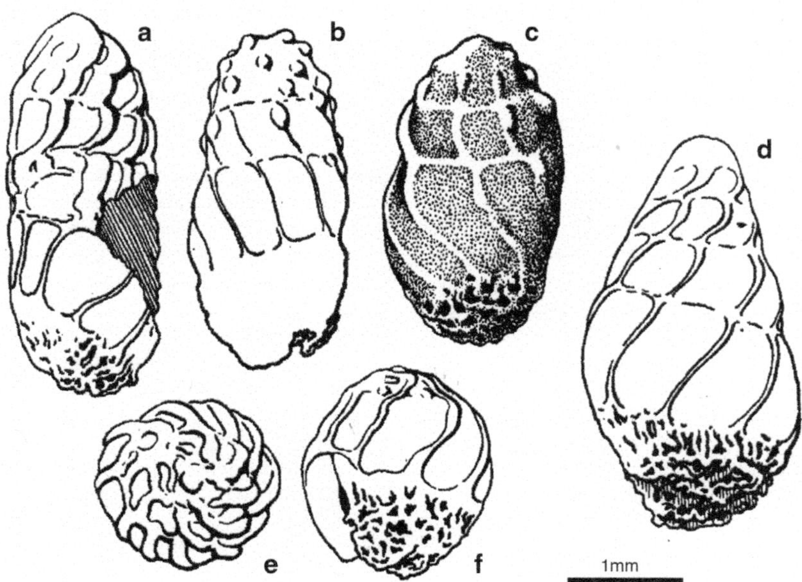

Fig. 5.3 Reproduction of Ruggieri's (1950) illustrations of selected Sakesaria species from Somalia, at standard enlargements according to the indications of the author. (**A**) *Sakesaria costulata* Ruggieri, 1950, lateral view. (**B**) *Sakesaria cylindrica* Ruggieri, 1950, lateral view. (**C**) *Sakesaria cylindrica* Ruggeri, 1950, juvenile specimen, lateral view; this specimen was selected for illustration to demonstrate the distribution of pores. (**D**) *Sakesaria pyrum* Ruggieri, 1950, lateral view. (**E, F**) *Sakesaria somalica* Ruggieri, 1950; **E**, apical view; **F**, lateral view

Remarks: The short, barrel-shaped shells have strikingly large megalospheres that are visible in the sections given by Smout (1954) for *Sakesaria dukhani*. The umbilicus is very narrow and filled with a severely restricted number of piles. The synonymy with *S. somalica* and *S. teretra* is based only on the similitude in shape of the shell.

Sakesaria trichilata Sander, 1962; Plate 5.25, Fig. 6.

1962 *Sakesaria trichilata*—Sander, p. 29, pl. 5, figs. 38–41.

Remarks: The pear-shaped shell exhibits a very strong ornamentation that rises not only from the chamber and whorl sutures and therefore must represent a coarsely cancellate structure. There was only a single specimen available for investigation.

"*Sakesaria*" *eocaenica* (Sacal and Debourle 1957), described for the first time under the generic name *Asanoina* from the lower Lutetian of Aquitaine (France) and attributed to *Sakesaria* by Sztràkos (2000), does not belong to the Paleocene and Early Eocene lockhartiid *Sakesaria*, nor does it to the Miocene-Pliocene pseudorotaliid genus *Asanoina*. A detailed description of the species *eocaenica* is given below in Chapter 9.

References

Accordi G, Carbone F, Pignatti J (1998) Depositional history of a Paleogene ramp (Western Cephalonia, Ionian islands, Greece). Geol Romana 34:131–205

Afzal J, Williams M, Aldbridge R (2009) Revised stratigraphy of the lower Cenozoic succession of the Grater Indus Basin in Pakistan. J Micropaleontol 28:7–23

Akhtar M, Butt AA (1999a) Lower Tertiary biostratigraphy of the Kala Chitta Range, Northern Pakistan. Rev Paléobiol 18:123–146

Akhtar M, Butt AA (1999b) Microfacies and foraminiferal assemblages from the early Tertiary rocks of the Kala Chitta Range (Northern Pakistan). Géol Mediterr 26(3–4):185–201

Allemann F (1979) Time of emplacement of the Zhob Valley Ophiolites and Bela Ophiolites, Baluchistan (Preliminary Report). In: Farah A, DeJong K (eds)

Geodynamics of Pakistan. Geol Surv Pakistan, Quetta, pp 215–249

Billman H, Hottinger L, Oesterle H (1980) Neogene to Recent Rotaliid foraminifera from the Indopacific Ocean; their canal system, their classification and their stratigraphic use. Schweiz Paläontol Abh 101:71–113, 27 text figs, 39 pls

BouDagher-Fadel MK (2008) Evolution and geological significance of larger benthic foraminifera. Developments in palaeontology & stratigraphy 21, 540 pp. Springer, Amsterdam

Carbonnel JP, Blondeau A (1977) Le groupe paléogène de Kerghana (Afghanistan du SW). Implications paleogéographique et structurale. Ann Soc Géol Nord 97:107–114

Carpenter WB, Parker WK, Jones R (1862) Introduction to the study of the foraminifera. Ray Society, Robert Hardwicke, London, 319 pp, 22 pls

Carter HJ (1861) Further observations on the structure of Foraminifera and on the larger fossilized forms of Sind, etc., including a new genus and species. J Bombay Branch Royal Asiatic Soc 6:31–96

Davies LM (1924) Notes on the geology of Kohat, with reference to the homotaxial position of the Salt Marl at Barhadur Khel. J Asiatic Soc Bengal Bombay (ns) 20: 207–224, 2 pls

Davies LM (1926) Remarks on Carter's genus *Conulites–Dictyoconoides* Nuttall, with description of some new species from the Eocene of North West India. Rec Geol Surv India Calcutta 59(2):237–257, 16–20 pls

Davies LM (1927) The Ranikot beds at Thai (North-West Frontier Provinces of India). Quart J Geol Soc 83:260–290, pls. 17–22, London

Davies LM (1930) The fossil fauna of the Samana range and some neighbouring areas: Part 6. The Palaeocene Foraminifera, Mem Geol Surv India (Palaeontol Indica) Calcutta (ns) 15:67–79, 10 pls

Davies LM (1932) The genera *Dictyoconoides* Nuttall, *Lockhartia* nov. and *Rotalia* Lamarck: their type species, generic differences and fundamental distinction from the *Dictyoconus* Group of forms. Trans R Soc Edinb 57:397–428, 4 pls

Davies LM, Pinfold ES (1937) The Eocene beds of the Punjab Salt range, Palaeontol Indica Calcutta 24(1):79 pp, 7 pls

Drooger CW (1960a) Microfauna and age of the Basses Plaines Formation of French Guyana I & II. Proc K Ned Akad Wet Amsterdam, Ser B 63(4):449–468

Drooger CW (1960b) Some Early Rotaliid Foraminifera I-III. Proc K Ned Akad Wet Amsterdam, Ser B 63 (2):287–334

Gill WD (1948) On the foraminifer *Dictyoconoides cooki* (Carter). Proc Leeds Phil Lit Soc (Sci Sect) 5 (2):174–182

Hasson PF (1985) New observations on the biostratigraphy of the Saudi Arabian Umm er Radhuma Formation (Paleogene) and its correlation with neighboring regions. Micropaleontology 31(4):335–364

Ho Y, Zhang B, Hu L, Shang J (1976) Mesozoic and Cenozoic foraminifera from the Mount Jolmo Lungma Region. In: A report of the scientific expedition in the Mount Jolmo Lungma 1966–1968. Paleontology (Beijing), Special publication 2, pp 1–76, 36 pls (in Chinese)

Hottinger L (2006) The "face" of benthic foraminifera. Boll Soc Paleontol Ital 45:75–89

Hottinger L (2009) The Paleocene and earliest Eocene foraminiferal family Miscellaneidae: neither nummulites nor rotaliids. Carnets Géol Article 2009/06, CG2009_A06

Ismail AA, Boukhary M (2008) Larger foraminifera from the Early Eocene of Shabwa area, Southeastern Yemen. Rev Paléobiol 27:89–97

Kaever M (1970) Die alttertiären Grossforaminiferen Südost Afghanistans unter besonderer Berücksichtigung der Nummulitiden-Morphologie, Taxonomie und Biostratigraphie. Münst Forsch Geol Paläontol 16(17):400 pp, 19 pls

Loeblich AR, Tappan H (1987) Foraminiferal genera and their classification. Van Nostrand Reinhold, New York, 1, 970 pp; 2, 212 pp, 847 pls

Müller-Merz E (1980) Strukturanalyse ausgewählter rotaloider Foraminiferen. Schweiz Paläontol Abh 101:5–68, 15 pls

Nicora A, Garzanti E, Fois E (1987) Evolution of the Tethys Himalaya continental shelf during Maastrichtian to Paleocene. Riv Ital Paleontol Strat 92:439–496

Nuttall WL (1926) The larger Foraminifera of the Upper Ranikot Series (Lower Eocene) of Sind (India). Geol Mag 63(2):112–121

Nuttall WL, Brighton AG (1931) Larger Foraminifera from the Tertiary of Somaliland. Geol Mag 63:49–65, 4 pls

Ovey CD (1947) A new Eocene species of *Lockhartia* Davies from British Somaliland. Ann Mag Nat Hist Lond, Ser 11, 13:571–576, 2 pls

Pignatti J, Matteucci R, Parlow T, Fantozzi L (1998) Larger foraminiferal biostratigraphy of the Maastrichtian-Ypresian Wadi Mashib succession (South Hadramawt Arch, SE Yemen). Z Geol Wiss 26(5–6):609–635

Rahaghi A (1978) Paleogene biostratigraphy of some parts of Iran. Nat Iranian Oil Comp Geol Lab 7:82 pp, 41 pls, Teheran

Rahaghi A (1983) Stratigraphy and faunal assemblage of Paleocene-Lower Eocene in Iran. Nat Iranian Oil Comp Geol Lab 10:73 pp, 49 pls, Teheran

Reiss Z, Merling P (1958) Structure of some Rotaliidea. Bull Geol Surv Israel 21:1–19

Revets SA (2001) The genus *Rotorbinella* Bandy, 1944 and its classification. J Foram Res 31:315–318

Rijsinge van C (1930) Some remarks on *Dictyoconoides* Nuttall (= *Conulites* Carter = *Rotalia* Lamarck). Ann Mag Nat Hist Ser 10, 5:116–137, 2 pls

Ruggieri G (1950) Foraminiferi del genere *Sakesaria* nel Paleocene della Migiurtinia. Giorn Geol ser 2 (11):94–98

Sacal V, Debourle A (1957) Foraminifères d'Aquitaine. 2e partie – Peneroplidae à Victoriellidae. Mém Soc Géol Fr (n s) 78(36):1–87, 35 pls

Samanta BK, Bandopadhyay KP (1994) Foraminiferal genus *Lockhartia* Davies from the Eocene succession of Cutch, Gujarat, western India. Indian J Geol 66 (3):165–189

Sander NJ (1962) Aperçu paléontologique et stratigraphique du Paléogène en Arabie Séoudite orientale. Rev Micropaleontol 5:3–40

Schwager C (1863) Die Foraminiferen aus den Eocaenablagerungen der lybischen Wüste und Aegyptens. Palaeontographica 30:79–154, 6 pls

Singh P (1970) Larger foraminifera from the Subathus of Beragua-Jangalgali Area, Jammù and Kashmir State. J Geol Soc India 11:34–44, 4 pls

Sirel E (1972) Systematic study of new species of the genera *Fabularia* and *Kathina* from Paleocene. Türk Jeol Kur Bül 15:277–249, 8 pls

Smout AH (1954) Lower Tertiary foraminifera of the Qatar peninsula. Brit Mus (Nat Hist), 96 pp, 44 figs, 15 pls

Smout AH, Haque M (1956) A note of the larger foraminifera and ostracoda of the Ranikot from the Nammal Gorge, Salt Range, Pakistan. Rec Geol Surv Pakistan 8 (2):49–60, 3 pls

Sztràkos K (2000) Les foraminifères de l'Eocène du Bassin de l'Adour (Aquitaine, France): biostratigraphie et taxinomie. Rev Micropaléontol 43(1–2):71–172, 23 pls

Tambareau Y (1972) Thanétien supérieur et Ilerdien inférieur des Petites Pyrénées, du Plantaurel et des Chaînons audois. Trav Lab Géol Pétrol Univ P Sabatier, 383 pp, 20 pls

Torre M (1966) Alcuni foraminiferi del Cretacico superiore della Peninsula Sorrentina. Boll Soc Natur Napoli 75:409–431

Van den Bold WA (1946) Contribution to the study of Ostracoda with special reference to the Tertiary and Cretaceous microfauna of the Caribbean region. Proefschrift Rijks-Univ. Utrecht, 167 pp, 18 pls

Vecchio E, Hottinger L (2007) Agglutinated conical foraminifera from the Lower-Middle Eocene of the Trentinara Formation (southern Italy). Facies 43:509–533

Wan X (1991) Paleocene larger foraminifera from Southern Tibet. Rev Esp Micropaleontol 23:7–28

Weiss W (1993) Age assignments of larger Foraminiferal assemblages to Eocene Age in Northern Pakistan. Zitteliana 20:223–252

New Subfamily Kathininae

6

Abstract
The new subfamily Kathininae encompasses rotaliid species with a solid umbilical mass pierced by radial slits, row of funnels or open but feathered interlocular space. Three genera (*Kathina*, *Dictyokathina*, *Plumokathina* n. gen.) represent this group and nine species (*K. aquitanica* n. sp., *K. selveri*, *K. pernavuti*, *K. delseota*, *K. major*, *D. simplex*, *P. dienii* n. sp., *P. lenticula* n. sp., *P. subsphaerica*) are here described and illustrated.

This new subfamily groups all rotaliid shells with a solid umbilical mass pierced by radial slits, rows of funnels or open but feathered interlocular spaces. The rotaliid umbilical structure with folium, foliar aperture and spiral umbilical space is reduced to minimal size and complexity.

6.1 *Kathina* Smout, 1954

Type species: *Kathina delseota* Smout, 1954

Description: The bilmellar-perforate shells have a lenticular shape with a sharp but unkeeled periphery. The lens formed by the shell presents always a similar convexity of both, ventral and dorsal sides of the test. The dorsal side is evolute and smooth. An interiomarginal foramen forms a low arch, positioned immediately below the dorsal chamber wall. The ventral side of the shell lacks any ornaments and is dominated by a solid umbilical mass that is pierced by numerous radial slits or parallel funnels. The folia are very small, bent sharply backward and fused one to another by their oblique umbilical tips. A spiral umbilical space seems to be absent or very small, restricting the umbilical communications to radial slits or rows of funnels from the intraseptal interlocular space in early whorls to the ambient environment, i.e. to the substrate on the ventral side of the lens. Where the sediment has not penetrated the canal system prior to the final deposition of the shell on the sea bottom, early carbonate cements may much reduce the optical contrast between wall and cavity in the shell and obscure the patterns of the umbilical canal system.

Remarks: The species of this genus must be attributed to at least two more or less coeval phyletic lineages: the most complete one is characterized by funnels evenly distributed over the shell's face. This lineage starts with *Kathina aquitanica* in SBZ 3, followed by *K. pernavuti* and *K. delseota* reaching SBZ 4 with considerable overlaps. The second lineage has a heavy umbilical filling, forming umbos of various extensions. The funnels are few and with narrow diameter that may be responsible for their sometimes poor preservation. The successive species

L. Hottinger, *Paleogene larger rotaliid foraminifera from the western and central Neotethys*,
DOI 10.1007/978-3-319-02853-8_6, © Springer International Publishing Switzerland 2014

K. selveri and *K. major* range apparently from SBZ 3 to SBZ 5 with overlaps. *Dictyokathina simplex* might represent a third lineage closer to the *K. aquitanica* group by the high number of funnels with even distribution but the stratigraphic range starting also in SBZ 3, whereas the multiple spiral indicates rather an independent separate group with poor phenotypic response to the "creative" times of the Late Paleocene.

The relationship of the Paleocene Tethyan species with the Late Cretaceous Caribbean ones, and in particular with *K. jamaicensis* Cushman and Jarvis, 1931 (see Butterlin and Fourcade 1989) and *K. bermudezi* remains obscure until the Caribbean faunas are thoroughly revised. The Tethyan Late Cretaceous *Orbitokathina* Hottinger, 1966 that has a megalospheric generation similar to *Kathina* (=*K. "bermudezi"*? see Boix et al. 2009) is much too large and complex to represent a Cretaceous root of the Paleocene Kathina. The nature of "*Orbitokathina*" *sarayensis* Sirel et al., 1983 from the Thanetian of the Van area, Eastern Turkey (see Sirel 2004, pls. 10–11), remains obscure until better preserved material permits to analyse the structures in detail.

***Kathina aquitanica* n. sp.**; Plate 4.4, Figs. 1–14; Plate 6.2, Figs. 1–7.

1998 ?*Kathina pernavuti* Sirel—Accordi et al., p. 200, pl. 16, fig. 2.

2006 *Kathina* sp.—Hottinger, p. 87, pl. 2, figs. 17–19.

Holotype: Specimen illustrated in Plate 4.4, Fig. 1; syntypes figured in Plate 4.4, Figs. 2–14.

Type locality and type level: Lafarge Quarry, Western Aquitaine, southern France; Paleocene (SBZ 3).

Derivation of name: Aquitania, region in south-western France, locations the type locality.

Diagnosis: The bilamellar-perforate shells form small-sized, smooth lenses of equal convexity on the ventral and the dorsal side of the test. The chambers are dorsally evolute and ventrally involute. There are 14–18 chambers in adult last whorls and a sharp but unkeeled periphery. The ventral face bears radial slits as orifices of the canal system. They correspond to the septa in the last whorl. In places, the radial slits are complemented by a second branch forming with the radial slit a kind of Y. The shorter branch reflects the interlocular space in the foliar suture. This is a major argument to assign this and similar species of *Kathina* to the rotaliids, in spite of the almost total fusion of the folia in the rest of the umbilical filling. Around the axial area of the shell, there is a circular furrow communicating with the canal system. It delimits an umbo that is a single axial pile free of funnels. The umbilical plate is small and short, delimiting a tiny spiral canal difficult to find in sections perpendicular to the coiling axis. The proloculi are small, reaching about 0.05 mm. We have not observed any dimorphism.

Remarks: *Kathina aquitanica* is the smallest species of this genus known so far. Its simple interlocular canal system with single slits as orifices reminds us of early stages of growth in *Kathina major*. Therefore, *K. aquitanica* is tentatively considered here as precursor species of *K. major*.

Kathina selveri Smout, 1954; Plate 6.1, Figs. 6–7, 9–14; Plate 6.2, Figs. 12–18.

1954 *Kathina selveri*—Smout, p. 62, pl. 6, figs. 11–13.

?1956 *Kathina nammalensis* Smout—Smout and Haque, p. 56, pl. 11, figs. 7–9.

1987 unidentifyed rotaliid—Nicora et al., pl. 34, fig. 5.

1998 unidentified specimens—Accordi et al., p. 180, pl. 6, fig. d, lower center.

2009 *Kathina selveri* Smout—Afzal et al., p. 17, pl. 1, fig. 8.

Description: The bilamellar-perforate shells are smooth on both sides, lenticular to cone-shaped, almost triangular in axial section. The chambers are dorsally evolute, ventrally involute. The dorsal side of the shell is flattened or slightly convex, the ventral side highly convex. The chambers are a little higher than long, separated by straight, radial septa that incline backward only in the immediate vicinity of their junction with the dorsal chamber wall. There are about 24 chambers in the last adult whorl. The umbilicus is occupied by a strong, undivided umbonal pile projecting around the coiling axis of the shell.

Remarks: *Kathina selveri* is distinguished from *K. major* Smout, 1954 by its shell proportions: the equatorial to axial diameter ratio is larger in *K. major* (Plate 6.5, Figs. 1–6) than in *K. selveri* (Plate 6.2, Figs. 9–10, 14). This is due to an extension of the umbilical filling towards the periphery of the shell.

Kathina nammalensis Smout and Haque, 1956 is in this respect similar to *K. selveri*. The original description of *K. nammalensis* is rather poor and does not allow defining clearcut differences with *K. selveri* that are not even discussed in Smout and Haque's paper. Until a number of well-preserved topotypes will be available, we have chosen to treat *K. nammalensis* as junior synonym of *K. selveri*.

Kathina pernavuti Sirel, 1972; Plate 6.4, Figs. 1–22.

1972 *Kathina pernavuti*—Sirel, p. 289, pl. 5, fig. 7.

Description: As stated in the original description, this species is a small-sized "version" of *Kathina delseota* Smout, 1954 with a somewhat simpler umbilical architecture. The bilamellar-perforate, low-trochospiral shells rarely exceed 1 mm in equatorial diameter. Adult whorls exhibit 14–18 chambers. The evolute, dorsal side of the shell is decorated by pustules that are denser at the shell apex and provide for a slightly projecting axial area. On the slightly flattened ventral side of the shell, instead of the many funnels that are observed in *K. delseota*, the umbilicus of *K. pernavuti* is filled with a central umbo. The rest of the umbilical area is covered by fused foliar walls, as in all *Kathina* species. The intraseptal interlocular spaces open as simple radial slits to the substrate of the animal (Plate 6.4, Figs. 4–7). They may be subdivided into a short row of funnels (Plate 6.4, Fig. 22). In spite of the small size of the shell, the folia are clearly visible (Plate 6.3, Fig. 19). They are bent backwards over the whole breadth of the previous chamber, but have a short extension in the axial direction of the shell. The foramen is an interiomarginal arch that extends over the dorsal half of the septal face. The proloculus is too small to be accurately measured in thin sections, about 0.04 mm in diameter. A dimorphism of generations was not observed.

Kathina delseota Smout, 1954; Plate 6.1, Figs. 15–17; Plate 6.2, Figs. 1 (bottom), 8–11; Plate 6.5, Figs. 1–18.

1954 *Kathina delseota*—Smout, p. 61, pl. 7, figs. 1–8.

2008 *Kathina delseota* Smout—BouDagher-Fadel, p. 260, pl. 628, fig. 18.

Description: A low-trochospiral chamber arrangement produces a lenticular shell with a broadly rounded periphery. The ratio of equatorial to axial diameter is about 2.5 in adult tests. The chambers are almost isometric, with a slight inclination of the straight septa. There are 17–19 chambers in an adult whorl. The dorsal side of the shell is faintly decorated with limbate spiral sutures. The interiomarginal foramen is positioned in the umbilical half of the septal face. The diameter of the megalosphere is about 0.4 mm if the section passes through the center of the proloculus (Plate 6.4, Fig. 4). The umbilical face on the ventral side of the shell is covered by numerous funnel orifices. On the peripheral part of the umbilicus the funnel orifices may be replaced by short radial slits. Some of them show the Y-shapes due to the foliar suture below the umbilical face (Plate 6.5, Fig. 18 bottom). Some specimens may show a kind of umbo in the coiling axis of the shell where the funnels are missing (Plate 6.5, Fig. 17).

Kathina major Smout, 1954; Plate 6.1, Fig. 19; Plate 6.2, Figs. 1–5, 8; Plate 6.5, Figs. 1–15.

1954 *Kathina major*—Smout, p. 63–64, pl. 6, figs. 1–10.

2008 *Kathina major* Smout—BouDagher-Fadel, p. 260, pl. 6.28, fig. 18.

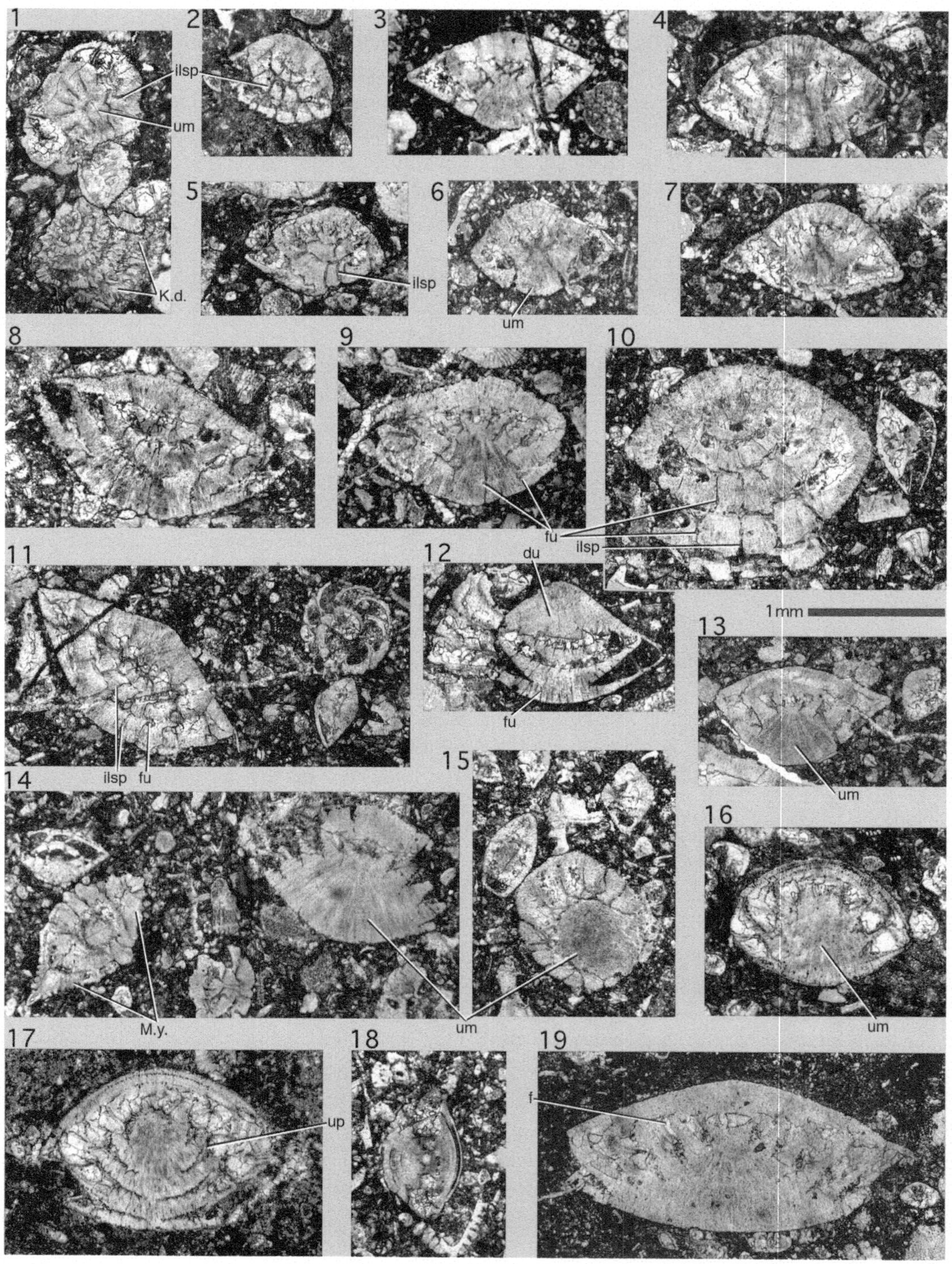

Plate 6.1 (1–5, 7) *Kathina aquitanica* n. sp.; from cemented rock, sample Kar 13, Kuh-e-Kargan, Kermanshah, Iranian Zagros ranges, collected by J. Braud; associated with *Miscellanites iranicus* (Rahaghi, 1983), emend. Hottinger, 2009 and associated with the usual Thanetian larger foraminifera such

Description: This species has a lenticular shape with a sharp periphery and equal convexity on both sides of the shell. The surface of the lens is smooth, without conspicuous ornaments. The umbilicus is filled with a solid mass of fused lamellae that are pierced by a restricted number of funnels, avoiding the axial zone without forming a projecting umbo. The dorsal parts of the totally evolute chambers are isometric or slightly higher than long, separated by straight septa that decrease their inclination from the dorsal chamber wall towards the equatorial plane. Megalospheric specimens have a spherical proloculus of 0.08–0.1 mm. Adult whorls of the shell number 24–26 spiral chambers.

The microspheric generation produces significantly larger shells. Adult spiral chambers are about twice as high in radial direction as their length in the direction of growth; there are 40–44 chambers per whorl. The interlocular spaces consist of a single, radial slit per septum that tends to get subdivided, in late stages of growth and in short distance from the umbilical shell surface, into a row of few, elongate funnels (Plate 6.2, Fig. 3).

6.2 *Dictyokathina* Smout, 1954

Type species: *Dictyokathina simplex* Smout, 1954

Description: Bilamellar discoidal shells with the basic architecture of kathinines. The evolute dorsal side of the shell is decorated by limbate sutures. The chambers are arranged in multiple spirals. There is a clear dimorphism: the microspheric generation, twice the size of the megalospheric shells, are often deformed to cup-like or plate-like shapes. It is the ventral side with its funnel orifices that becomes convex.

Dictyokathina simplex Smout, 1954; Plate 6.6, Figs. 1–13; Plate 6.7, Figs. 1–13.

1954 *Dictyokathina simplex*—Smout, p. 65, pl. 8, figs. 1–11.
1983 *Orbitokathina hottingeri*—Rahaghi, p. 56, pl. 37, figs. 1–6, 9–10.
1983 *Planorbulina* sp.—Rahaghi, no text, pl. 37, figs. 7–8.
1983 *Orbitokathina sarayensis* Sirel—Gündüz and Acar, p. 149, text-fig. 2, pl. 1, figs. 1–14; pl. 2, figs. 1–9; pl. 3, fig. 3.
1985 *Dictyokathina simplex* Smout—Hasson, p. 360, pl. 6, figs. 4–5.
1988 *Dictyokathina vanica*—Sirel, p. 481, pl. 3, figs. 1–3; pl. 4, figs. 1–6; pl. 5, figs. 1–9.
non: 1993 *Dictyokathina simplex* Smout—Weiss, p. 250, pl. 1, fig. 5.
1998 *Dictyokathina vanica* Sirel—Sirel, p. 83, pl. 42, figs. 1–2 (microspheric), figs. 3–5 (megalospheric).
1998 *Orbitokathina sarayensis*—Sirel et al. Sirel, p. 84, pl. 43, figs. 3–4, ?non pl. 43, figs. 1–2.
2004 *Dictyokathina vanica* Sirel—Sirel, p. 16, pl. 12, figs. 1–5; pl. 13, figs. 1–6; pl. 14, figs. 1–9.

Plate 6.1 (continued) as *Glomalveolina primaeva* (Reichel, 1937), *Fallotella alavensis* Mangin, 1954 and "*Taberina*" *daviesi* Henson, 1950; Paleocene (SBZ 3). (**1**) (*top*) Section perpendicular to coiling axis showing the massive ventral umbo. (**2, 5**) Oblique sections. (**3**) Axial section. (**4, 7**) Subaxial section. (**8–9, 11**) *Kathina delseota* Smout, 1954. (**8, 9, 11**) Subaxial sections. (**1**) (*bottom*), (**10**) Oblique section inclined for about 60° in respect to the shell's coiling axis. (**12–18**) *Kathina selveri* Smout, 1954; sample All 77233, Zhob Valley, Baluchistan, Pakistan, collected by F. Allemann; Paleocene (SBZ 3–4). (**12**) Oblique section; note the large dorsal umbo exaggerated by the position of the oblique section. (**13, 18**) Axial section. (**14**) Oblique section (*right*) associated to *Miscellanea yvettae* Leppig, 1988, axial section. (**15**) Section perpendicular to coiling axis showing the massive ventral umbo that might have a kind of skipjack function. (**16–17**) Oblique sections. (**19**) *Kathina major* Smout, 1954; sample All77233, Zhob Valley, Baluchistan, Northwestern Pakistan, collected by F. Allemann. (**19**) Axial section. Abbreviations: *f* foramen, *um* umbo, *ilsp* intraseptal interlocular space, *fu* funnel, *up* umbilical plate, *du* dorsal umbo

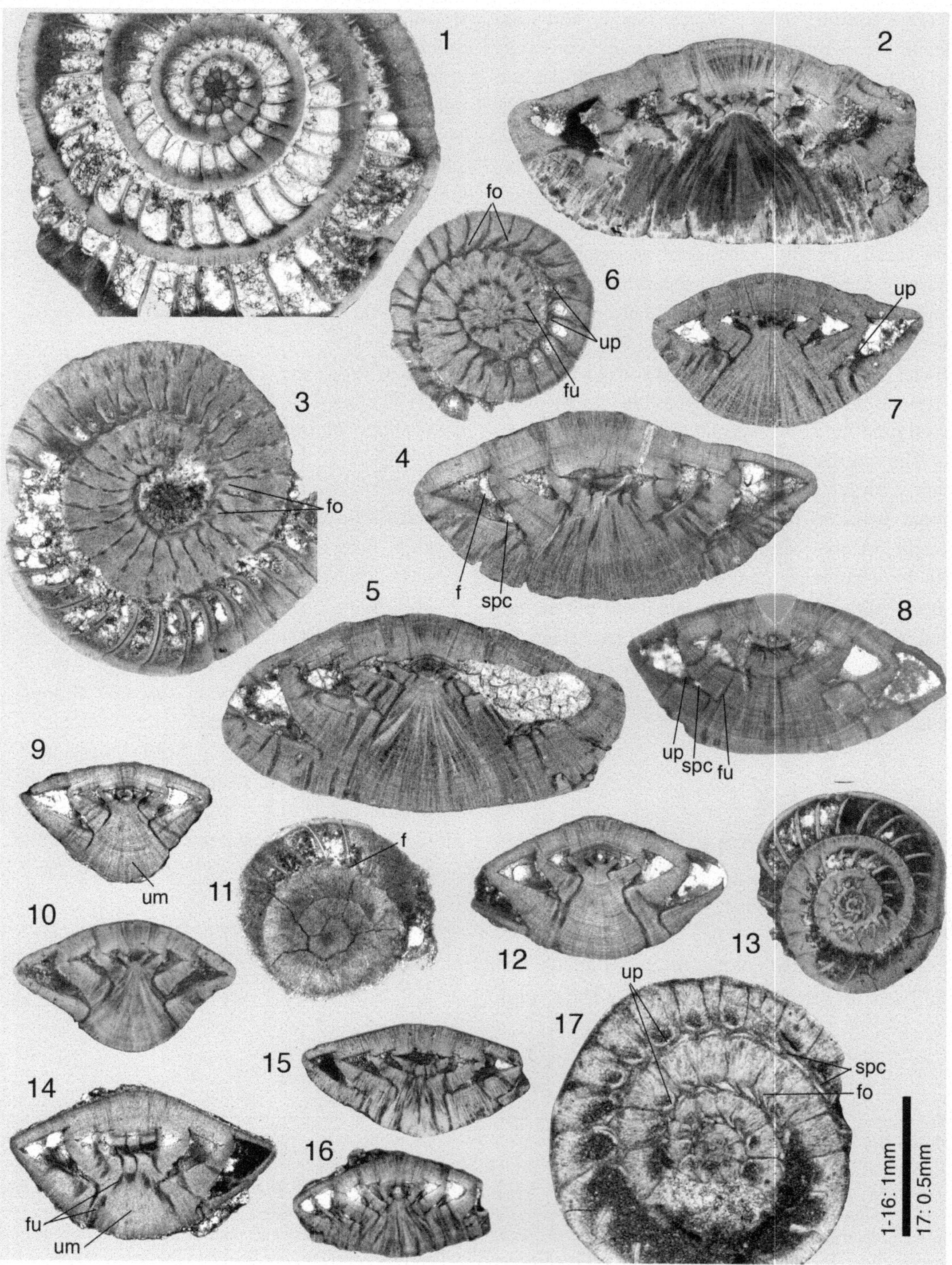

Plate 6.2 (**1–5**, **8**) *Kathina major* Smout, 1954. (**1**) Microspheric specimen, equatorial section. (**2**, **5**) Microspheric specimens, axial sections. (**3**) Section perpendicular to coiling axis showing distribution of narrow intraseptal interlocular spaces. (**4**, **8**) Megalospheric specimens, axial sections. (**6**, **9–14**) *Kathina selveri*

2004 partim *Orbitokathina sarayensis* Sirel—Gündüz and Acar. Sirel, p. 15, pl. 11, figs. 3–4, 6–8; non pl. 10, figs. 1–3, 10–13; pl. 11, figs. 1–2.

2008 *Dictyokathina simplex* Smout—BouDagher-Fadel, p. 260, pl. 6.28, fig. 21.

Description: The megalospheric generation has a discoidal shell with a broadly rounded periphery. The dorsal side is always moderately convex, the ventral side may be flattened, sometimes a little concave, eventually with a faint umbo where the number of funnels is more or less reduced. There are three supplementary spirals starting early in ontogeny, at a shell diameter of about 1 mm. The first chambers of each intercalated spiral (arrows on Plate 6.6, Fig. 7) are fed from the interlocular space of adjacent chambers in the previous whorl. The chambers exhibit approximately isometric equatorial sections with slightly inclined, straight septa. There are 19–23 chambers in half a whorl at a shell diameter of 1 mm. As pointed out already by Smout (1954), axial and oblique sections of megalospheric specimens are difficult to distinguish from the ones of *Kathina delseota*, because they do not show the presence or absence of multiple spirals.

The microspheric specimens reach twice the size of the megalospheric generation, with an equatorial to axial diameter ratio of 13–14.5. No specimens are available to produce any significant sections in equatorial direction.

Remarks: There seems to be difficulties with the enlargements given by Smout (1954) for *Dictyokathina simplex* and as well by Sirel (1988, 2004) for *D. vanica*. Smout (1954) indicates the enlargement of the holotype and one paratype of *D. simplex* as 12 magnifications (pl. 8, figs. 2–3) whereas the similar sized specimen in Fig. 4 is indicated by 5 magnifications (×5) and the obviously much less enlarged Fig. 9 as 10 magnifications. Sirel's specimen of *D. vanica* occupies the total largeness of a plate with an enlargement of 15 magnifications.

Sirel et al. (1983) and Rahaghi (1983) erected a new Paleocene species of the Cretaceous genus *Orbitokathina* Hottinger, 1966 under two different names. This "species" is mainly characterized by a systematic deformation of the discoïdal shell to take the shape of a sombrero (Plate 6.7, Fig. 1, 4–5). The axial or nearly axial sections are in my view identical with Smout's *Dictyokathina simplex* from 1954 (pl. 8, fig. 11). The horizontal sections given for *Orbitokathina sarayiensis* by Sirel et al. (1983) can not belong to the same taxon because the solid umbonal area, perforated by numerous funnels characteristic for all Kathininae, does not appear in tangential sectors of the horizontal sections. These call for a detailed structural analysis of the chambers (or chamberlets?) and of the patterns of their connections by foramina or stolons. Well preserved and isolated specimens would be necessary to perform such an analysis. For the time being, and considering the contemporary stratigraphical distribution of all taxa involved, I recommend their synonymization as presented in the synonymy list.

Plate 6.2 (continued) Smout, 1954; megalospheric specimens. (**6**) Section perpendicular to the coiling axis showing distribution of the interlocular spaces around the adaxial ventral umbilical umbo. (**7**, **9–10**, **12**) Axial sections. (**11**, **13**) Sections perpendicular to the coiling axis showing the number and the inclination of the chamber septa below the dorsal surface of the shell. (**14**) Oblique section inclined for little more than 45° in respect to the coiling axis of the shell. (**15–17**) *Kathina delseota* Smout, 1954; megalospheric specimens. (**15–16**) Subaxial sections. (**17**) Section perpendicular to coiling axis showing the tiny rotaliid structure, characteristic for the genus, with its backwards inclined folia, the umbilical plate and the narrow spiral canal. (**1–5**, **8**) from sample M 5681; (**6–7**, **9–14**) from sample M 5623; (**15–17**) from sample M 2710; Natural History Museum (London), Arabian Peninsula. Abbreviations: *f* foramen, *up* umbilical plate, *spc* spiral canal, *fo* folium, *fu* funnel

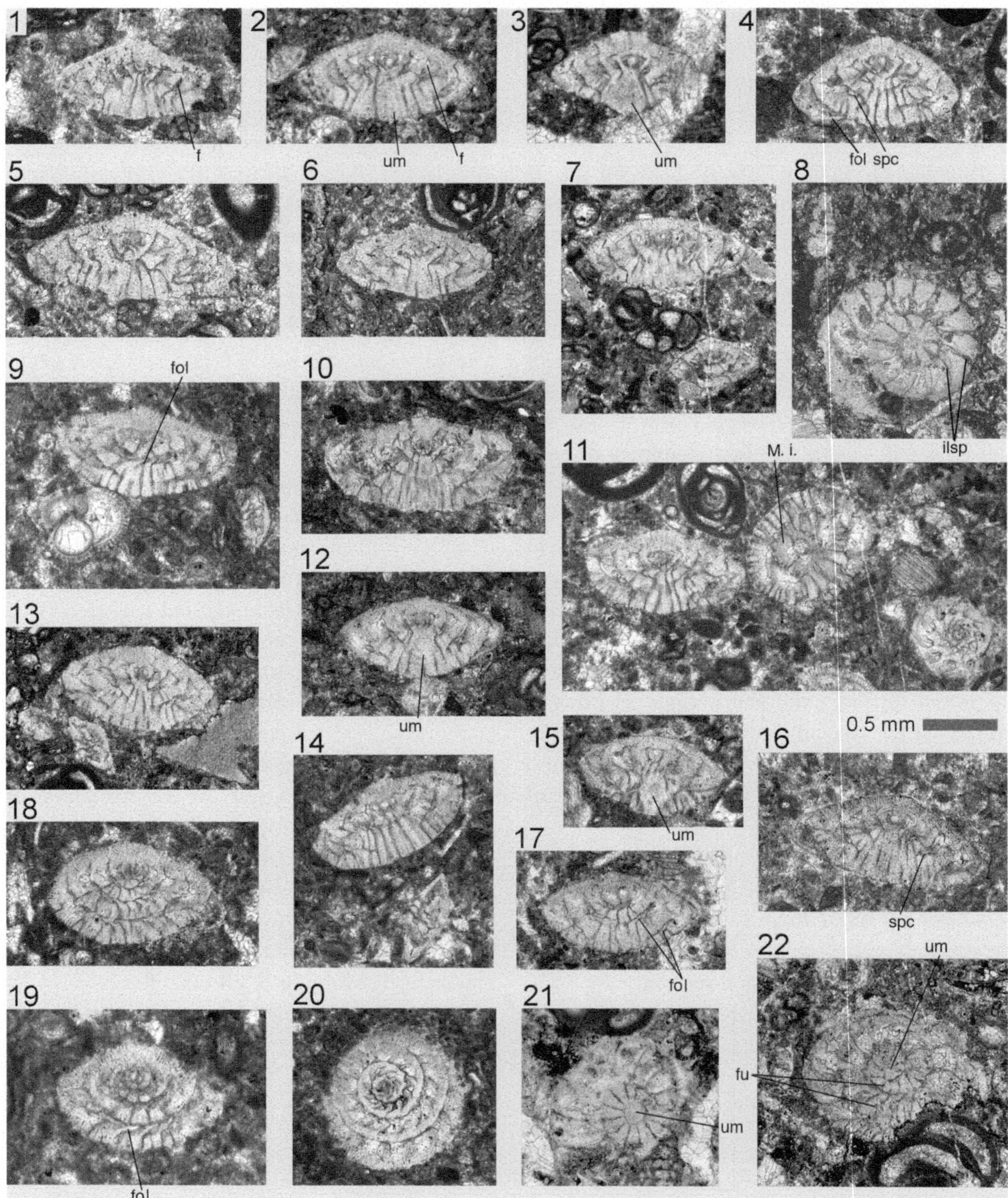

Plate 6.3 *Kathina pernavuti* Sirel, 1972; sample Kar 11, Kuh-e-Kargan, Kermanshah, Iranian Zagros, collected by J. Braud; Paleocene (SBZ 3). (**1–6**, **12**, **14**, **15**) Axial and subaxial sections; note in strictly axial sections (such as **6** or **12**) the small axial umbo free of funnels; this documents the close relationship with *Kathina aquitanica* n. sp. (**7, 9–11**) (*left*), (**13**, **17–18**) Oblique sections with inclinations below 45° in respect to the coiling axis of the shell. (**8, 11**) (*right*), (**19–22**) Oblique sections with inclinations above 45° in respect to the coiling axis; note in (**11**) (*center*) the oblique section of (M. i.) *Miscellanites iranicus* (Rahaghi, 1983). Abbreviations: *f* foramen, *um* umbo, *fol* folia, *spc* spiral canal, *ilsp* intraseptal interlocular space, *fu* funnel

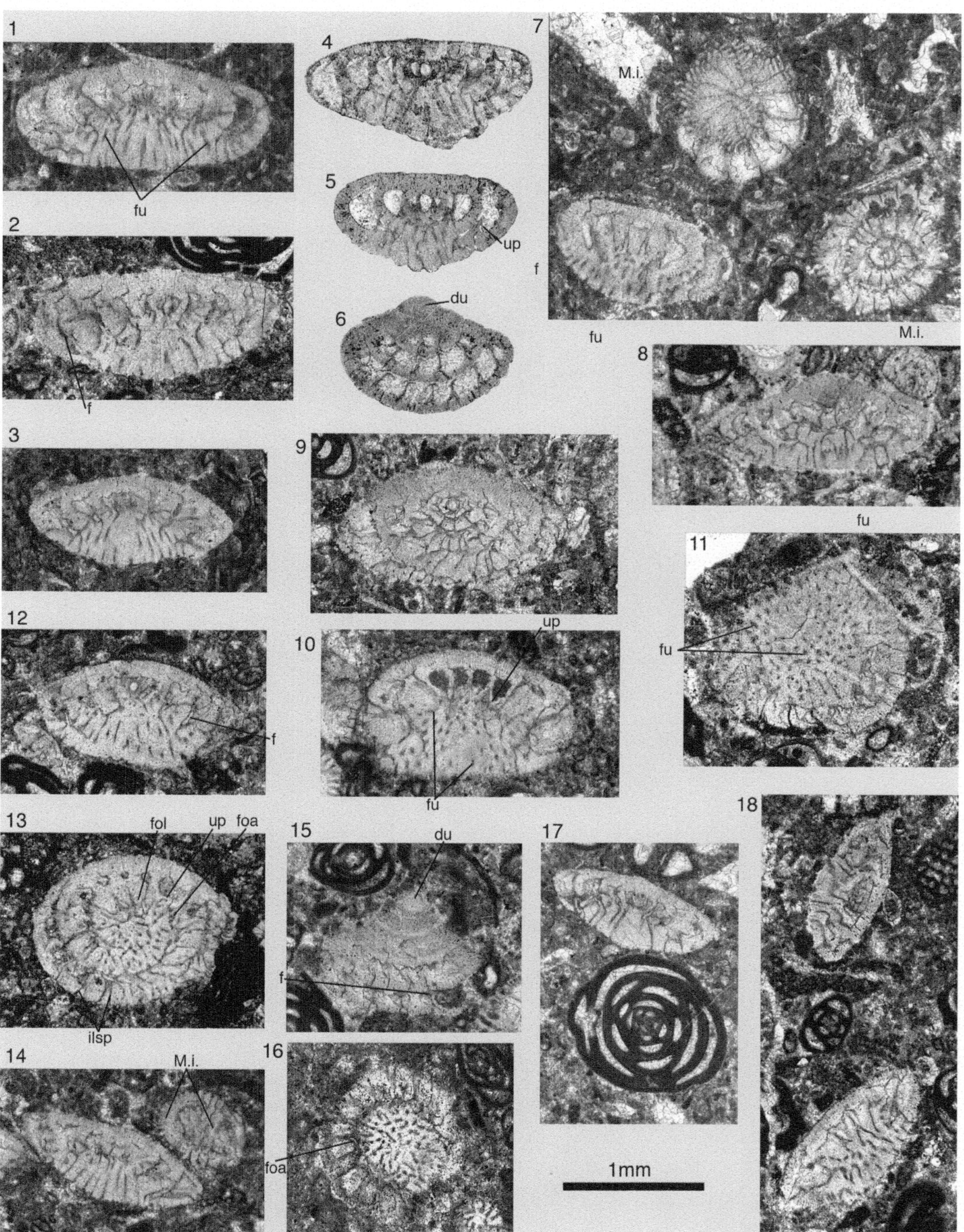

Plate 6.4 *Kathina delseota* Smout, 1954; megalospheric specimens; sample Kar 12, Kuh-e-Kargan, Kermanshah, Iranian Zagros, collected by J. Braud; Paleocene (SBZ 3). (**1**, **3**, **14**, **17**) Axial sections. (**2**, **4**, **5**, **7**) (*left*), (**8**, **12**, **18**) (*top*) Oblique sections inclined for less than 45° in respect to the coiling axis of the shell. (**6**, **9–10**, **13**) Oblique sections inclined for more than 45° in respect to the coiling axis. (**11**, **16**) Sections tangential to the shell face showing the distribution of the numerous funnels. (**15**) Section tangential to the dorsal surface of the shell showing the inclination of the septa. (**18**) Transverse sections. Note in (**7**) (*center* and *right*) and in (**14**) (*right*) the presence of (M. i.) *Miscellanites iranicus* (Rahaghi, 1983). Abbreviations: *f* foramen, *fol* folia, *ilsp* intraseptal interlocular space, *fu* funnel, *up* umbilical plate, *du* dorsal umbo, *foa* foliar aperture

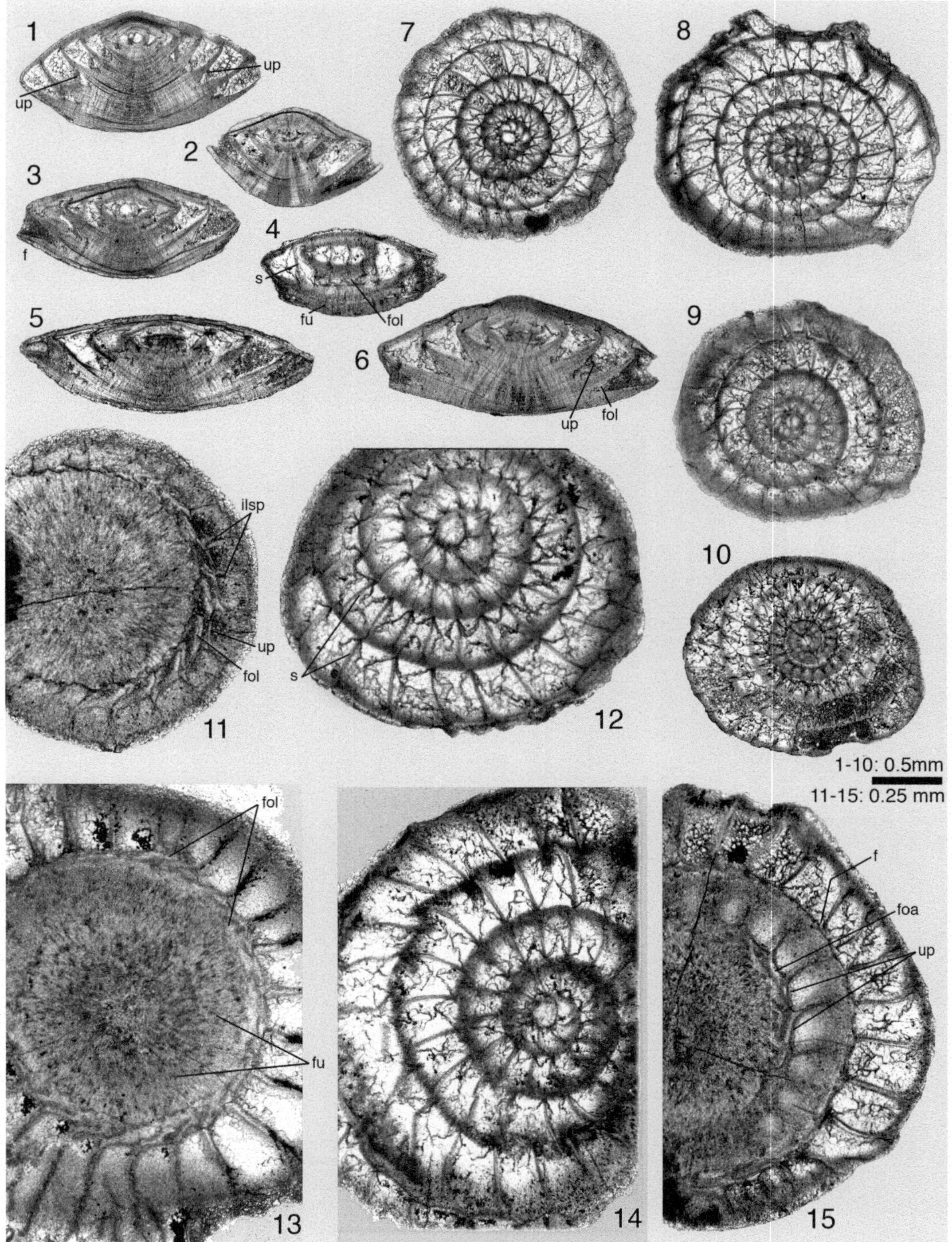

Plate 6.5 *Kathina major* Smout, 1954; megalospheric specimens; sample 95108, Dhak Pass, Salt Range, North-western Pakistan, base of Patala Formation; associated with *Ranikothalia nuttalli* (Davies, 1927); Paleocene (SBZ 4). (**1–3**) Axial sections. (**4–6**) Transverse sections parallel to the shell's axial plane. (**7–10**) Equatorial sections. (**11, 13, 15**) Sections perpendicular to the coiling axis showing the generotypic details of the shell's structure. (**12, 14**) Equatorial section sections showing the details of the bilocular nepionic stage of growth. Abbreviations: *f* foramen, *up* umbilical plate, *s* septum, *fol* folia, *ilsp* intraseptal interlocular space, *fu* funnel, *foa* foliar aperture

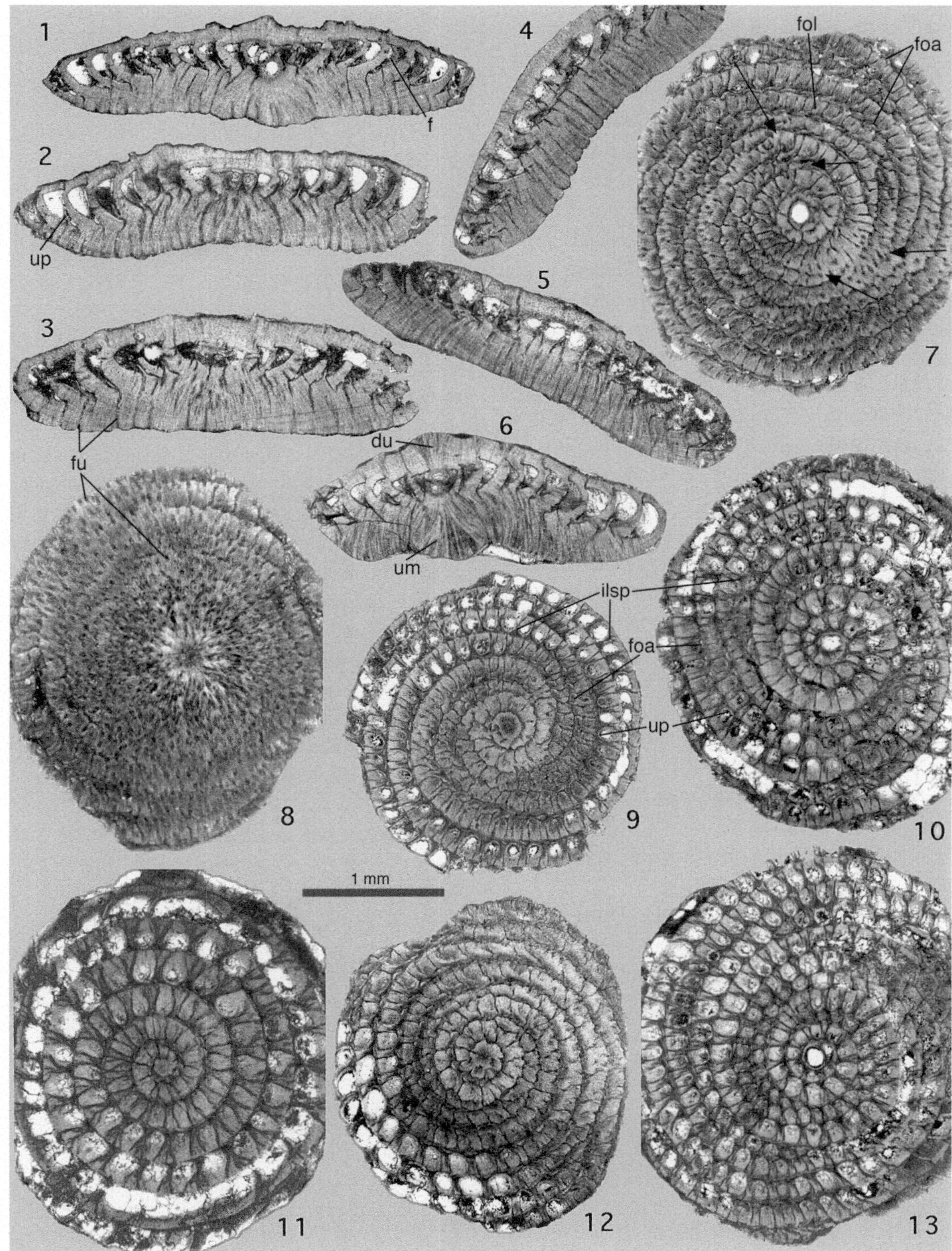

Plate 6.6 *Dictyokathina simplex* Smout, 1954; isolated specimens, megalospheric generation; specimens (**1–3**, **7**, **9**) from sample P 40174 Natural History Museum (London), paratypes from Qatar, see Smout (1954, p. 66); specimens (**4–6**, **8**, **10–13**) from DK 5 well, 400', Qatar; Late Paleocene or Early Eocene (SBZ 4 or 5). (**1–6**) Axial and subaxial sections; note the faint ventral umbo in axial position. (**7–8**) Sections perpendicular to

6.3 *Plumokathina* n. gen.

Type species: *Plumokathina dienii* n. sp.

Description: The bilamellar-perforate shells are trochospiral, evolute on the dorsal side, involute on the ventral side. The septa are inclined backwards in the dorsal and radial in the ventral parts of the spiral. The dorsal shell surface is smooth, the ventral surface dominated by an umbo that fills the umbilical area almost completely. The intraseptal interlocular space remains uncovered but is restricted in space by the spurs between the grooves of the coarse and deep feathering. This latter feature is the main character of the genus. The foramen is a low but long interiomarginal arch or slit extending over almost the whole length of the interiomargin of the spiral chamber. The umbilical plate separates a very narrow spiral interlocular space from the chamber lumen that communicates with a very low foliar chamberlet by a large gap. The folium is bent backwards in relation to the radial axis of the chambers.

***Plumokathina dienii* n. sp.**; Figs. 3.5J, 6.1A–N; Plate 6.8, Figs. 1–21.

1998 *Plumokathina* sp.—Accordi et al., p. 196, pl. 14, fig. 6.

1998 unidentified rotaliid—Accordi et al., p. 190, pl. 11, fig. 8.

2000 "*Plumokathina dienii*"—Peybernès et al., p. 44, fig. 6/5.

Holotype: megalospheric specimen figured in Fig. 6.1D.

Type locality and type level: Cuccuru 'e Flores Conglomerate, Orosei, Sardinia, Italy; Paleocene (SBZ 2).

Derivation of name: in honor of Iginio Dieni (University of Padova, Italy) and his work carried out in Sardinia.

Diagnosis: The equatorial diameter of adult shells reaches 1.2 mm. The convexity of the dorsal and ventral sides of the lenticular shell is about equal or favouring the ventral side with its progressive umbo. The ratio of equatorial to axial diameter of the shells varies from 1.6 to 2.0. At a diameter of 1 mm, the last whorl counts 10–12 spiral chambers. The septa are straight, inclined backwards in the dorsal and strictly radial in the ventral parts of the chambered spiral. The slight inflation of the chambers is obscured in the penultimate half whorl and earlier stages of growth by the feathering of the interlocular space that invades the periphery from the ventral side of the test. The periphery of the shell is without keel but very sharp and slightly upturned in dorsal direction. The umbilical architecture corresponds to the generic diagnosis. There is a modest dimorphism of generations concerning beside the proloculus mainly the shell size. The diameter of the megalosphere reaches 0.06 mm.

***Plumokathina lenticula* n. sp.**; Plate 6.9, Figs. 1–6; Plate 6.10, Figs. 1–12; Plate 6.11, Figs. 1–31.

1972 *Rotalia hensoni* Smout—Samuel et al., p. 79, pl. 42, figs. 1–2; pl. 43, figs. 1–2; Plate 6.5, Figs. 1–3.

Plate 6.6 (continued) coiling axis showing the distribution of the funnels in the ventral part of the shell and over its face; the start of supplemental spirals is marked in (**7**) by *arrows*. (**9–10**, **12–13**) Sections perpendicular to the coiling axis in the dorsal layer of the shell showing the multiple spirals of the megalospheric generation. (**11**) Section perpendicular to the coiling axis in the dorsal part of the shell; unique specimen with an absence of multiple spirals and a larger diameter of the chambers as compared to the specimens with intercalated supplemental spirals; we do not know whether this is a mere variant of *D. simplex* or may represent a new species of larger dimensions and close to *Kathina delseota*, including the axial section illustrated in 6. Abbreviations: *f* foramen, *fol* folia, *ilsp* intraseptal interlocular space, *fu* funnel, *up* umbilical plate, *du* dorsal umbo, *foa* foliar aperture, *um* umbo

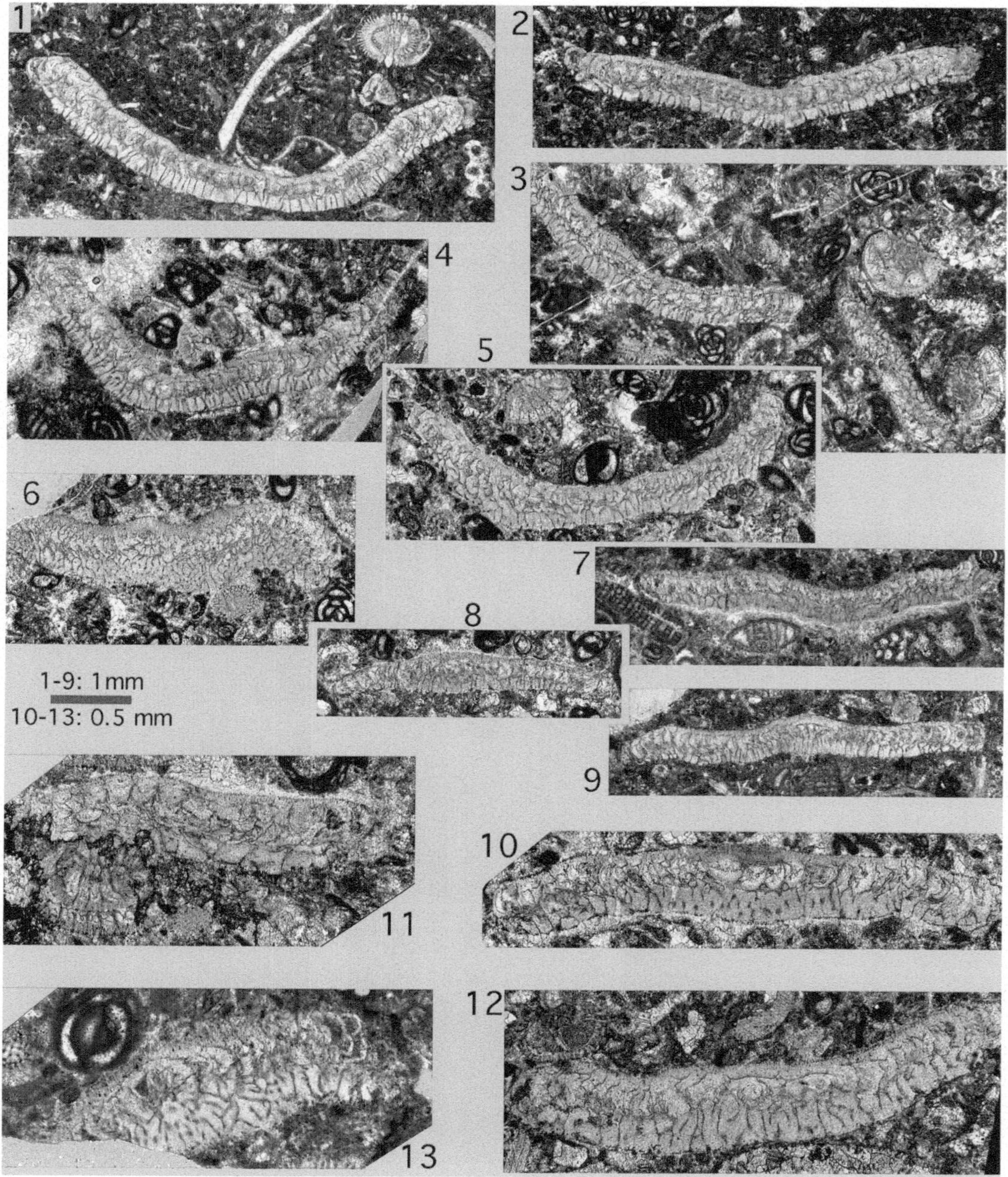

Plate 6.7 *Dictyokathina simplex* Smout, 1954; specimens of both generations in cemented rock; sample Kar 12, Kuh-e-Kargan, Kermanshah, Iranian Zagros, collected by J. Braud; Paleocene (SBZ 3). (**1–5**) Microspheric specimens, sections parallel but with various distances to coiling axis; note the cup-shaped deformation of the discoidal shells; the convex side represents the shell's face. (**6, 13**) Oblique section inclined for more than 45° in respect to the coiling axis; specimen 1 (*top right*) is associated with *Miscellanites minutus* (Rahaghi, 1983). (**7–12**) Megalospheric generation. (**9**) Axial sections, all others oblique sections inclined for less than 45° in respect to coiling axis

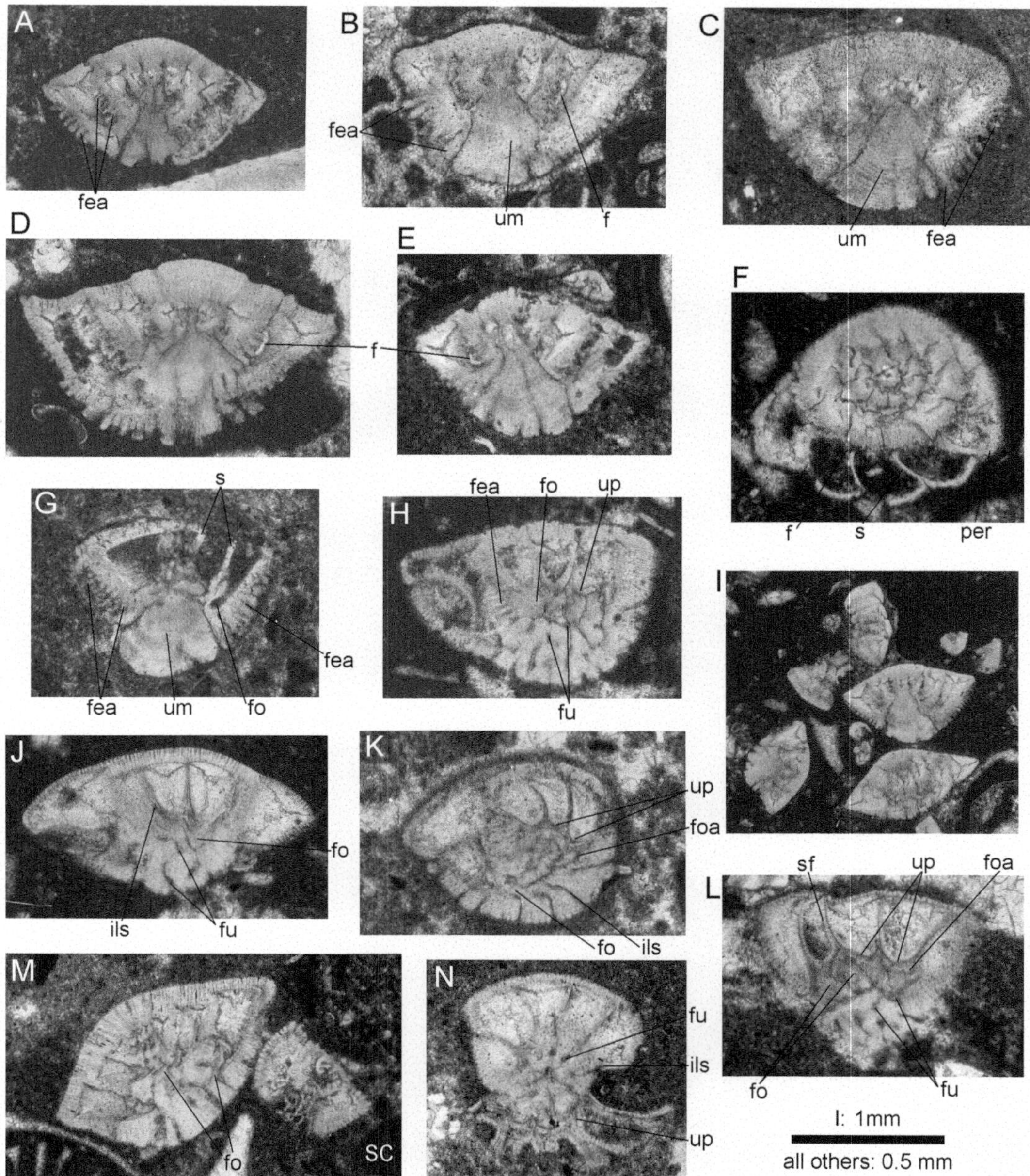

Fig. 6.1 *Plumokathina dienii* n. sp. Specimens from the type locality in Sardinia, Paleocene (SBZ 2). Note the feathering (*fea*) of the ventral cameral sutures. (**A–E**) Axial sections. (**F**) Oblique section almost perpendicular to the coiling axis. (**G–L, N**) Oblique sections with various inclinations. (**M**) Transverse section, parallel to coiling axis and oblique section of *Scarificatina reinholdi* (Pozaryska and Szczechura, 1970). Abbreviations: *s* septum, *f* foramen, *per* periphery, *sf* septal flap, *up* umbilical plate, *um* umbo, *fea* feathering of the intraseptal interlocular space, *fo* folium, *fu* funnel, *foa* foliar aperture, *ils* intraseptal interlocular space

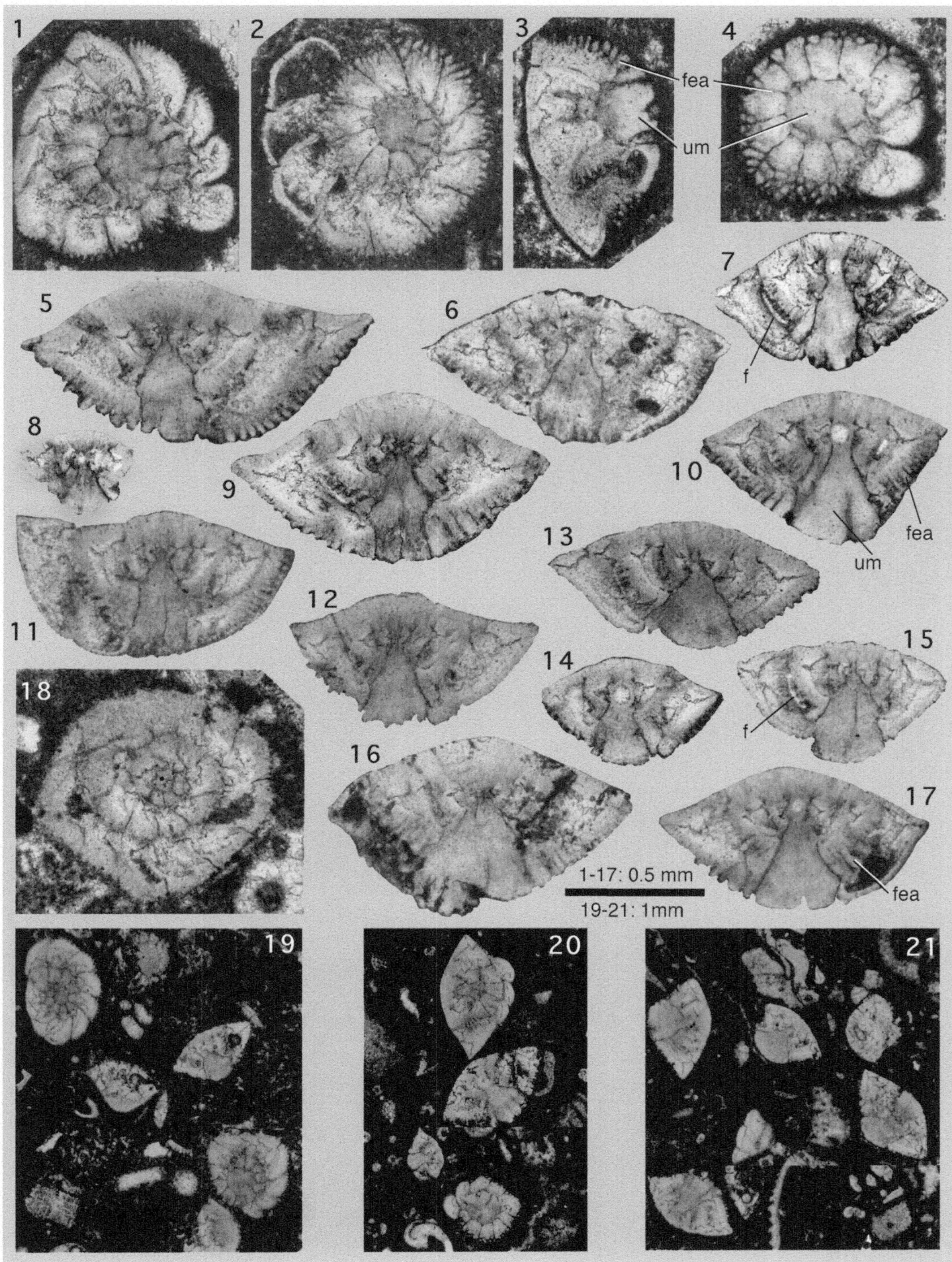

Plate 6.8 *Plumokathina dienii* n. sp.; sample F4a, pebbles of the Cuccuru'e Flores Conglomerate, Orosei, Sardinia (see Dieni et al. 1985); Paleocene (SBZ 2). (**1–2, 4, 18**) Sections more or less perpendicular to the coiling axis. (**3**) Oblique section showing the feathering of the shoulders of the interlocular space. (**5–17**) More or less perfect axial sections; note in (**7**) and (**15**) the long, a low arch in interiomarginal position representing the intercameral foramen. (**19–21**) Random sections in cemented rock produced by sediments that are dominated by micrites. Abbreviations: *f* foramen, *fea* feathering of interlocular space, *um* umbo

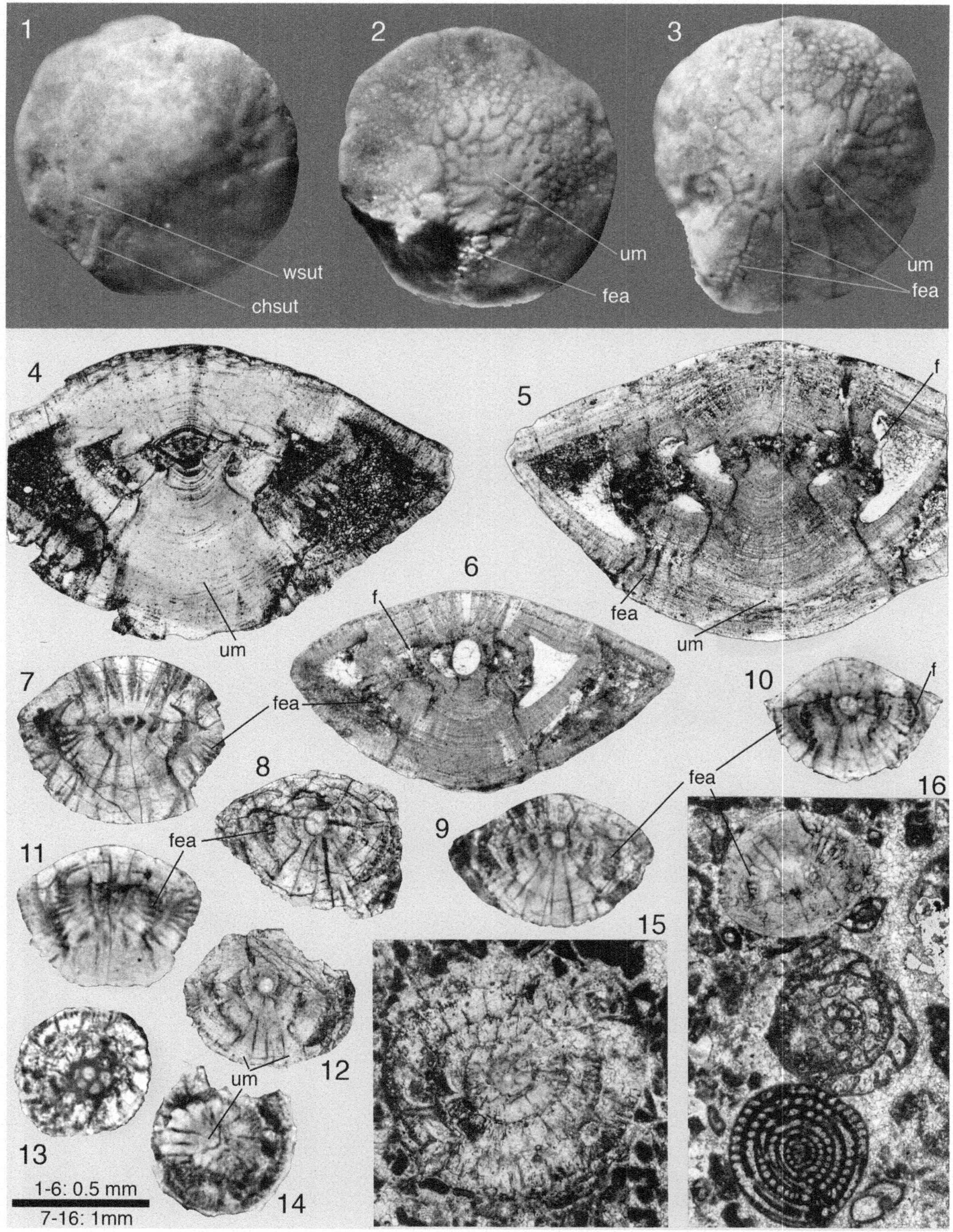

Plate 6.9 (**1–6**) *Plumokathina lenticula* n. sp.; isolated specimens; from Narp, Petites Pyrénées, Paleocene (SBZ 3). (**1–3**) Dorsal (**1**) and ventral (**2–3**) views. (**4–5**) Axial sections, microspheric forms. (**6**) Axial section, megalospheric form. (**7–16**) *Plumokathina subsphaerica* (Sirel, 1972); megalospheric forms; topotypes from Pernavut village, Turkey (Sirel 1972), Paleocene. (**7–10, 12**) Axial sections. (**11, 16**) Subaxial sections. (**13–15**) Sections perpendicular to coiling axis. Abbreviations: *f* foramen, *wsut* whorl suture, *chsut* chamber suture, *um* umbo, *fea* feathering of interlocular space

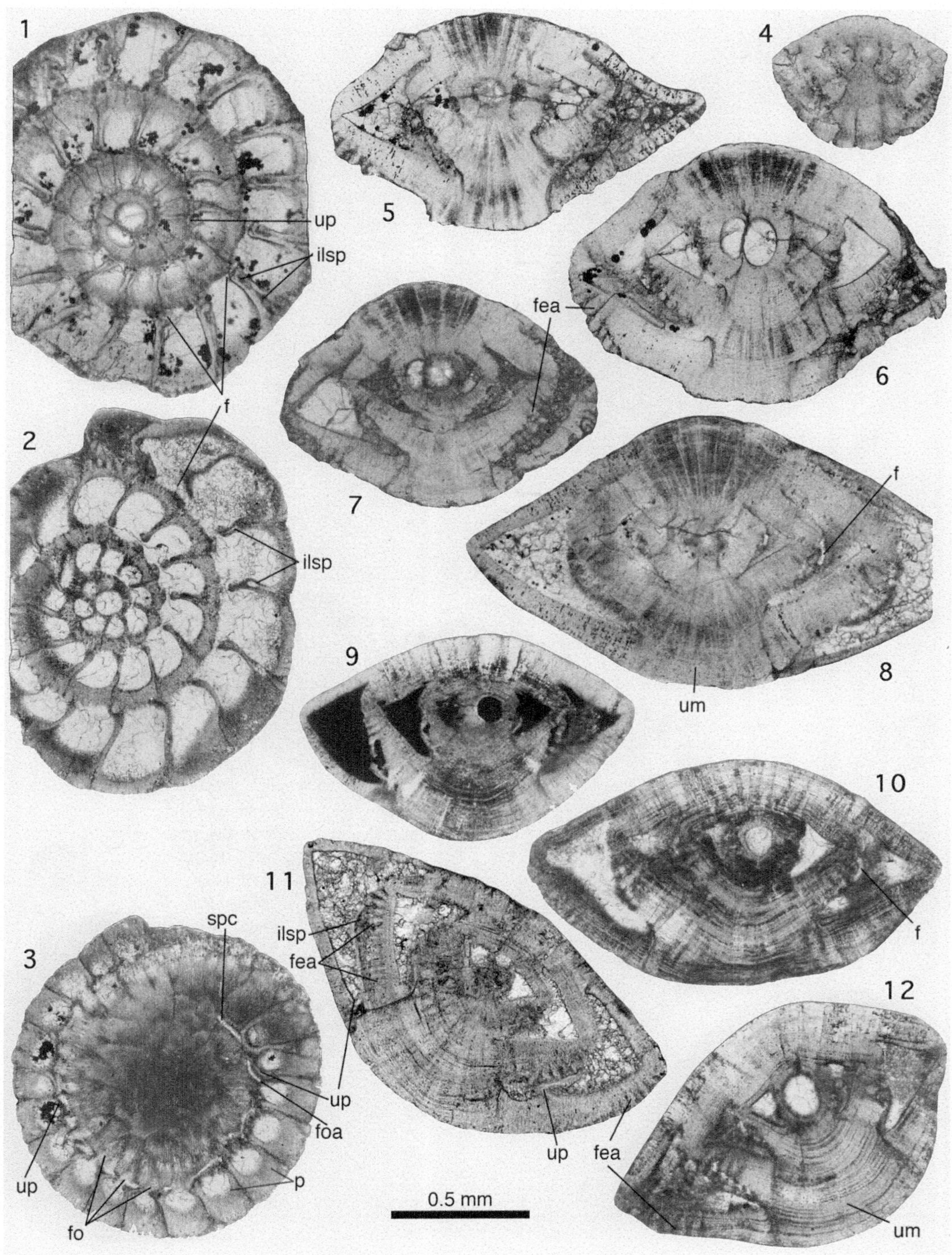

Plate 6.10 *Plumokathina lenticula* n. sp.; from Narp, Petites Pyrénées, Paleocene (SBZ 3). (**1–2**) Section perpendicular to coiling axis. (**3**) Section not perpendicular (uncentered) to coiling axis. (**4–10, 12**) Axial sections. (**11**) Subaxial sections. Abbreviations: *f* foramen, *up* umbilical plate, *ilsp* intraseptal interlocular space, *um* umbo, *fea* feathering of interlocular space, *spc* spiral canal, *foa* foliar aperture, *p* pores, *fo* folium

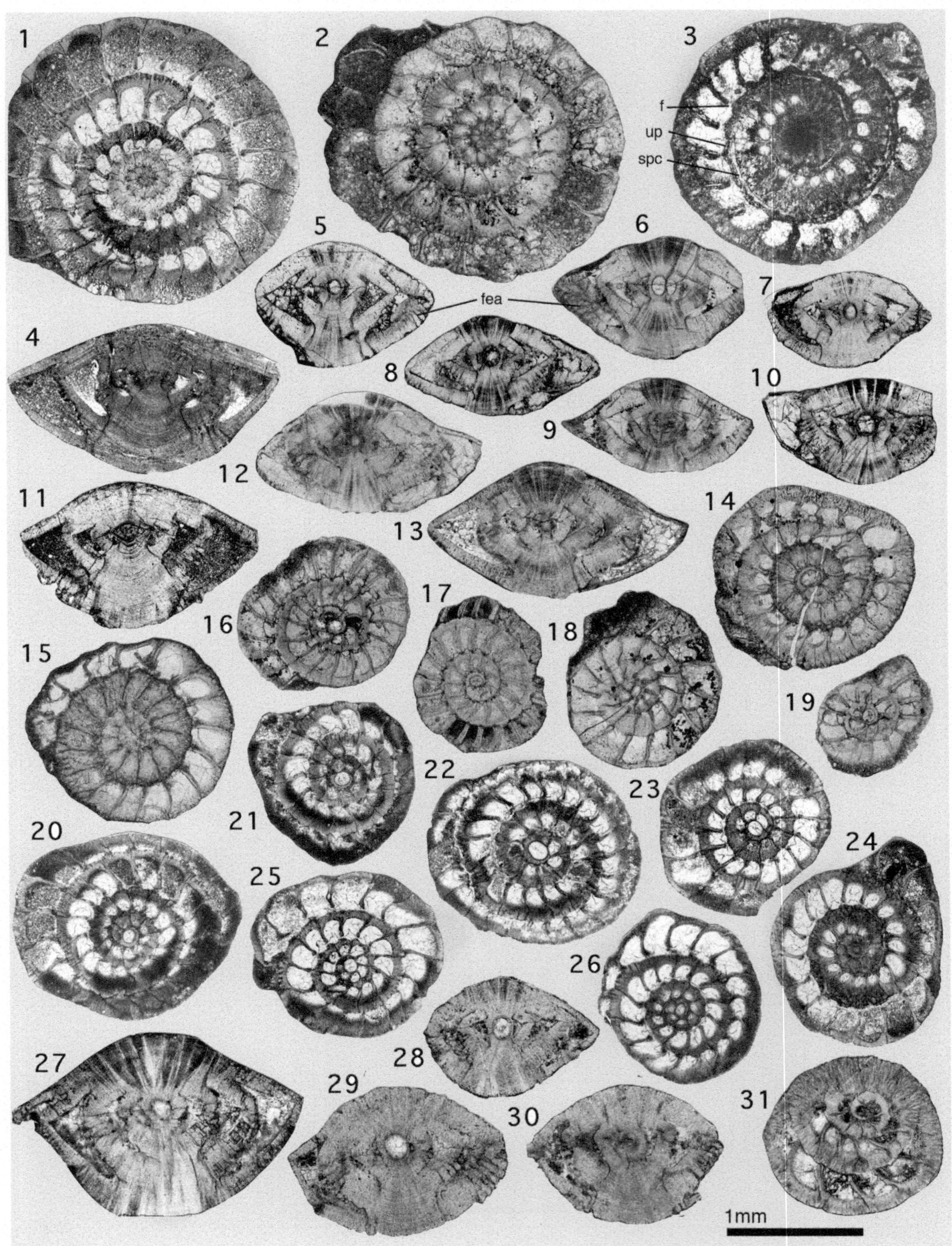

Plate 6.11 *Plumokathina lenticula* n. sp.; (**1–4**) microspheric forms; (**5–26**) megalospheric forms; from Narp, Petites Pyrénées; Aquitaine, southwestern France; Paleocene (SBZ 3). (**1–3**) Sections perpendicular to coiling axis. (**4**) Axial section. (**5–13**, **27–30**) Axial sections. (**14–26**) Sections perpendicular to coiling axis. (**31**) Oblique section. Abbreviations: *f* foramen, *up* umbilical plate, *fea* feathering of interlocular space, *spc* spiral canal

1972 *Rotalia* sp. 3 pars—Samuel et al., pl. 38, figs. 1–2.

1998 *Smoutina* ? sp.—Sirel, p. 79; pl. 37, figs. 1–4, 6–9, 11–16; pl. 68, figs. 4–20.

1998 *Plumokathina* sp.—Accordi et al., pl. 14, fig. 6.

Holotype: megalospheric specimen figured in Plate 6.9, Fig. 3.

Type locality and type level: Narp, Petites Pyrénées, France; Paleocene (SBZ 3).

Derivation of name: *plumokathina*, a *Kathina* with feathering instead of funnels; *lenticula*, lentil shaped.

Diagnosis: The trochospiral shell forms a lens with equal convexity on both sides and with an angular but unkeeled periphery. The equatorial diameter of the microspheric generation reaches 2.4 mm, of the megalospheric generation 2.0 mm. The ratio of equatorial to axial diameter of the lens varies from 1.8 to 2.0. Adult whorls count 16–19 chambers in the megalospheric, 18–22 in the somewhat larger microspheric generation. The diameter of the megalosphere reaches 0.12 mm. The umbilical architecture including the foramen is conform with the generic diagnosis but, as the external aspect of the holotype shows, the most proximal parts of the ventral chamber walls are welded to the large umbo leaving a ring of short, linear orifices of the interlocular space that are irregular-oblique to the shell radius. The umbilical plates are comparatively heavy and separate a large spiral canal from the chamber lumen.

Remarks: *Plumokathina lenticula* is a very frequent rotaliid from the middle Thanetian that has a wide distribution in the Western Neotethys. Its morphology permits an easy identification in random thin sections of cemented carbonates and attracted therefore the attention for many years. It was cited under the name "*Plumokathina*" *subsphaerica* (Sirel, 1972). It is formally described here for the first time. However, it will be necessary to separate the lenticular forms from the subspherical ones, keeping in mind that the two species occur respectively in SBZ 3 and 4.

The name *Smoutina* was also used for plumokathinas, but *Smoutina* Drooger, 1960 is a Caribbean form from Jamaica with numerous funnel orifices distributed all over the umbilical area, close or even identical with *Kathina jamaicensis* (Cushman and Jarvis, 1931) in Butterlin and Fourcade (1989).

Plumokathina subsphaerica (Sirel, 1972); Plate 6.9, Figs. 7–16.

1972 *Kathina subsphaerica*—Sirel, p. 287, pl. 5, figs. 1–5.

1972 *Kathina selveri* Smout—Sirel, p. 290, pl. 5, fig. 6.

1983 *Kathina* aff. *subsphaerica* Sirel—Rahaghi, pl. 36, figs. 17–18 (the same specimen).

1983 *Kathina* group *selveri* Smout—Rahaghi, pl. 36, figs. 12–15.

1988 *Kathina selveri* Smout—Drobne et al., pl. 26, fig. 9.

1998 ?*Kathina subsphaerica* Sirel—Accordi et al., p. 200, pl. 16, figs. 1, 3.

1998 *Smoutina* ? *subsphaerica* (Sirel)—Sirel, p. 78, pl. 36, fig. 1–20.

2008 "*Plumokathina*" sp.—Pignatti et al., p. 134, pl. 2, figs. 7–10; pl. 4, figs. 5–6.

Description: The bilamellar-perforate, trochospiral shell has a globular to ovoid shape with a rounded, unkeeled periphery. The ventral side, dominated by a large, composite umbo, has a hemispherical shape, The dorsal side is a little less convex. The chamber walls are comparatively thick. Taking into account the relatively low number of chambers per whorl, the individual lamellae must be thicker than in other *Plumokathina* species. The ratio equatorial to axial diameter of the shells always remains below 1.5. At an equatorial diameter of 2 mm, there are 20, at a diameter of only 1 mm, there are 13 chambers in the last whorl. The intraseptal interlocular space remains uncovered but is partially subdivided by the coarsest spurs of the feathering system known in the subfamily. The feathered space reaches the periphery and sometimes invades the dorsal side of the test over a very short distance. The umbo has a lobed contour that hints to the composite nature of the umbilical fill. Few radial slits in the umbo form orifices of funnels remaining as canal system from earlier whorls. No microspheric specimens have been found. The diameter of the megalosphere reaches 0.16 mm.

Remarks: *Plumokathina subspherica* is the largest of the *Plumokathina* species and occurs within a characteristic association of larger fotaminifera in SBZ 4. The phylogenetic trend in this group to more sphericity in the shell is opposed to what is observed in other groups of larger benthics and must have a functional meaning. *Ammonia pila* Billman et al., 1980 from the Plio-Pleistocene of the Mahakam Delta (Kalimantan, Borneo) presents similar proportions in a subspherical shell. The heavy umbo in this form has been tentatively interpreted as weight of a skipjack mechanism helping to stabilize the animal immerged in soft, fine-grained substrates (Hottinger 2006) (Plate 6.9).

References

Accordi G, Carbone F, Pignatti J (1998) Depositional history of a Paleogene ramp (Western Cephalonia, Ionian islands, Greece). Geol Romana 34:131–205

Afzal J, Williams M, Aldbridge R (2009) Revised stratigraphy of the lower Cenozoic succession of the Grater Indus Basin in Pakistan. J Micropaleontol 28:7–23

Billman H, Hottinger L, Oesterle H (1980) Neogene to recent Rotaliid foraminifera from the Indopacific Ocean; their canal system, their classification and their stratigraphic use. Schweiz Paläontol Abh 101: 71–113, 27 text figs, 39 pls

Boix C, Villalonga R, Caus E, Hottinger L (2009) Late Cretaceous Rotaliids (Foraminiferida) from the Western Tethys. Neues Jb Geol Paläontol Abh 253(3): 197–227

BouDagher-Fadel MK (2008) Evolution and geological significance of larger benthic foraminifera. Developments in palaeontology & stratigraphy, 21. Springer, Amsterdam, 540 pp

Butterlin J, Fourcade F (1989) Extension stratigraphique et distribution géographique du genre *Lockhartia* Davies, 1932 (Foraminifères, Rotaliidae). Rev Micropaleontol 31(4):225–242

Davies LM (1927) The Ranikot beds at Thai (North-West Frontier Provinces of India). Quart J Geol Soc 83:260–290, pls. 17–22, London

Dieni I, Massari F, Radoicic R (1985) Marine Paleocene pebbles included in the Cuccuru'e Flores Conglomerate (Post-Cuisian) of Orosei (Sardinia). In: 19th European micropaleontological colloquium excursion guide. AGIP Mineraria, San Donato Milanese, Milano, 213–214 pp

Drobne K, Ogorelec B, Plenicar M, Zucchi-Stolfa M-L, Turnsek D (1988) Maastrichtian, Danian and Thanetian Beds in Dolenjavas (NW Dinarides, Yugoslavia): microfacies, foraminifers, rudists and corals. S A Z U Ljubljana (4 Historia Naturalis) 29(6):147–224, 35 pls

Drooger CW (1960a) Microfauna and age of the Basses Plaines Formation of French Guyana I & II. Proc K Ned Akad Wet Amst Ser B 63(4):449–468

Drooger CW (1960b) Some Early Rotaliid Foraminifera I-III. Proc K Ned Akad Wet Amst B 63(2): 287–334

Hasson PF (1985) New observations on the biostratigraphy of the Saudi Arabian Umm er Radhuma formation (Paleogene) and its correlation with neighboring regions. Micropaleontology 31(4): 335–364

Henson FRS (1950) Middle East tertiary Peneroplidae (Foraminifera) with remarks on the phylogeny and taxonomy of the family. West Yorkshire Printing Co., Wakefield, ii+70 pp, 3 figs, 10 pls

Hottinger L (1966) Foraminifères rotaliformes et Orbitoides du Sénonien inférieur pyrénéen. Eclogae Geol Helv 59(1):277–301

Hottinger L (2006) The "face" of benthic foraminifera. Boll Soc Paleontol Ital 45:75–89

Hottinger L (2009) The Paleocene and earliest Eocene foraminiferal family Miscellaneidae: neither nummulites nor rotaliids. Carnets Géol Article 2009/06, CG2009_A06

Nicora A, Garzanti E, Fois E (1987) Evolution of the Tethys Himalaya continental shelf during Maastrichtian to Paleocene. Riv Ital Paleontol Strat 92:439–496

Peybernès B, Fondecave-Wallez M-J, Hottinger L, Eichène P, Segonzac G (2000) Limite Crétacé-Tertiaire et biozonation micropaléontologique du Danien-Sélandien dans le Béarn occidental et la Haute-Soule (Pyrénées Atlantiques). Geobios 33: 35–48

Pignatti J, Di Carlo M, Benedetti A, Bottino C, Brigulio A, Falconi M, Matteucci R, Perugini G, Ragusa M (2008) SBZ 2–6 larger foraminiferal assemblages from the Apulian and Pre-Apulian domains. Atti Mus Civ St Nat Trieste (Suppl 53):131–146, 7 pls

Pozaryska K, Szczechura J (1970) On some warm-water foraminifers from the Polish Montian. Acta Palaeontol Pol 15:95–113

Rahaghi A (1983) Stratigraphy and faunal assemblage of Paleocene-Lower Eocene in Iran. Nat Iranian Oil Comp Geol Lab 10, 73 pp, 49 pls

Reichel M (1937) Etude sur les Alvéolines. Deuxième partie. Mém Soc Paléontol Suisse 59(2):95–147, pls. 10–11

Samuel O, Borza K, Köhler E (1972) Microfauna and Lithostratigraphy of the Paleogene and adjacent Cretaceous of the Middle Vah valley (West Carpathians). Geol Ustav Dionysza Stura, 246 pp, 153 pls

Sirel E (1972) Systematic study of new species of the genera *Fabularia* and *Kathina* from Paleocene. Türk Jeol Kur Bül 15:277–249, 8 pls

Sirel E (1988) *Anatoliella*, a new foraminiferal genus and a new species of *Dictyokathina* from the Paleocene of the Van area (east Turkey). Rev Paléobiol 7(2):477–493

Sirel E (1998) Foraminiferal description and biostratigraphy of the Paleocene-Eocene shallow water limestones and discussion of the Cretaceous-Tertiary boundary in Turkey. Gen Dir Min Res Expl (MTA), Monograph ser 2, 117 pp, 68 pls

Sirel E (2004) New benthic foraminifera of the Mesozoic and Cenozoic of Turkey (in Turkish). Chamber of Geological Engineers of Turkey, Publication 84. Scientific Synthesis of Life Time Research, Special volume 1. TMMOB jeologji Mühendisleri Odasi (Ankara), 219 pp, 66 pls

Sirel E, Gündüz H, Açar S (1983) Sur la présence d'une nouvelle espèce d'*Orbithokathina* Hottinger dans le Thanétien de Van (Est de la Turquie). Rev Paléobiol 2: 149–159

Smout AH (1954) Lower Tertiary foraminifera of the Qatar peninsula. Brit Mus (Nat Hist), 96 pp, 44 figs, 15 pls

Smout AH, Haque M (1956) A note of the larger foraminifera and ostracoda of the Ranikot from the Nammal Gorge, Salt Range, Pakistan. Rec Geol Sur Pak 8(2): 49–60, 3 pls

Weiss W (1993) Age assignments of larger Foraminiferal Assemblages of Maastrichtian to Eocene age in Northern Pakistan. Zitteliana 20:223–252

7 New Subfamily Daviesininae

Abstract

This new subfamily is characterised by bilamellar-perforate, heavily ornate shells which have a low trochospiral arrangement of the chambers. The intraseptal canal system produces orifices on both sides of the test and may develop in an enveloping canal system forming a supplemental skeleton of considerable thickness. Eleven species of the genus *Daviesina* (*D. fleuriausi*, *D. labaneae*, *D. khatiyahi*, *D. danieli*, *D. langhami*, *D. intermedia*, *D. salsa*, *D. praegarumnensis* n. sp., *D. garumnensis*, *D. tenuis*, *D. ruida*) are described and illustrated.

The bilamellar-perforate, heavily ornate shells have a low trochospiral arrangement of the chambers. The dorsal sides of the chambers are involute, sometimes a little less than on the ventral side. Their distinction is difficult if the umbilical structures are not preserved to perfection. The spiral widens more rapidly than in all other groups of the rotaliidae. In a group of particularly flat daviesines with a wide spire, the septal flap is folded; the folds subdivide the spiral main chamber into chamberlets like in a *Heterostegina* but there is no marginal cord. The intraseptal canal system produces orifices on both sides of the test and may develop in an enveloping canal system, forming a supplemental skeleton of considerable thickness. The interiomarginal foramen is in peripheral position and often bears a kind of tooth that has an unclear origin and nature. In most cases it is asymmetrical. There are two umbilical plates, usually of different length, smaller on the dorsal side and larger of the ventral side of the shell. Accordingly, there are also two wide-open spiral canals.

The daviesines are difficult to distinguish from the truly planispiral miscellaneids (Hottinger 2009) because their chamber arrangement is so close to a planispiral pattern.

7.1 *Daviesina* Smout, 1954

Type species: *Daviesina khatiyahi* Smout, 1954

The type species, *Daviesina khatiyahi*, is characterized by a particularly heavy ornamentation on both sides of the shell by large pustules. Otherwise the dorsal side of the shell is smooth, the chamber sutures barely visible. On the ventral side, the sutures are hidden at the bottom of the interlocular spaces between the proximal ends of the chamber alae. The intraseptal canal system is reduced to a row of sutural canals that deviate rarely and for only short distances into the lateral chamber wall. The nearest relative to the type species is *Daviesina salsa* that has the same generic features and, in particular, a similar

L. Hottinger, *Paleogene larger rotaliid foraminifera from the western and central Neotethys*,
DOI 10.1007/978-3-319-02853-8_7,

canal system. The ornamentation of the lateral chamber walls consists here of a very crowded growth of pustules that give a false impression of thickened walls.

A second group of species is characterized by a supplemental skeleton that is based on a feathered interlocular space, covered by subsequent outer lamellae, as in *Daviesina danieli* and *D. langhami*. The resulting supplemental skeleton exhibits an enveloping canal system with numerous orifices in between spikes and piles forming a heavy ornamentation. In this group a doubling of the periphery is particularly frequent and adds to the confusion with miscellaneids.

Two additional species have an even more pronounced operculiniform habit in an almost perfectly planispiral, wide-spired, evolute shell: *Daviesina intermedia* have a row of stolons supplementing the main foramen whereas *Daviesina salsa* exhibits a kind of coarse trabeculae. Today these structural differences are treated as of specific rank, and, therefore, the two species are incorporated here in the group of *Daviesina langhami*.

The third group is characterized by narrowly coiled, involute early stages, followed by an adult last whorl tending to get evolute and wide-spired. The early species in this group, *D. praegarumnensis* and *D. garumnensis*, have an involute, trochospiral early stage of growth that is heavily ornate by pustules and piles. Their aperture is an areal slit, starting ventrally from an interiomarginal position and reaching in dorsal direction obliquely into the septal face. This is not a typical daviesinid foramen, but all other morphological elements favour a classification of these species in the group of *D. tenuis*. The long septa of the wide-spired adult stages of this group's advanced species are covered by a septal flap that gets folded. *Daviesina fleuriausi* from the Late Cretaceous might belong to the same group. However, the conspicuous kink in the septal sutures of this species has never been observed in any other species of *Daviesina*. *D. fleuriausi* and *D. labanae* are the only representatives of this genus in the Cretaceous. Both occur in the type Maestrichtian. *D. fleuriausi* at least seems to be wide-spread in the Neotethys; it has been observed in the Spanish Pyrenees, the Betics but also in Madagascar.

Daviesina fleuriausi (d'Orbigny, 1826); Plate 7.1, Figs. 1–9.

1826 *Amphistegina fleuriausi* d'Orbigny, p. 304, (name only, *nomen nudum*), *fide* Reuss.

1951 *Operculina fleuriausi* (d'Orbigny)—Visser, p. 251, pl. 1, fig. 17; pl. 10, figs. 1, 6, with synonymy.

2002 *Daviesina fleuriausi* (d'Orbigny)—Abramovich et al., p. 61, pl. 4, figs. 5–10.

Remarks: This species is one of the Late Cretaceous rotaliid precursors. The bilamellar-perforate shell is small, with an operculinid habit where the spire opens rapidly in late stages of growth. Early whorls are involute, late whorls evolute. In the early whorls, the dorsal alar prolongations are longer and more clearly involute than the ventral prolongations that bump against a large, solid and progressive umbo. The chambers are near to planispiral but, in axial sections, the asymmetry of the foramen and the presence of an umbilical plate, restricted to one side of the test, reveal its trochospiral original nature. The triangular foramen and the umbilical plate permit to assign this species to the genus *Daviesina*. Adult spiral chambers are marked by a sudden kink in the dorsal septa at about half their radial extension. This knee is not known in any other *Daviesina* species rising after the K-T event. Therefore, it is difficult to assign *D. fleuriausi* to a particular phylogenetic lineage within the daviesines.

Daviesina labanae (Visser, 1951); Plate 7.1, Figs. 10–14.

1951 *Operculina labanae*—Visser, p. 253, pl. 1, fig. 18; pl. 10, figs. 2–3.

Remarks: In association with *Daviesina fleuriausi*, F. Liebau has found another "*Operculina*" with a much tighter coil, a triangular foramen with a kind of tooth and an umbilical plate that obliges to classify *D. labanae* as rotaliid with affinities to *Daviesina*. The shell is

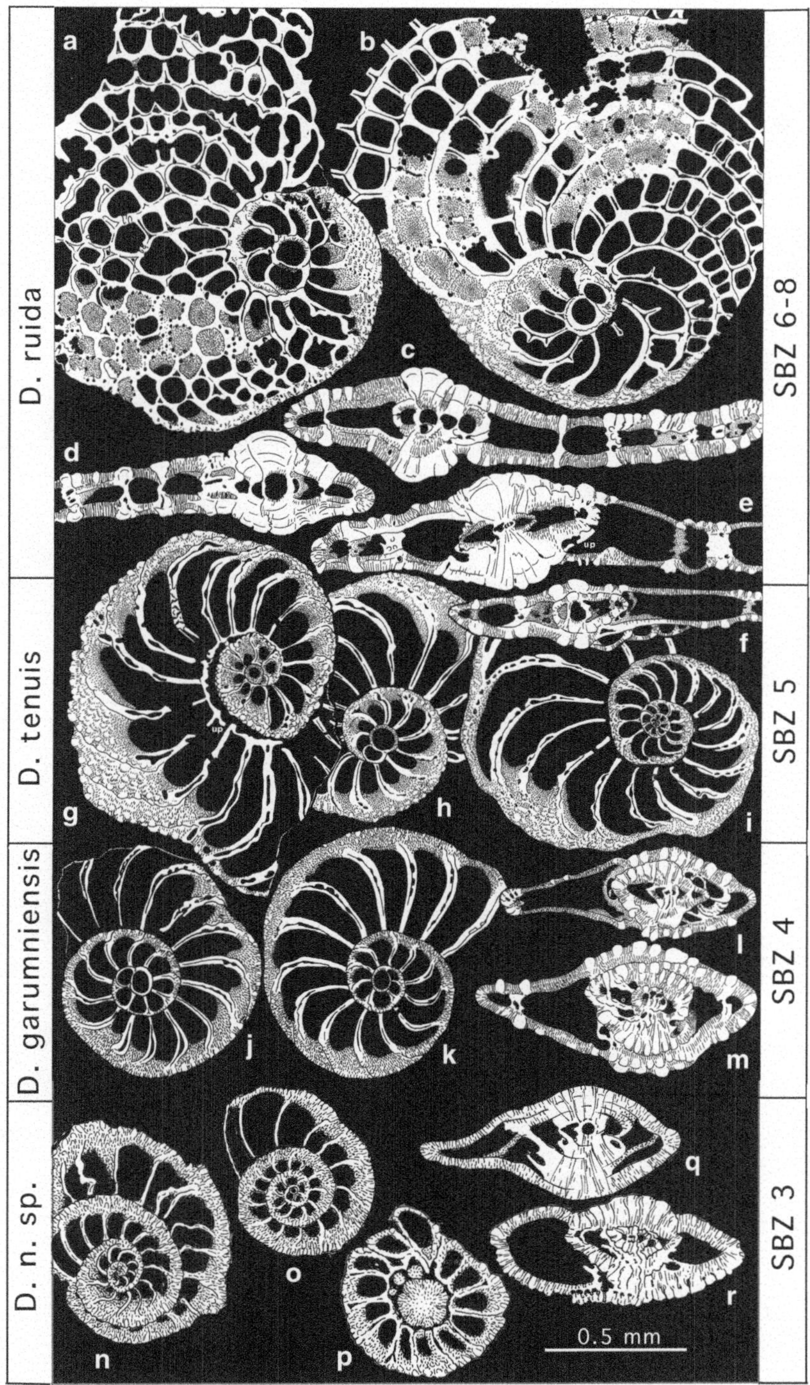

Fig. 7.1 Biostratigraphic distribution of the operculiniform Daviesininae; drawings made from camera lucida

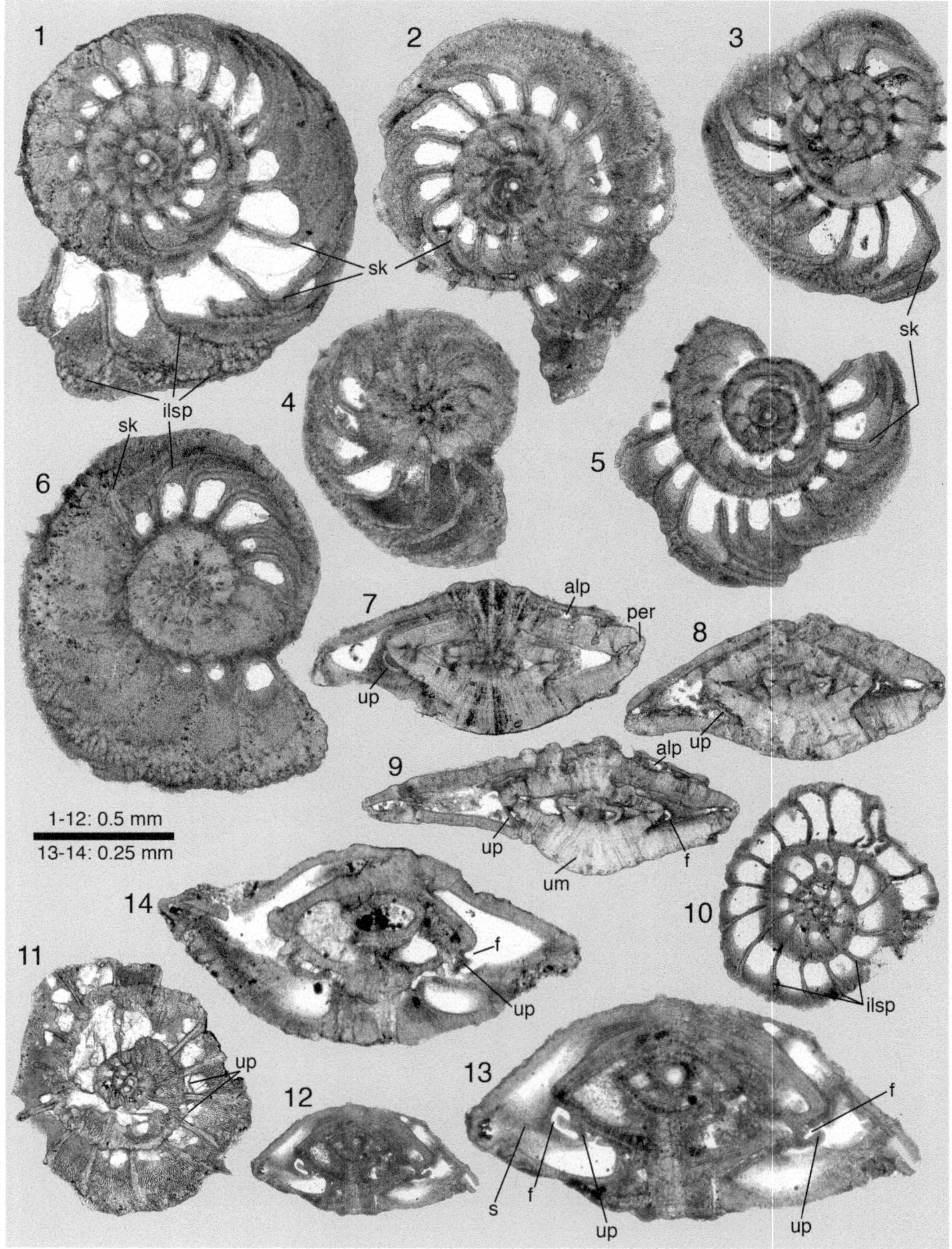

Plate 7.1 (**1–9**) *Daviesina fleuriausi* (d'Orbigny, 1826); megalospheric specimens; from Isona, Tremp Basin, Lleida Province, Spanish Pyrenees, collected by F. Liebau; Maastrichtian. (**1–6**) Sections perpendicular to the coiling axis; note the kink (sk) of the septum that is one of the characteristics of this species. (**7–9**) Axial sections showing the triangular foramen and the asymmetric umbilical plates. (**10–14**) *Daviesina labanae* (Visser, 1951); from Isona, Tremp Basin, Lleida Province, Spanish Pyrenees, collected by F. Liebau; Maastrichtian. (**10–11**) Equatorial sections. (**12–14**) Axial sections; note the foramen with a "tooth" in 13. Abbreviations: *f* foramen, *s* septum, *up* umbilical plate, *ilsp* intraseptal interlocular space, *um* umbo, *alp* alar prolongation, *per* periphery, *sk* septal kink

almost planispiral, with comparatively voluminous alar prolongations on both sides of the shell.

Daviesina khatiyahi Smout, 1954; Plate 7.2, Figs. 1–16; Plate 7.3, Figs. 1–15.

1954 *Daviesina khatiyahi*—Smout, p. 67, pl. 12, figs. 1–11; pl. 14, Fig. 7.
1962 *Miscellaneoides bramkampi*—Sander, p. 14, pl. 2, figs. 1–16.
1980 *Daviesina khatiyahi* Smout—Caus et al., p. 1054, text-Fig. 6G, H; pl. 1, figs. 1–3.
1980 *Daviesina khatiyahi* Smout—Gaetani et al., p. 144, pl. 16, figs. 2–4; pl. 17, Fig. 3.
1985 *Daviesina khatiyahi* Smout—Hasson, p. 360, pl. 5, figs. 4–6.
1991 *Daviesina khatiyahi* Smout—Wan, 15, pl. 2, figs. 15–16.
1993 *Miscellanea miscella* (d'Archiac & Haime)—Weiss, p. 252, pl. 1, fig. 6.
1998 *Daviesina khatiyahi* Smout—Pignatti, et al., p. 621, text-fig. 8.1–8.
2008 *Daviesina khatiyahi* Smout—Ismail and Boukhary, p. 90, pl. 1, figs. 3–6, non Fig. 7.
2008 *Daviesina khatiyahi* Smout—BouDagher-Fadel, p. 362, pl. 6.28, fig. 20.

Description: The bilamellar-perforate shells are low-trochospiral to almost planispiral. The chambers are evolute on their dorsal side and ventrally somewhat involute. The dorsal exterior of the shell is decorated with few heavy beads located at the junction of the chamber sutures with the whorl suture. They produce thus a spiral ornament on the otherwise smooth dorsal face. In the area around the coiling axis, the ventral face is covered by a closely packed group of heavy piles that obscure the radial pattern of the ventral chamber sutures. These are visible only in the last whorl as deep furrows. The sutural furrows communicate with the rest of the interlocular space by a single row of short sutural tubiform passages (Plate 7.3, Fig. 15). The intercameral foramen is formed as low but long interiomarginal slit positioned over the periphery of the previous whorl. In some places, the daviesinid asymmetric-triangular foramen can be discovered (Plate 7.3, Fig. 5). The umbilical plate is small, restricted to the ventral side of the test and delimits a minute foliar chamberlet.

The microspheric generation exhibits a chamber shape that has at least twice the radial height in relation to its length in the direction of growth. The septa deviate from the radius of the shell by a strong falciform inclination. Adult whorls count 26–32 chambers.

The megalospheric generation is about a third smaller, has a tighter spiral with 24–30 septa in the last adult whorl. The diameter of the megalosphere is 0.16–0.24 mm. The first two chambers form a dyad.

Daviesina danieli Smout, 1954; Plate 7.4, Figs. 1–21.

1954 *Daviesina danieli*—Smout, p. 69, pl. 7, figs. 15–17.
1972 *Miscellanea* sp.—Bizon et al., pl. 1, fig. 2.
1980 *Daviesina danieli* Smout—Caus et al., p. 1056, text-figs. 6A–C; pl. 2, Figs. 5–7.
1980 *Daviesina danieli* Smout—Gaetani, et al., p. 144, pl. 15, figs. 1, 3, 6.
1983 *Daviesina shirazensis*—Rahaghi, p. 59, pl. 40, figs. 1–16.
non 1987 *Daviesina danieli*—Nicora et al., pl. 34, fig. 3.

Description: Bilamellar-perforate, spiral chambers construct a trochospiral shell that has a low-convex to flattened, more or less evolute dorsal aspect and a completely involute conical axial intersection on the ventral side of the test. The shell is small, 1–2 mm in equatorial diameter, with a ratio of equatorial diameter to axial thickness of 1.5–2.0. At an equatorial diameter of 1 mm the last whorl counts 13–15 chambers.

The dorsal side is decorated by beads along the sutures and some smaller papillae on the dorsal chamber wall. Around the axis of coiling, an apical area is occupied by a group of piles that admit among them some canal orifices fed from dorsal funnels. The latter spring from the intraseptal canal system at the dorsal interiomarginal angle of the septal face. The conical venter of the shell is marked at its axial center by a central pile surrounded by a ring of smaller piles.

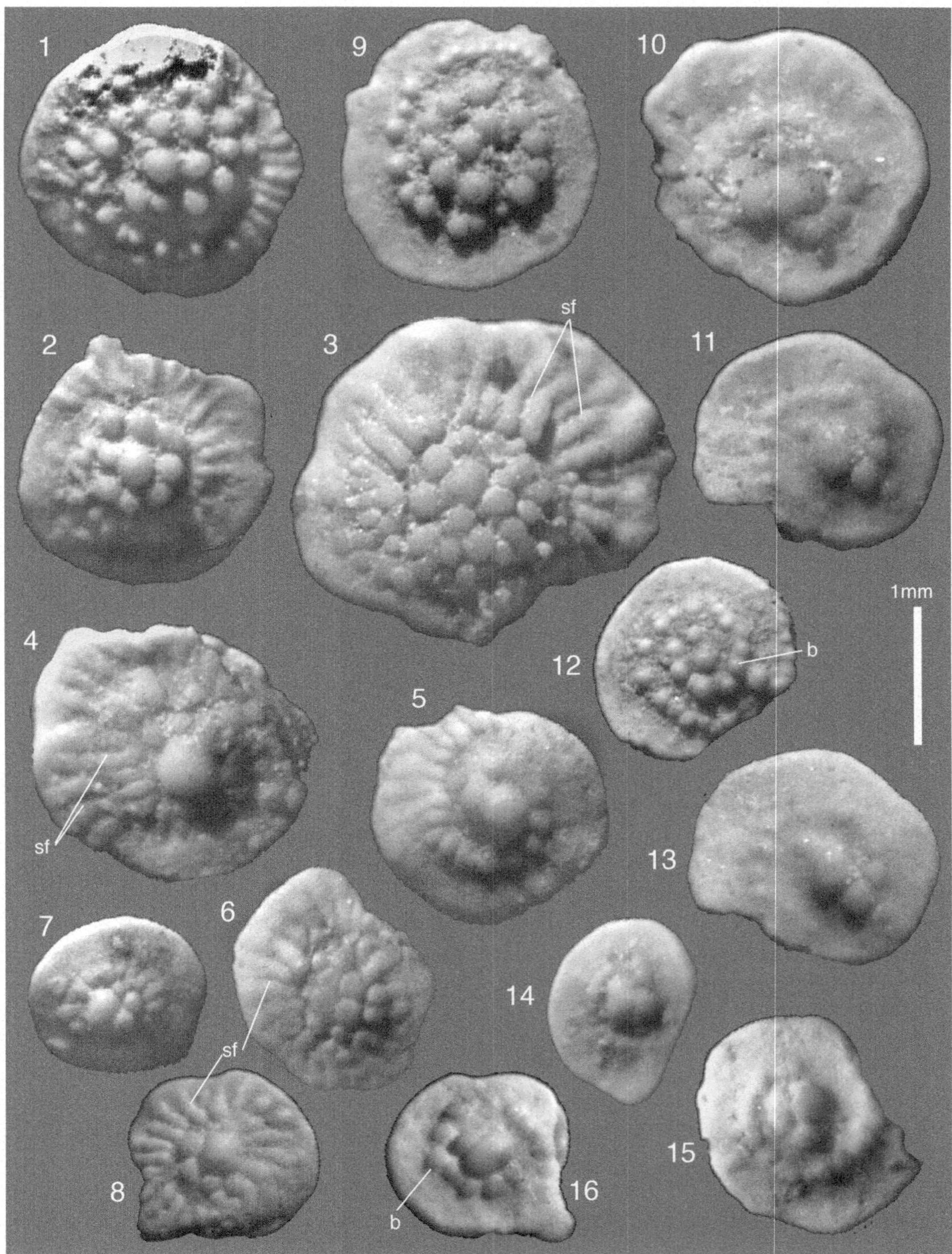

Plate 7.2 *Daviesina khatiyahi* Smout, 1954; isolated specimens; sample AQ 6, Qatar, Arabian Peninsula, collected by M. Chatton. (**1–8**) Ventral views. (**9–16**) Dorsal views; note the spiral arrangement of the dorsal ornaments, mostly beads. Abbreviations: *sf* septal face, *b* bead

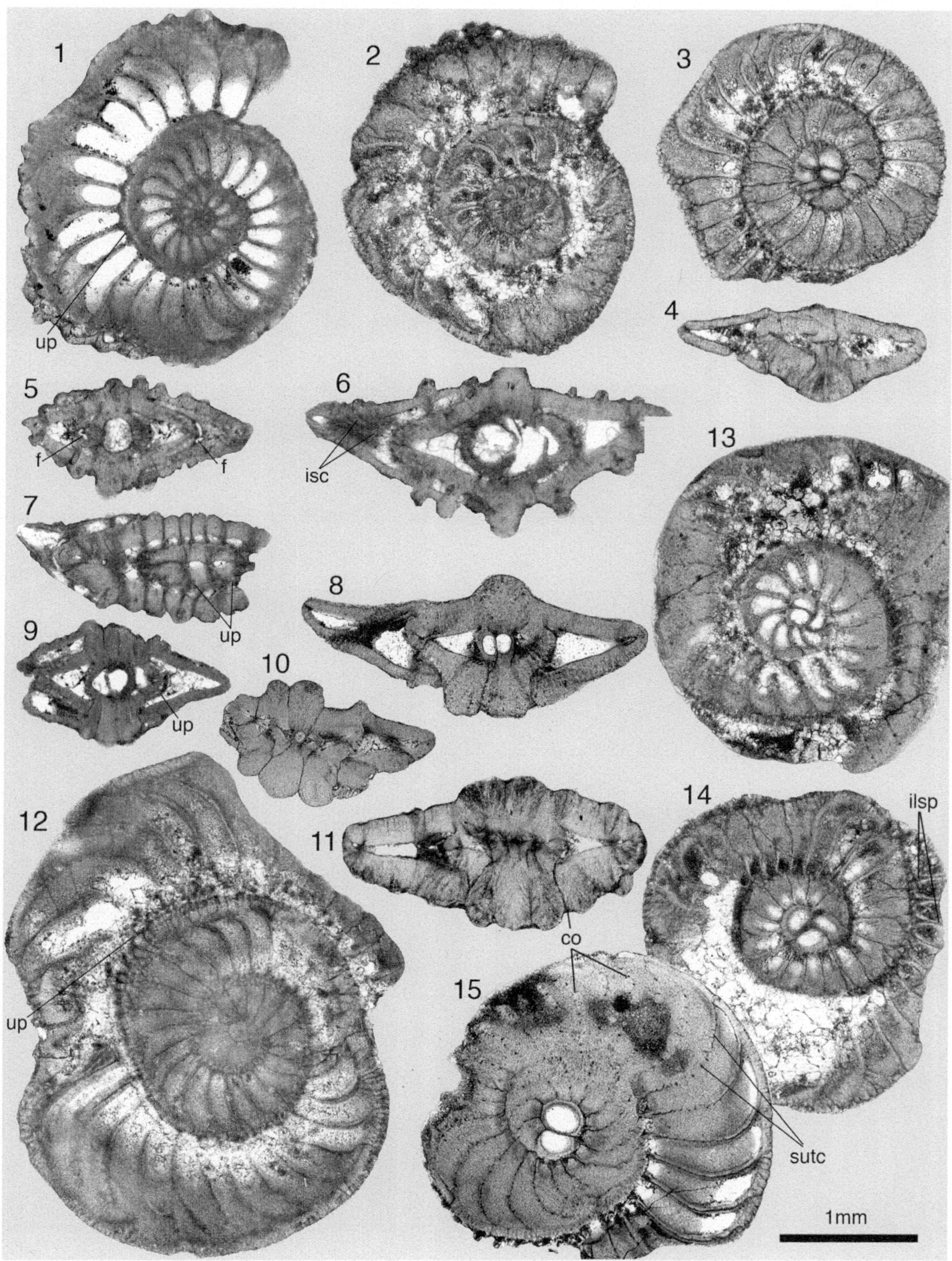

Plate 7.3 *Daviesina khatiyahi* Smout, 1954; sample AQ 6, Qatar, Arabian Peninsula, collected by M. Chatton. (**1–3**, **12–15**) Equatorial sections. (**4–11**) Axial sections. (**1–2**, **12**) Microspheric specimens, all others belong to the megalospheric generation. Abbreviations: *f* foramen, *up* umbilical plate, *isc* intraseptal canal system, *co* canal orifice, *sutc* sutural canal, *ilsp* intraseptal interlocular space

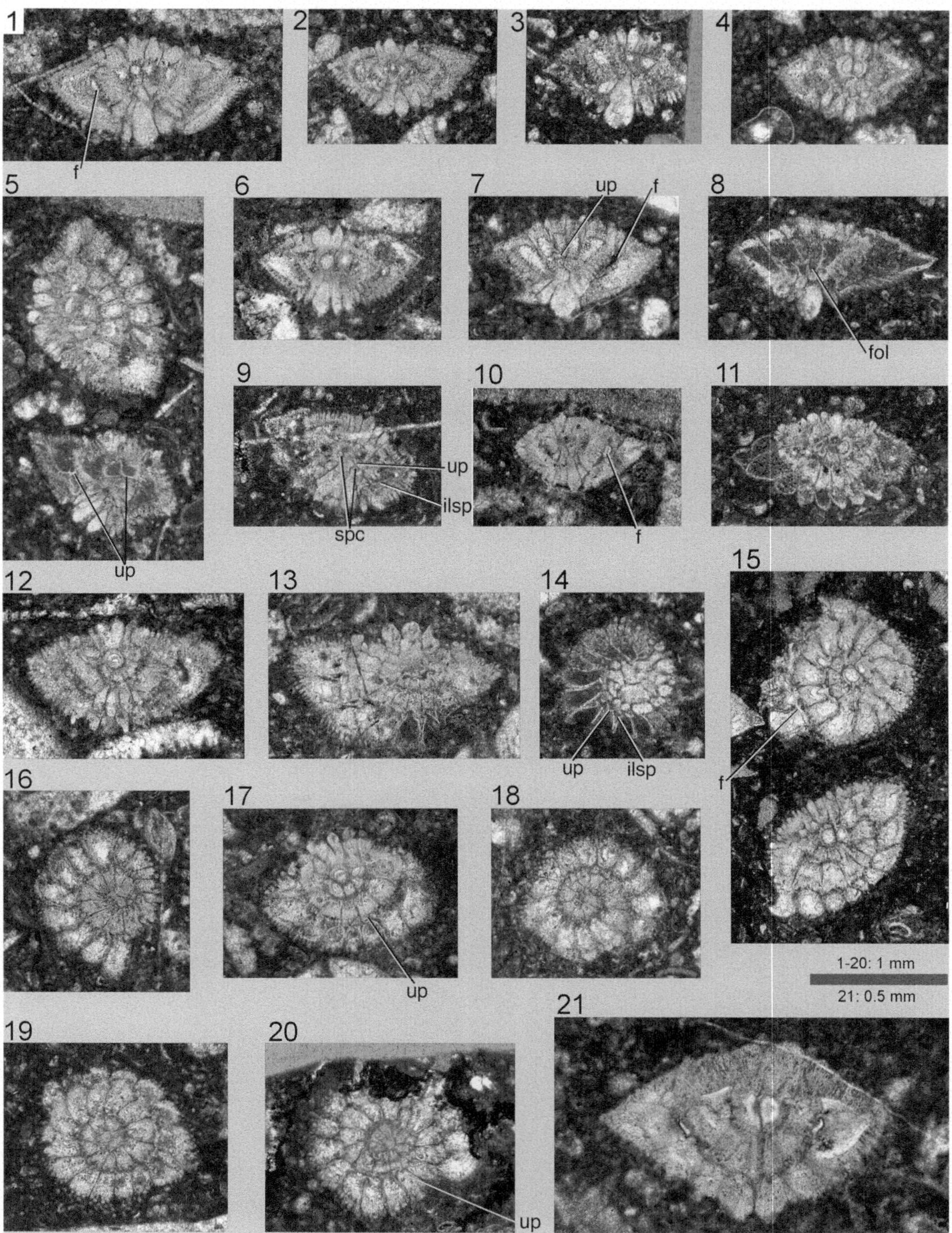

Plate 7.4 *Daviesina danieli* Smout, 1954; sample Kar 9, Kuh-e-Kargan, Kermanshah, Iranian Zagros, collected by A. Braud. Paleocene (SBZ 3). (**1–4**, **6**, **9–10**, **12**) Axial sections. (**5**, **7–8**, **11**, **13**, **17**) Oblique sections. (**14**) Tangential section perpendicular to coiling axis showing the umbilical architecture. (**15**) Equatorial and oblique-centered sections showing dimensions of the megalosphere. (**16**, **18–20**) Sections perpendicular to the coiling axis showing numbers and shape of chambers in the outer whorls. (**21**) Axial section, details showing foramina in the last whorl. Abbreviations: *f* foramen, *up* umbilical plate, *fol* folia, *ilsp* intraseptal interlocular space, *spc* spiral canal

The intercameral foramina form a high, interiomarginal arch at an equatorial level defined by the location of the periphery of the previous whorl. Below the foramen, at the ventral umbilical chamber end, the intraseptal space is enlarged to form a triangular space delimited by the septum, the septal flap and the umbilical plate. Thus, the umbilical architecture has a rotaliid nature, in spite of the pararotaliid overall aspect of the shell's ornamentation at the exterior (Hottinger et al. 1991).

The proloculus has a diameter of 0.04–0.12 mm and probably marks the megalospheric generation. No microsphere has been observed.

Daviesina langhami Smout, 1954; Plate 7.5, Figs. 1–14, 16–17; Plate 7.6, Figs. 1–5; Plate 7.8, Figs. 1–10; Plate 7.9, Figs. 1–11; Plate 7.10, Figs. 1–4.

1954 *Daviesina langhami*—Smout, p. 68, pl. 11, figs. 1–11.

1956 *Daviesina langhami* Smout—Smout and Haque, p. 52, Pl. 10, figs. 7–10, ?11.

1962 *Miscellanoides pruvosti*—Sander, p. 15, pl. 3, figs. 1–12.

non 1980 *Daviesina langhami* Smout—Gaetani et al., p. 145, pl. 17, figs. 1–2, 4 (= *Miscellanea dukhani* Smout—Hottinger, 2009, p. 6, pl. 6, figs. 1–7; pl. 7, figs. 1–8).

1983 *Daviesina persica*—Rahaghi, p. 38, pl. 39, figs. 1–13.

1985 *Daviesina langhami* Smout—Hasson, p. 360, pl. 5, fig. 7.

1987 *Daviesina langhami* Smout—Nicora et al. p. 464, pl. 33, fig. 1.

1987 *Daviesina danieli* Smout—Nicora et al., pl. 34, fig. 3.

1987 *Daviesina kathyahi* Smout—Nicora et al., pl. 34, fig. 2.

1991 *Daviesina langhami* Smout—Wan, p. 15, pl. 2, figs. 17–18.

1993 *Miscellanea miscella* (d'Archiac and Haime)—Weiss, p. 252, pl. 3, figs. 3–6.

1998 unidentified rotaliid—Accordi et al., p. 182, pl. 7, fig. a, upper left.

2008 *Daviesina persica* Rahaghi—Ismail and Boukhary, p. 93, pl. 1, fig. 8.

2008 *Daviesina langhami* Smout—BouDagher-Fadel, p. 361, pl. 6.27, figs. 14–15.

2009 *Miscellanea miscella* (d'Archiac and Haime)—Afzal et al., p. 17, pl. 1, figs. 3, 5, ? non fig. 1.

2009 *Daviesina langhami* Smout—Hottinger, text-fig. 2.

Description: The very low-trochospiral and semi-involute shell exhibits a supplemental skeleton that provides the lateral chamber walls on both sides of the test with a heavy, dense and deep pustular ornamentation. The supplemental skeleton is generated by the feather grooves perpendicular to the chamber sutures in the primary chamber wall, covered by secondary lamellation. Thus, an enveloping canal system separates the supplemental skeleton from the primary, bilamellar and perforate chamber wall (Plate 7.10, Figs. 1, 4). The periphery is angular to sharp, without keel, but covered with pustules like the rest of the shell's surface. Often, the periphery doubles in the last whorl by lacing the shell in equatorial direction. The frequency of this phenomenon seems to be a diagnostic feature at species level, as well as in the genus *Miscellanea* (Hottinger, 2009). To discern dorsal and ventral is very difficult, often impossible from the outside aspect of the shell.

The intercameral foramen is triangular and asymmetric in respect of the equatorial plane. A kind of tooth at the shell's equator seems to subdivide the apertural arch into two unequal parts. The origin of the "tooth" is unclear as long as the three-dimensional extension, possibly into the umbilical plate, has not been observed. An umbilical plate is present in both umbilical regions: a larger one on the ventral side, a smaller one on the dorsal side of the shell. A dorsal folium has not been found.

The microspheric forms reach an equatorial diameter of at least 4.4 mm and an axial diameter of 1 mm. At this stage of growth the last whorl counts 34–37 chambers. They have an equatorial

Plate 7.5 (**1–14**, **16–17**) *Daviesina langhami* Smout, 1954; megalospheric specimens, external, spiral views. (**1–10**) Sample 95114, lower Lockhart Limestone, Dhak Pass, Salt Range, Pakistan; Paleocene (SBZ 3). (**11–14**) Sample 95119, Lockhard Limestone, Dhak Pass, Salt Range, Pakistan; Paleocene (SBZ 3). (**16–17**) Microspheric specimens, sample 95119 as in Figs. 11–14. (**15**) *Miscellanea miscella* (d'Archiac and Haime, 1853), sample 95117, top Lockhart Limestone, Dhak Pass, Salt Range, Pakistan; Paleocene (SBZ 4). Abbreviations: *b* beads

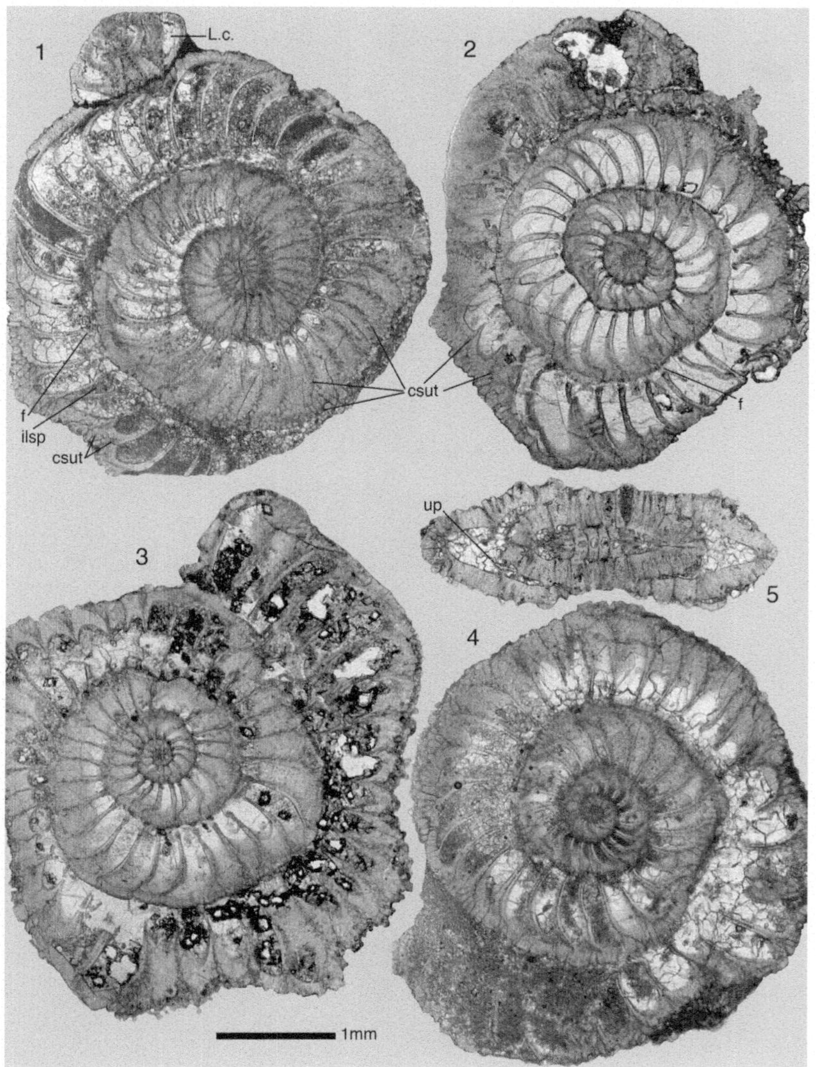

Plate 7.6 *Daviesina langhami* Smout, 1954; microspheric specimens; sample 92010a, Base Patala Formation, Dhak Pass, Salt Range, Pakistan; Paleocene (SBZ 4). (**1**) Equatorial section associated (*at the top*) with (L. c.) *Lockhartia conditi* (Nuttall, 1926). (**2–4**) Equatorial sections. (**5**) Axial section. Abbreviations: *f* foramen, *up* umbilical plate, *csut* chamber suture, *ilsp* intraseptal interlocular space

extension in radial direction that is four times their extension in the direction of growth.

The megalospheric generation reaches an equatorial diameter of 3.5 mm, with 26–28 chambers in the last whorl. The diameter of the megalosphere varies from about 0.16–0.24 mm. In some specimens, the first two chambers form a dyad.

Both generations are difficult to distinguish from *Miscellanea dukhani* Smout, 1954 that is comparatively little known (Hottinger 2009, pl. 6, figs. 1–7; pl. 7, fig. 1–8). Additional microspheric specimens are provided here on Plate 7.7, Figs. 1–6 to facilitate the comparison. Note the involute character of the shell that is visible in the axial sections, the symmetrical foramina (Plate 7.7, Fig. 3), the symmetrical disposition and size of the umbilical plates (Plate 7.7, Fig. 4) and the comparatively wide coils of the late spiral chambers.

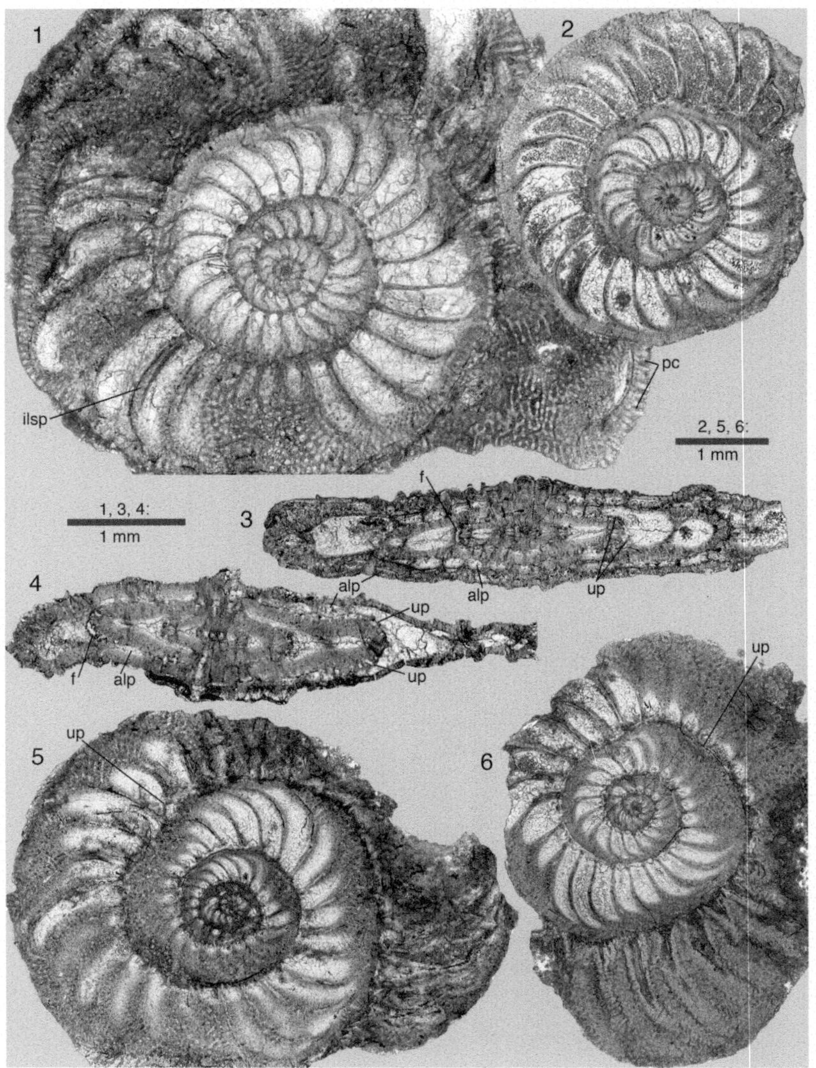

Plate 7.7 *Miscellanea dukhani* Smout, 1954; microspheric specimens. (**1–2**, **5–6**) Equatorial sections. (**3–4**) Axial sections. (**1**) Sample 93525, Patala Formation, Dhak Pass; (**2–6**) samples 95108 and 95109, Patala Formation, about 10 m above the hardground marking the top of the Lockhart Limestone, Dhak Pass, Salt Range, Pakistan; Paleocene (SBZ 4). Abbreviations: *f* foramen, *up* umbilical plate, *alp* alar prolongation, *ilsp* intraseptal interlocular space, *pc* peripheral canals

Daviesina intermedia Smout and Haque, 1956; Plate 7.11, Figs. 1–9; Plate 7.12, Figs. 1–14; Plate 7.13, Figs. 1–10.

1956 *Daviesina intermedia*—Smout and Haque, p. 54, pl. 10, figs. 4–6.

1983 *Daviesina iranica*—Rahaghi, p. 60, pl. 41, figs. 1–18.

1991 *Daviesina intermedia* Smout and Haque—Wan, p. 14, pl. 2, figs. 9–11.

Description: The bilamellar-perforate test is almost perfectly planispiral, evolute and operculiniform in shape, flattened (equatorial diameter reaching eight times the axial diameter), but with a characteristic axial inflation by a dense group of coarse piles. In most cases, the axial swelling is more pronounced on the ventral side of the test. The septal sutures are marked by a string of minute beads; the whorl sutures are unmarked and often

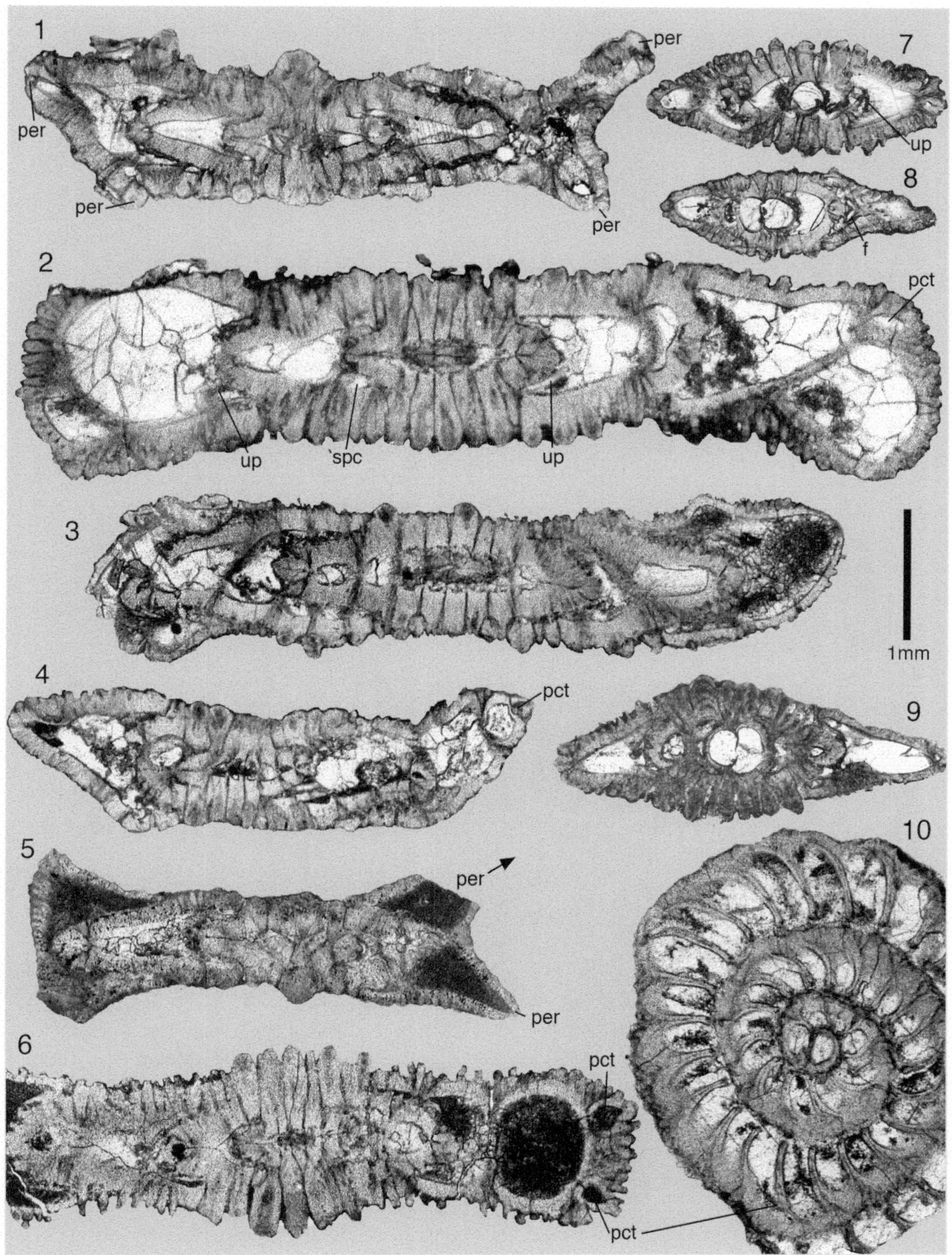

Plate 7.8 *Daviesina langhami* Smout, 1954; sample 92010a, base Patala Formation, Dhak Pass, Salt Range, Pakistan, except Figs. 9–10 from sample 93503, Patala Formation immediately above the hardground marking the top of the Lockhart Limestone, Nammal Gorge, Salt Range, Pakistan; Paleocene (SBZ 4). (**1–6**) Axial sections, microspheric specimens. (**7–9**) Axial sections, megalospheric specimens; note in (**1**, **4–6**) the asymmetric, double angular periphery in late whorls. (**10**) Equatorial section, megalospheric specimen. Abbreviations: *f* foramen, *up* umbilical plate, *per* periphery, *spc* spiral canal, *pct* peripheral chamber tip

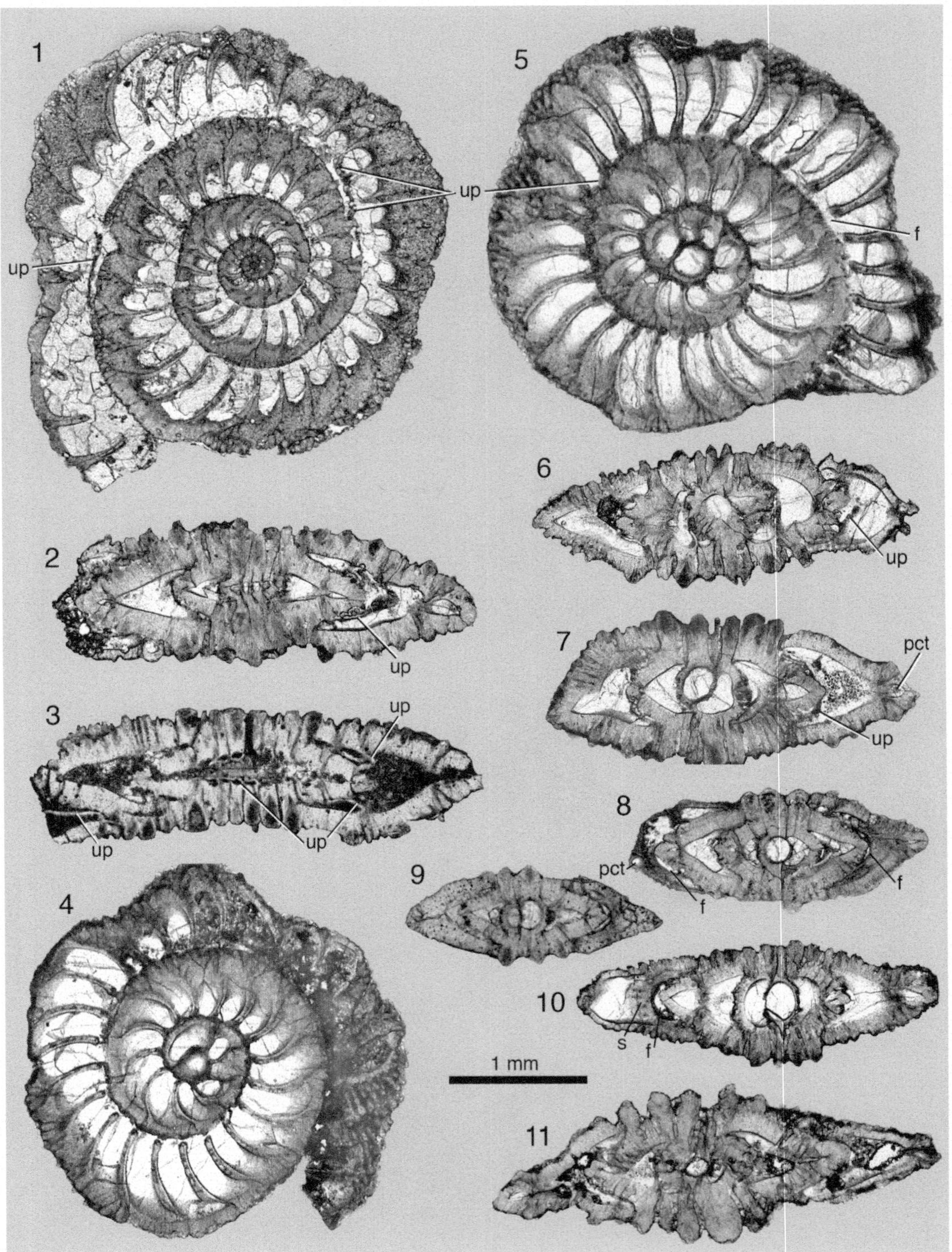

Plate 7.9 *Daviesina langhami* Smout, 1954; sample 92010a, base Patala Formation, Dhak Pass, Salt Range, Pakistan; Paleocene (SBZ 4). (**1–3**) Small microspheric specimens. (**4–11**) Megalospheric specimens. (**1**, **4–5**) Equatorial sections. (**2–3**, **6–11**) Axial sections; note the umbilical plates in (**1–3. 11**) Specimen with crushed chambers in the last whorl due to the compaction of the encasing sediment. Abbreviations: *f* foramen, *up* umbilical plate, *pct* peripheral chamber tip, *s* septum

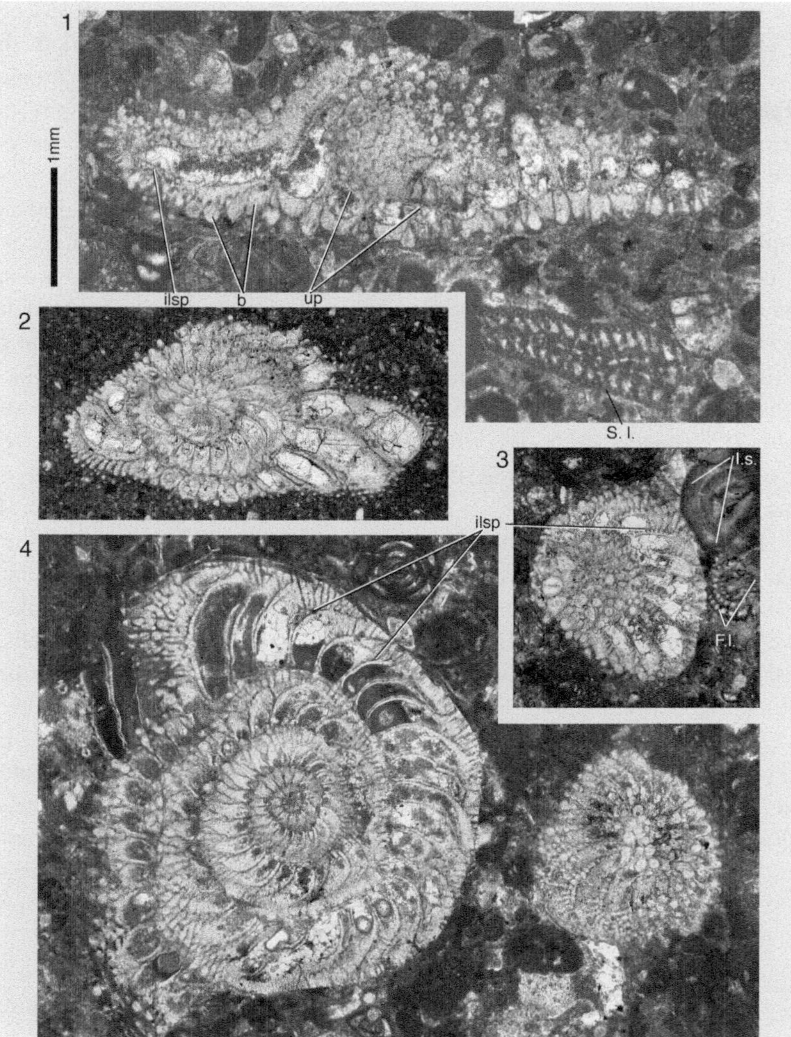

Plate 7.10 *Daviesina langhami* Smout, 1954; random sections in cemented rock; samples Kar 2 and Kar 3, Kuh-e-Kargan, Kermanshah, Iranian Zagros, collected by A. Braud; Early Eocene, Ilerdian (SBZ 5). (**1**) Oblique section of a microspheric specimen, inclined for about 45° in respect to the shell's coiling axis, associated (*at the bottom*) with (S. l.) *Saudia labyrinthica* Grimsdale, 1952. (**2**) Oblique section of megalospheric specimen, inclined for about 60° in respect to the coiling axis. (**3**) Oblique-tangential section of megalospheric specimen, associated with (I. s.) *Idalina sinjarica* Grimsdale, 1952 (*top right*) and (F. l.) *Fabularia liburnica* Drobne, 1974 (*right*). (**4**) Subequatorial section of microspheric specimen and tangential section of megalospheric specimen. Abbreviations: *b* beads, *up* umbilical plate, *ilsp* intraseptal interlocular space

difficult to recognize. In between the sutures, a single row of papillae marks the median line of the lateral chamber wall. Sections tangential to the lateral chamber wall reveal the density of the enveloping canal system with its numerous orifices that are distributed over the whole surface of the test (Plate 7.12, Figs. 2–4). In late chambers with few outer lamellae, feather grooves betray the origin of the canal system by the cover of the drains with secondary lamellation (Plate 7.13, Figs. 7–8). I have not succeeded in observing and figuring shape and position of the intercameral foramen. However, the umbilical plate is very distinct in many specimens. The plates are confined to a single side of the shell and interpreted as ventral. There are no other means to distinguish ventral from dorsal sides of the shell without ambiguity.

Microspheric specimens reach an equatorial diameter of 5.6 mm with about 45 chambers in the very loose last whorl. A single, median row of stolons complements the foramen as a connection between successive chamber lumina. There is no communication between the stolons and the intraseptal interlocular space.

The megalospheric generation has a much tighter spiral that reaches an equatorial diameter of 2.3 mm, with 28 chambers in the last whorl. The diameter of the megalosphere varies from 0.16 to 0.24 mm.

Remarks: The original description of *Daviesina intermedia* by Smout and Haque (1956) is based on a few megalospheric specimens that are deprived of their adult stages. The slight undulation of the equatorial plane is responsible for the apparently tighter coiling of the early whorls of the types unlike the specimens figured here (compare with Plate 7.13, Fig. 8).

The figured specimens all come from sample 93563 of the Dandot village section, where they are associated with *Daviesina garumniensis* Tambareau, 1972 and *Miscellanea juliettae* Leppig, 1988 (see Hottinger 2009), Paleocene (SBZ 3).

Daviesina salsa (Davies and Pinfold, 1937); Plate 7.14, Figs. 1–5; Plate 7.15, Figs. 1–5; Plate 7.16, Figs. 1–10; Plate 7.17, Figs. 1–17.

1937 *Operculina salsa*—Davies and Pinfold, p. 37, pl. 5, figs. 1, 3, 7, 10, 15.

1937 *Operculina subsalsa*—Davies and Pinfold, p. 37, pl. 5, figs. 2, 4, 8–9.

1956 *Rotalia daviesinoides*—Smout and Haque, p. 55, pl. 11, figs. a–c.

non: 1993 *Operculina salsa* (Davies and Pinfold)—Weiss, p. 252, pl. 1, fig. 1 (= *Ranikotalia sindensis*).

1977 *Ranikothalia nuttalli* (Davies)—Hottinger, pars, pl. 18, figs. 15–17.

1991 *Operculina subsalsa* Davies—Wan, p. 17, pl. 3, figs. 20–22.

Remarks: Originally described under the generic designation *Operculina*, this species is characterized by its loose, operculiniform spiral in an evolute, almost planispiral test. A whorl suture is clearly to be seen at the exterior of the shell. A dense coat of papillae produces a conspicuous thickening of the lateral chamber walls. The position of the septal sutures is marked by a single row of beads. The area around the coiling axis of the shell is thickened by a dense group of piles that obscure the ornamental pattern of the early whorls. There seems to be no morphological feature on the exterior surface of the test, in order to distinguish a ventral from a dorsal side. In axial sections, only ventral umbilical plates can be recognized. In equatorial sections, umbilical plates and the spiral canal are seen only where the section is slightly displaced from the equatorial plane to a level cutting many lateral chamber walls.

The interlocular space is very peculiar in this species: the intraseptal canals are of a coarser calibre than usual, form a single-layered fan in the septum and deviate forward and backward in alternating order into the thickened lateral walls of both neighboring chambers to emerge between the papillae all over the lateral surface of the test. In the genus *Nummulites*, similar structures are called trabeculae (Hottinger 1977). However, in this genus the trabecular canals have a much narrower calibre and are bifurcating much more frequently. The regular backward-foreward alternation of the canals in *D. salsa* hint to a construction of the canals by covering the grooves of the feathering by secondary lamellation. Some of the grooves may be seen in tangential sections of very late spiral chambers where the secondary lamellation does not yet obscure the sutural features (Plate 7.15, Figs. 2, 4). The narrow, rounded periphery of the test is covered by enveloping canals that are forming a closely meshed envelope of the chamber wall in continuation with the intraseptal canal system. This peripheral structure is not a marginal cord characteristic for all Nummulitidae, because all canal elements parallel to the shell's periphery, including a sulcus, are missing.

Plate 7.11 *Daviesina intermedia* Smout and Haque, 1956; sample 93563, Patala Formation, below coal seam, Dandot village, Salt Range, Pakistan; Paleocene (SBZ 4). (**1–9**) External views; the distinction of dorsal and ventral shell surfaces is difficult as shown in (**9**) with the views of both sides in the same specimen; views where the spiral of late whorls is faintly visible represent probably the shell's dorsal surface. Abbreviations: *bd* beads, *pap* papillae

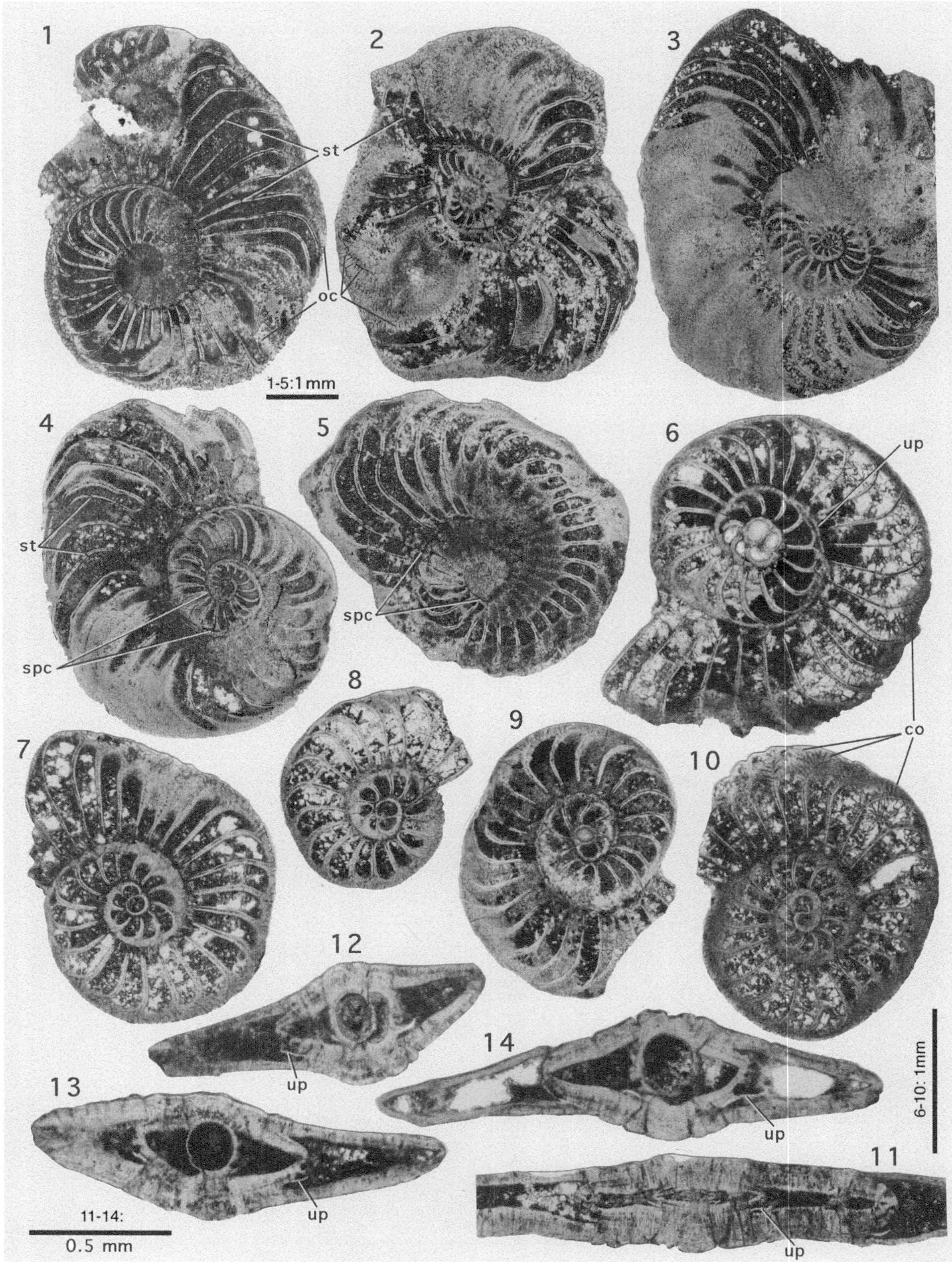

Plate 7.12 *Daviesina intermedia* Smout and Haque, 1956; sample 93563, Patala Formation, below coal seam, Dandot village, Salt Range, Pakistan; Paleocene (SBZ 4). (**1–5**) Equatorial sections of microspheric specimens; note the stolons in the late septa. (**6–10**) Equatorial sections of megalospheric specimens. (**12–14**) Axial sections of megalospheric specimens; details showing the umbilical plates on only one, the ventral side of the shell. (**11**) Axial section of microspheric specimen with broken outer whorl. Abbreviations: *st* stolons, *up* umbilical plate, *spc* spiral canal, *co* canal orifices

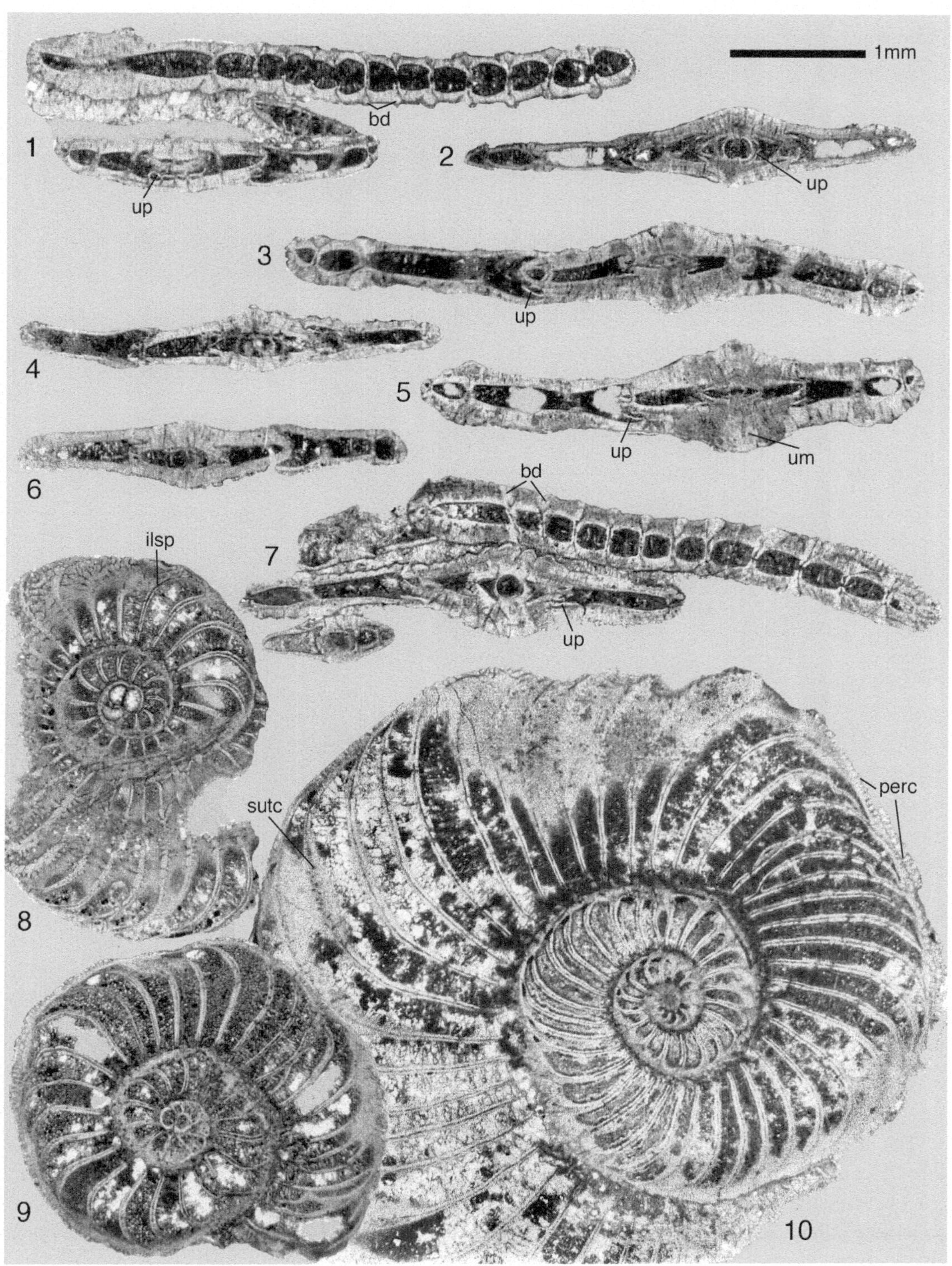

Plate 7.13 *Daviesina intermedia* Smout and Haque, 1956; sample 93563, Patala Formation, Dandot village, Salt Range Pakistan, Paleocene (SBZ 4). (**1**) Transverse sections of micro-and megalospheric specimens showing stolons supplementing the single, interiomarginal foramina. (**2**, **4**, **6–7**) Axial sections of megalospheric specimens. (**3**, **5**) Axial sections of microspheric specimens. (**8–9**) Equatorial sections of megalospheric specimen. (**10**) Equatorial section of microspheric specimen; note the foramina, the stolons and the umbilical plates. Abbreviations: *bd* beads, *up* umbilical plate, *um* umbo, *ilsp* intraseptal interlocular space, *sutc* sutural canal, *perc* peripheral canal

Plate 7.14 *Daviesina salsa* (Davies and Pinfold, 1937); samples 93504 and 93525, Patala Formation, about 5 m above the hardground marking the top of the Lockhart Limestone, Nammal Gorge, Salt Range, Pakistan; Paleocene–Early Eocene (SBZ 4–SBZ 5). Lateral, external views of free microspheric (**1**) and megalospheric (**2–5**) specimens. In both generations the distinction of dorsal from ventral sides is very difficult on outside aspects. Abbreviations: *wsut* whorl suture, *bd* beads, *pap* papillae

The microspheric generation reaches an equatorial diameter of 5.6 mm and counts 25–35 chambers in the last whorl, the megalospheric generation 3.5 mm in diameter and 24–30 chambers in the last whorl. The megalosphere diameter varies from 0.16 to 0.24 mm.

Remarks: *Daviesina salsa* has much the same proportions as "*Operculina*" *heberti* Munier-Chalmas, 1884 (see Hottinger 1977) and exhibits also trabeculae. These have, however, a much larger bore and are much less numerous as compared to "*Operculina*" *heberti*. The latter is

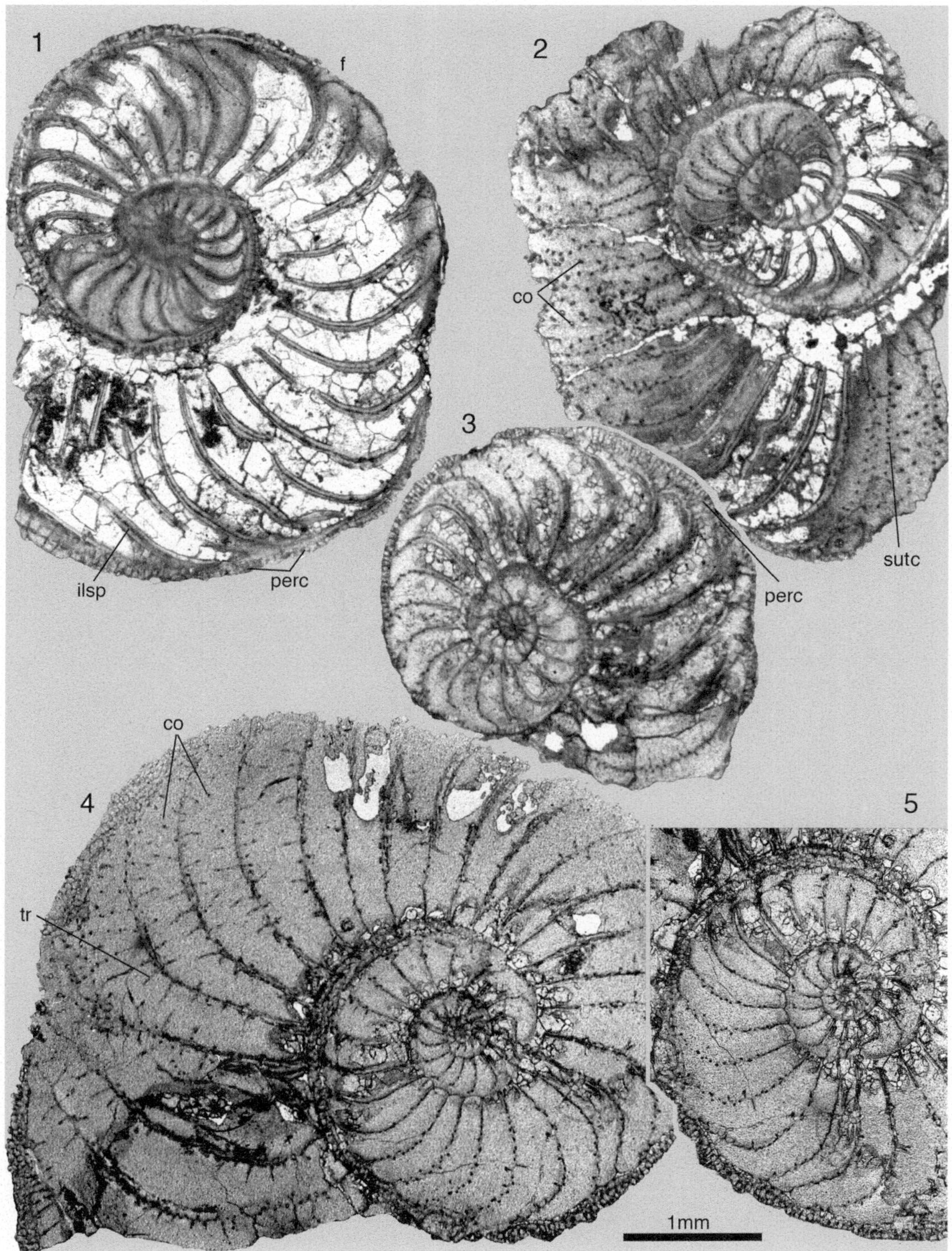

Plate 7.15 *Daviesina salsa* (Davies and Pinfold, 1937); samples 93504 and 93525, Patala Formation, Nammal Gorge, Salt Range, Pakistan; Paleocene–Early Eocene (SBZ 4–SBZ 5). (**1–4**) Equatorial sections of microspheric specimens; note the coarse trabecules and the radial structure of the shell's periphery due to a simple enveloping canal system. Abbreviations: *f* foramen, *co* canal orifices, *tr* trabeculae, *ilsp* intraseptal interlocular space, *sutc* sutural canal, *perc* peripheral canal

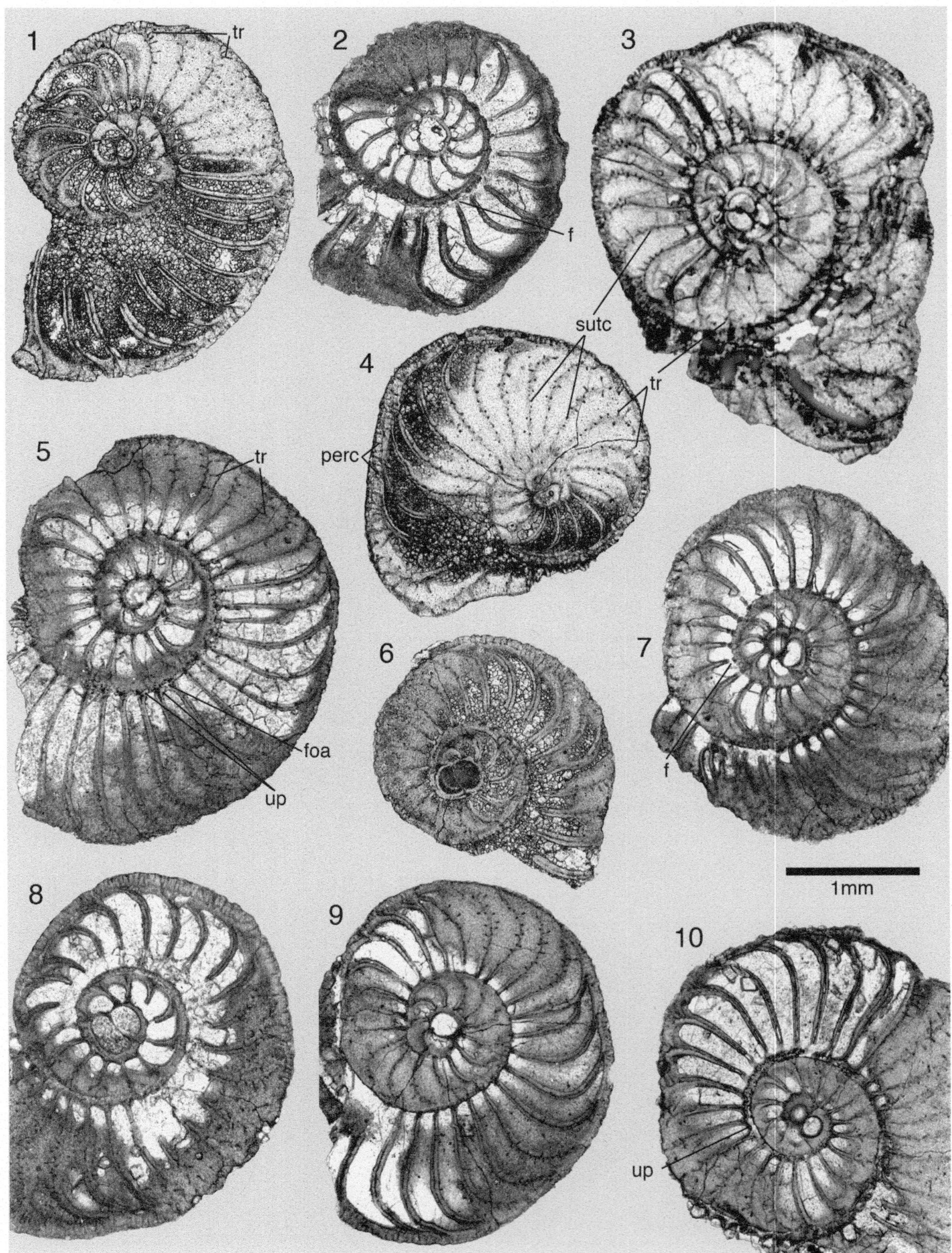

Plate 7.16 *Daviesina salsa* (Davies and Pinfold, 1937); samples 93504 and 93525, Patala Formation, Nammal Gorge, Salt Range, Pakistan; Paleocene–Early Eocene (SBZ 4–SBZ 5). (**1–10**) Equatorial and subequatorial sections of megalospheric shells. Abbreviations: *f* foramen, *up* umbilical plate, *tr* trabeculae, *foa* foliar aperture, *sutc* sutural canal, *perc* peripheral canal

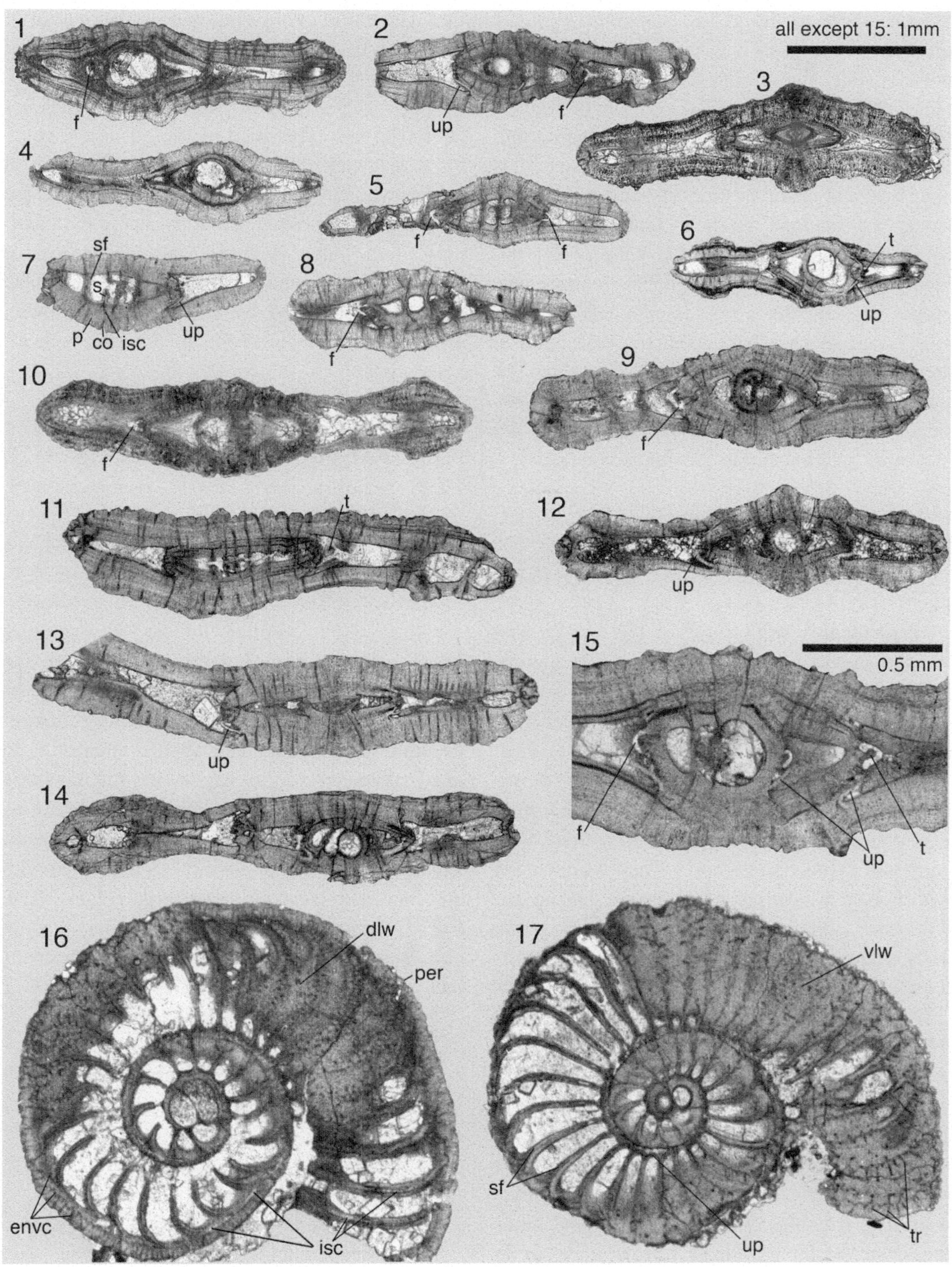

Plate 7.17 *Daviesina salsa* (Davies and Pinfold, 1937); samples 93504 and 93525, Patala Formation, Nammal Gorge, Salt Range, Pakistan; Paleocene–Early Eocene (SBZ 4–SBZ 5). (**1–15**) Axial sections. (**16–17**) Equatorial sections. (**3**, **13**) Microspheric specimens, all others are megalospheric. (**15**) Detail showing the foramina. Abbreviations: *f* foramen, *up* umbilical plate, *p* pores, *s* septum, *co* canal orifice, *isc* intraseptal canal system, *sf* septal flap, *t* tooth, *tr* trabeculae, *envc* enveloping canals, *dlw* dorsal wall, *vlw* ventral wall, *per* periphery

assigned here to the genus *Nummulites* by the presence of a marginal cord in combination with typically nummulite trabeculae. This is the most ancient species (SBZ 3) of the genus *Nummulites* that is, however, difficult to interpret as phylogenetic root for all Nummulitidae by it is huge size and chamber proportions. On the other hand, "*Operculina*" *salsa* is tentatively considered as senior synonym of *Rotalia daviesinoides* Smout and Haque, 1956 because the proportions of the compressed shell with its loose spire of chambers seem to correspond to the megalospheric generation of *D. salsa*. The triangular, tooth-bearing foramen and the one-sided umbilical plate (Plate 7.17, Fig. 15) point to a close relationship to the genus *Daviesina*.

Daviesina praegarumnensis n. sp.; Fig. 7.1N–R; Plate 7.18, Figs. 1–34.

Holotype: Microspheric specimen figured in Plate 7.18, Fig. 3.

Type locality and type level: Lafarge Quarry, Western Aquitaine, southern France; Thanetian, Paleocene (SBZ 3).

Derivation of name: precursor species of *Daviesina garumnensis*.

Diagnosis: There are a number of comparatively small sized *Daviesina* species that have in common an operculiniform habit with involute early and evolute late whorls. The involute early whorls bear a heavy ornamentation. During the Paleocene and the Early Eocene, the involute early stage is reduced in favour of progressively flattened late whorls with a widening spiral (Fig. 7.1). *Daviesina praegarumnensis* n. sp. is the earliest representative of this group. The last whorl of this species is just faintly looser and the whole shell is heavily decorated by numerous piles. The ventral side of the shell is marked by a heavy, progressive umbo. It is surrounded by a row of smaller piles arising each from a foliar wall. The irregular contour that the axial umbo produced by an umbilical furrow hints to its composite nature. The shell's equatorial to axial diameter ratio varies from 1.75 to 2.25.

The foramen is a short slit in interiomarginal position below the periphery of the previous whorl. Both sides of the apertural slit bear minute lips. The umbilical lip continues in the interior of the respective chamber as part of the umbilical plate.

The megalosphere has a diameter of 0.04–0.06 mm. The shells of about 1 mm in equatorial diameter exhibit 14–15 spiral chambers in the last whorl. Size and proportions of the shell are very similar in both generations except the thickness of the outer chamber walls depending, of course, on the number of secondary lamellae per whorl produced by every step of growth. These are more numerous in the microspheric generation.

Daviesina garumnensis Tambareau, 1972; Fig. 7.1J–M; Fig. 7.2A–I; Plate 7.19, Figs. 1–21.

1972 *Daviesina garumnensis*—Tambareau, p. 208, pl. 12, figs. 2–5.

1980 *Daviesina garumnensis* Tambareau—Caus et al., p. 1058; text-fig. 7A–D; pl. 3, figs. 4–6.

Remarks: The general habit of *Daviesina garumnensis* is similar to the new species *D. praegarumnensis*, but the last whorl in the microspheric specimens is looser, the dimorphism more pronounced and the shells get larger in both generations. The shell's equatorial to axial diameter ratio varies in microspheric forms from 1.5 to 2.8, in megalospheric forms from 1.7 to 2.1. The foramen is similar, a straight slit in interiomarginal position. A specimen with an incipient second periphery (Plate 7.19, Fig. 16) shows two slits as foramina, one on each side of the shell. Microspheric specimens have 18–20, megalospheric specimens 10–12 spiral chambers in adult growth stages.

Daviesina tenuis (Tambareau, 1967); Fig. 7.1F–I; Plate 7.20, Figs. 1–14; Plate 7.21, Figs. 1–11.

1967 *Operculina tenuis*—Tambareau, p. 425, text-figs. 1–3; pl. 1, figs. 1–19.

1972 *Daviesina tenuis* (Tambareau)—Tambareau, p. 211, pl. 12, figs. 6–7.

1980 *Daviesina tenuis* (Tambareau)—Caus et al., p. 1060, text-fig. 7E–H; pl. 3, figs. 1–3; pl. 4, figs. 1–2.

1980 *Daviesina khatyiahi* Smout, pars—Gaetani et al., pl. 17, fig. 3.

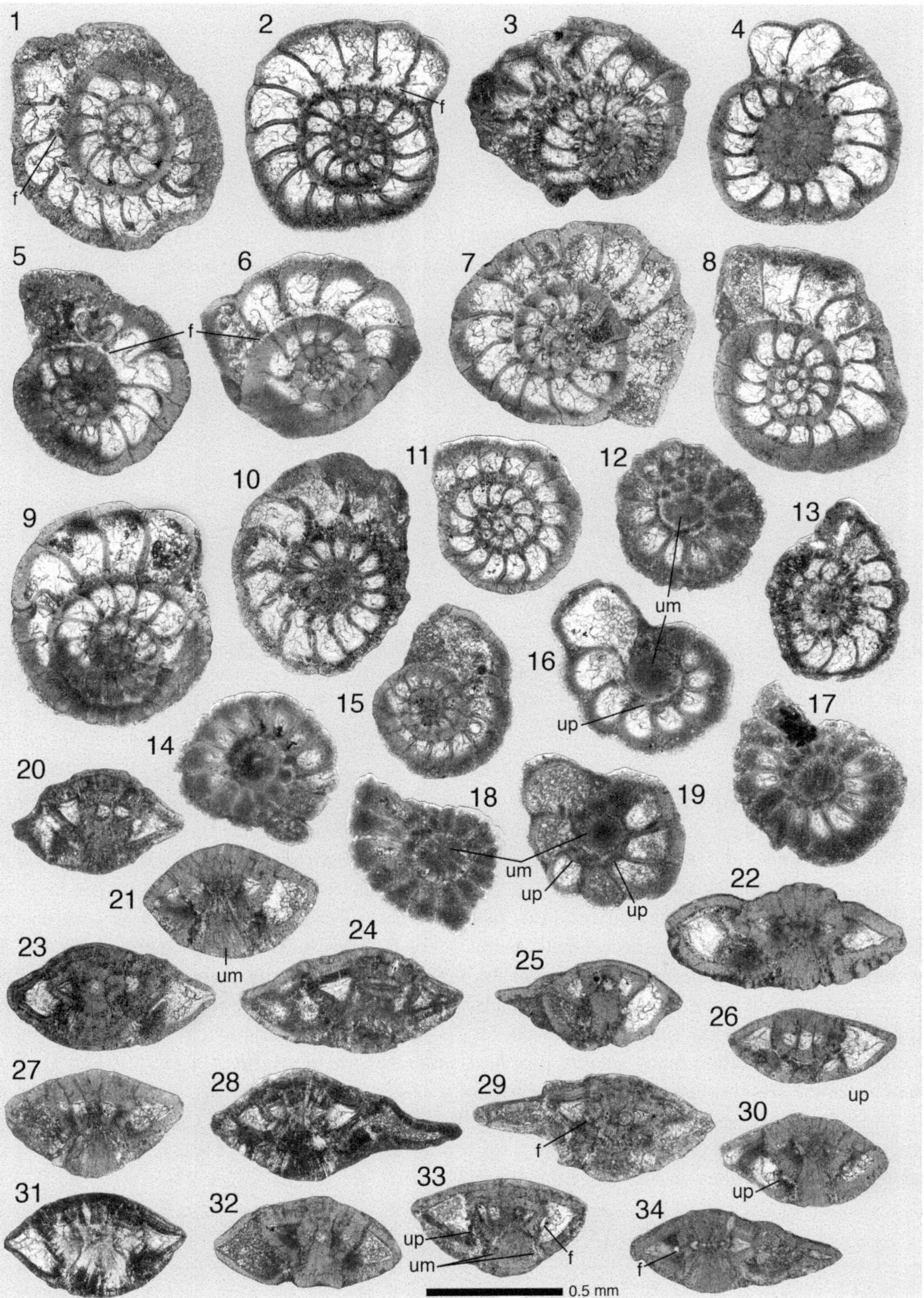

Plate 7.18 *Daviesina praegarumnensis* n. sp.; sample 9 from Lafargue Quarry, Western Aquitaine, Southern France; Thanetian, Paleocene (SBZ 3). (**1–11**, **13**) Equatorial sections. (**3**) Only microspheric specimen found; holotype. (**12**, **14–19**) Sections perpendicular to the coiling axis through the ventral part of the shell; note

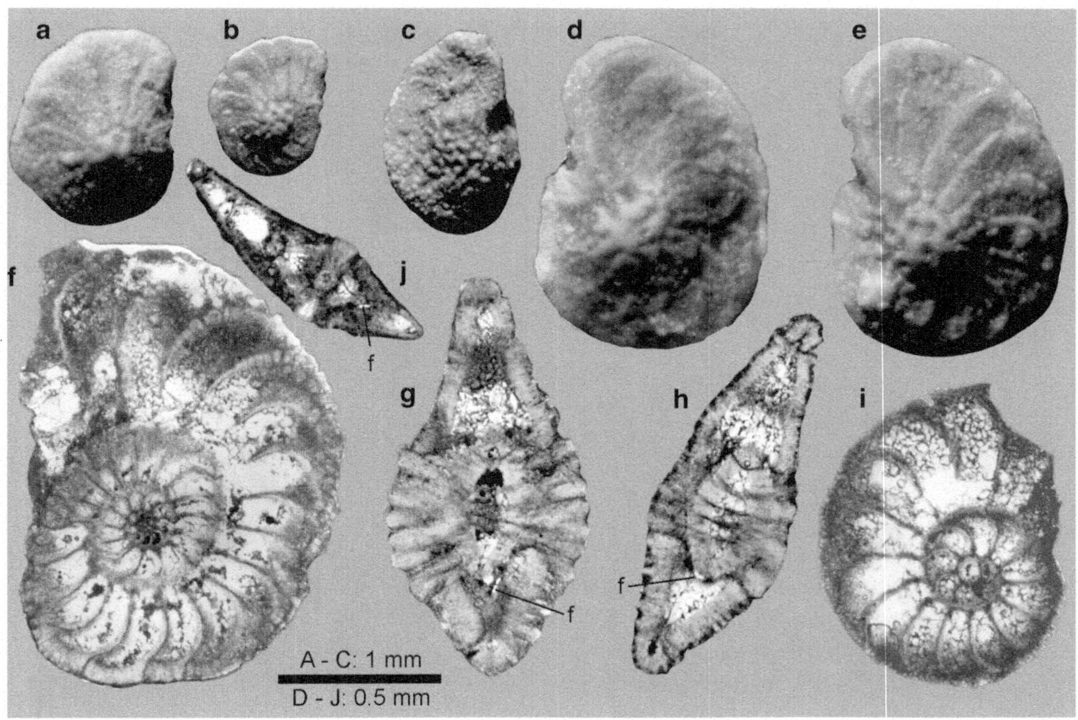

Fig. 7.2 *Daviesina garumniensis* Tambareau, 1972. (**A**, **D**) Dorsal view. (**B**, **C**, **E**) Ventral view; note the minor piles surrounding the main pile; microspheric forms. (**F**) Section perpendicular to coiling axis. (**G**) Axial section; megalospheric forms. (**H**, **J**) Axial sections. (**I**) Section perpendicular to coiling axis. Abbreviation: *f* foramen

1991 *Daviesina persica* Rahaghi—Wan, p. 15, pl. 1, figs. 12–14.

1998 *Daviesina* cf. *tenuis* (Tambareau)—Pignatti et al., p. 621, pl. 1, figs. 1–3.

Remarks: Many free specimens from the Salt Range permit to study this species with more accuracy than before. The ornamentation of the operculiniform shell is dominated by beads aligned along the septal sutures. Microspheric shells exhibit in addition many smaller pustules in between the septal sutures. The megalospheric specimens produce an adaxial ornament with a single, prominent axial pile surrounded by a spiral row of smaller piles arising from the foliar walls. These ornamental particularities are similar to the ornaments of *Daviesina garumnensis* and *D. praegarumnensis* and support the phylogenetic interpretation of the group given here.

Microspheric shells are almost completely evolute and have 13–15 chambers in adult whorls. The chambers are separated by septa that are bent backwards in a rounded pathway. The foramen is a triangular slit, at least in the microspheric generation (Plate 7.21, Fig. 5).

Daviesina ruida (Schwager, 1863); Fig. 19A–E; Plate 7.21, Figs. 12–16.

Plate 7.18 (Continued) in (**12**) and (**16**) the spiral canal (spc) around the axial umbilical pile and in (**12**), (**14**) and (**17**) the small piles (fop) arising from the foliar walls. (**20–34**) Subaxial and axial sections. Abbreviations: *f* foramen, *up* umbilical plate, *um* umbo

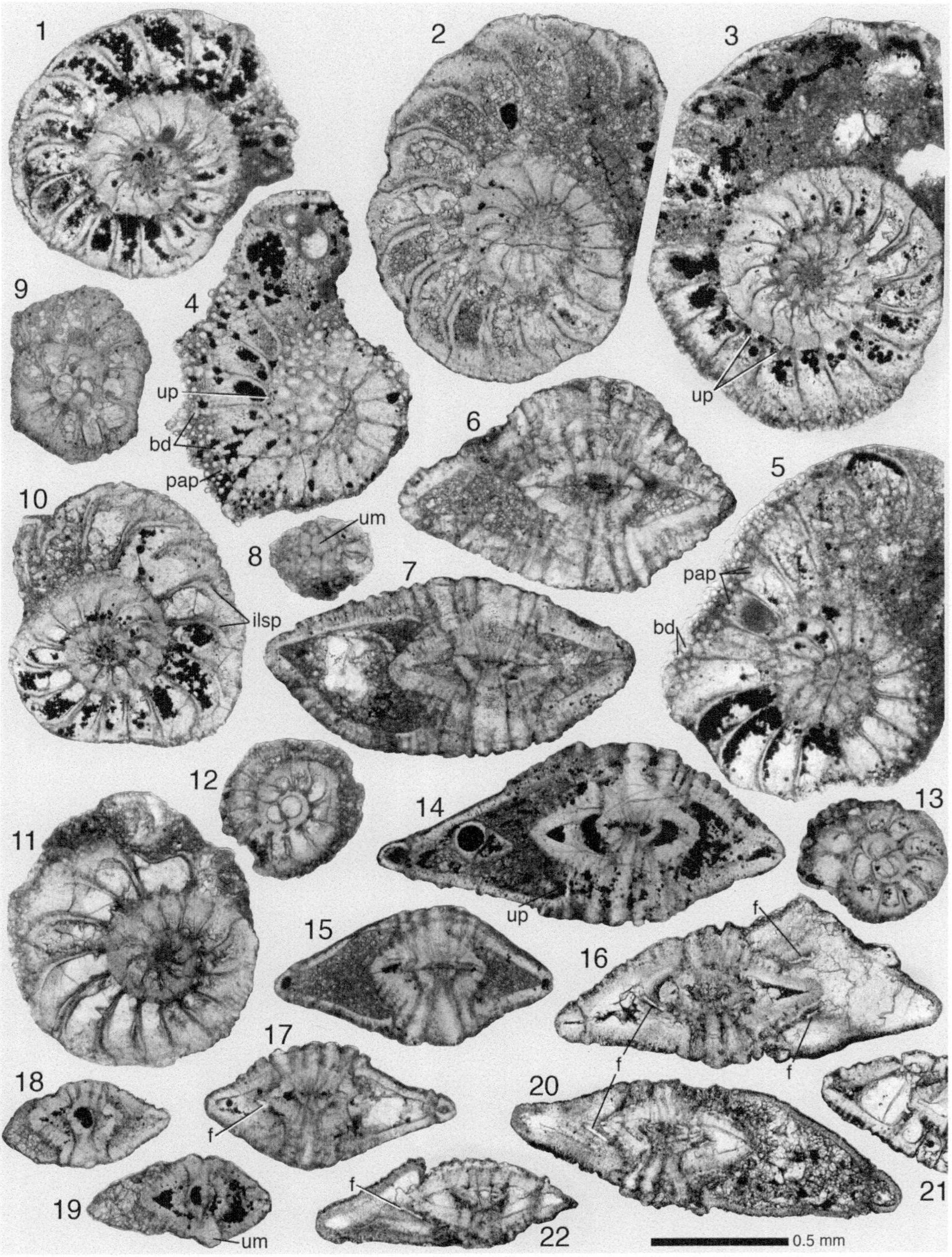

Plate 7.19 *Daviesina garumnensis* Tambareau, 1972; topotypes, sample 72008 collected by H. Schaub, deposited in the Museum of Natural History of Basel, Switzerland; Le Quillet, Petites Pyrénées, France; Paleocene (SBZ 4). Microspheric specimens figured in (**1–7**), (**10** and **11**, **14–17**, **20**); megalospheric specimens in (**8**). (**1–3**, **10–11**) Equatorial sections. (**4–5**) Sections perpendicular to the coiling axis showing positions of beads and pustules where the sections are tangential to the lateral surface of the shell. (**6–7**, **14–17**, **20**) Axial sections. (**8**) Section tangential to the ventral main pile and the surrounding minor piles standing on the foliar walls; note the similarity of this ornament with that of *Daviesina praegarumnensis* (Plate 7.18, Figs. 12, 14). (**9**, **12–13**) Equatorial sections. (**18–19**) Axial sections. Abbreviations: *f* foramen, *up* umbilical plate, *bd* beads, *um* umbo, *ilsp* intraseptal interlocular space, *pap* papillae

Plate 7.20 *Daviesina tenuis* (Tambareau, 1967); external, lateral views of isolated specimens, (**1–7**) sample 93520, (**8–14**) sample 95119; both samples are located in the *top* Hangu Formation and Lockhart Limestone, Dhak Pass, Salt Range, Pakistan; Paleocene–Early Eocene (SBZ 4–5). (**1–13**) Megalospheric specimens. (**1–2**), (**7**, **9–10**) Ventral view showing the characteristic prominent ornamentation of the umbilical zone (compare with axial sections Plate 7.21, Figs. 4–7, 10); all others may be dorsal views. (**14**) Possibly ventral view of a microspheric specimen. Abbreviations: *bd* beads, *pap* papillae

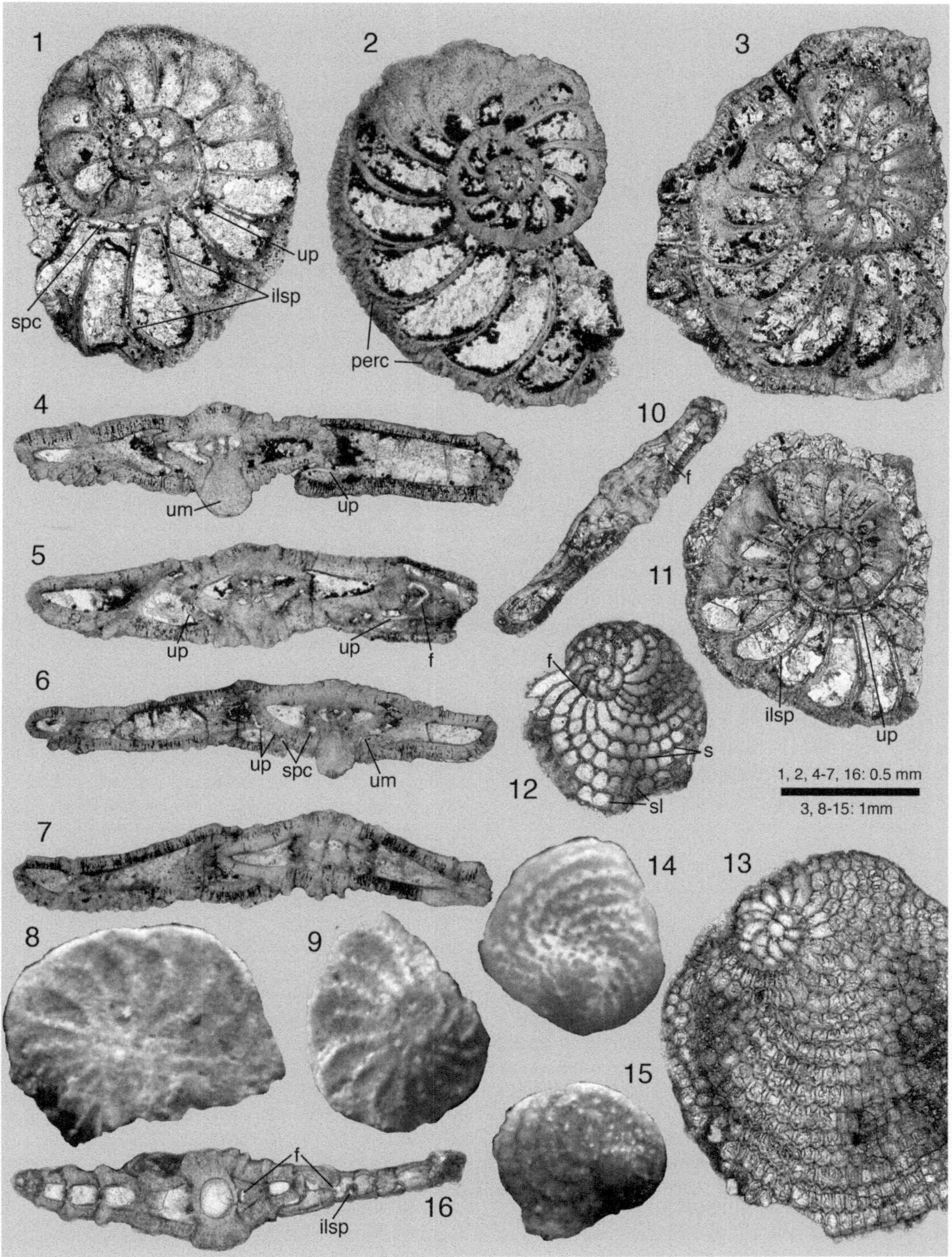

Plate 7.21 (**1–11**) *Daviesina tenuis* (Tambareau, 1967); sample 95115, top Hangu Formation, Dhak Pass, Salt Range, Pakistan; Paleocene–Early Eocene (SBZ 4–SBZ 5). (**1–3, 11**) Microspheric specimens, equatorial sections. (**4–7, 10**) Axial sections. (**4–5**) Possibly microspheric specimens. (**8–9**) External view of megalospheric specimens. (**12–16**) *Daviesina ruida* (Schwager, 1863); sample 93545, Alveolina beds below Sakesar Limestone, top Nammal Formation, Nilawahan Gorge, Nurpur, Salt Range, Pakistan; Eocene, Ilerdian (SBZ 7–SBZ 8). (**12–13**) Megalospheric specimens, equatorial section. (**14–15**) External view of lateral shell surface of megalospheric specimens. (**16**) Megalospheric specimen, axial section. Abbreviations: *s* septum, *sl* septulum, *f* foramen, *up* umbilical plate, *um* umbo, *spc* spiral canal, *perc* peripheral canal, *ilsp* intraseptal interlocular space

1863 *Heterostegina ruida*—Schwager, p. 145, pl. 29, fig. 6a–c.

1937 *Heterostegina* cf. *ruida* Schwager—Davies and Pinfold, p. 52, pl. 5, fig. 21.

1952 *Heterostegina* cf. *ruida* Schwager—Grimsdale, p. 237; pl. 24, figs. 3–8.

1967 *Heterostegina adamsi*—Eames and Clarke, p. 314, pl. 61, figs. 13–15.

1980 *Daviesina ruida* (Schwager)—Caus et al., p. 1060, text-fig. 5, 7I–M; pl. 4, figs. 1–2.

Remarks: The absence of a marginal cord obliges to classify this species as a *Daviesina* by its apertural characters in spite of the secondary subdivision of the chambers by folds of the septal flap. The outer aspect of the shell's lateral walls reveal the beaded chamber sutures. The beads are arising from the junctions of chamber with chamberlet sutures. The megalosphere has a diameter of 0.12 mm followed by 4 undivided chambers representing the nepiont. The insufficient preservation of the material available makes it difficult to analyse the stolon pattern in the adult stages. Below the junction of the septal with the septular sutures, a triangular space represents the interlocular realm. It is neither known whether this space opens between the beaded ornamental elements to the exterior, nor from where it might be fed.

References

Accordi G, Carbone F, Pignatti J (1998) Depositional history of a Paleogene ramp (Western Cephalonia, Ionian Islands, Greece). Geol Romana 34:131–205

Afzal J, Williams M, Aldbridge R (2009) Revised stratigraphy of the lower Cenozoic succession of the Grater Indus Basin in Pakistan. J Micropaleontol 28: 7–23

Bizon G, Bizon JJ, Ricou LE (1972) Étude stratigraphique et paléogéographique des formations Tertaires de la region de Neyriz (Fars interne, Zagros Iranien). Rev Inst Fr Pétrole 27(3):369–405

BouDagher-Fadel MK (2008) Evolution and geological significance of larger benthic foraminifera. Developments in palaeontology & stratigraphy 21, 540 pp. Springer, Amsterdam

Caus E, Hottinger L, Tambareau Y (1980) Plissements du septal flap et système de canaux chez *Daviesina*, foraminifères paléocènes. Eclogae geol Helv 73(3): 1045–1069

d'Archiac A, Haime J (1853) Description des animaux fossiles du groupe nummulitique des Indes, précédée d'un résumé géologique et d'une monographie des Nummulites, 373 pp, 36 pls, Gide & Baudry, Paris

D'Orbigny AD (1826) Tableau méthodique de la classe des céphalopodes. Ann Sci Nat Paris, Ser 1 7:96–314

Davies LM, Pinfold ES (1937) The Eocene beds of the Punjab Salt Range. Palaeontologia Indica Calcutta 24(1), 79 pp, 7 pls

Drobne K (1974) Les grandes Miliolidées des couches paléocènes de la Yougoslavie du Nord-Ouest. Razpr (4 razr) Slov Akad Znan Umetn 17(3):129–184, Ljubljana

Eames FE, Clarke WJ (1967) A Palaeocene *Heterostegina*. Palaeontology 10:314–316

Gaetani M, Nicora A, Premoli-Silva I (1980) Uppermost cretaceous and paleocene in the Zanskar range (Ladakh-Himalaya). Riv Ital Paleontol 86:127–166

Grimsdale TF (1952) Cretaceous and Tertiary Foraminifera from the Middle East. Bull Brit Mus (Nat Hist) 1: 223–247

Hasson PF (1985) New observations on the biostratigraphy of the Saudi Arabian Umm er Radhuma formation (Paleogene) and its correlation with neighboring regions. Micropaleontology 31(4): 335–364

Hottinger L (1977) Foraminifères operculiniformes, Mem Mus Nat Hist Nat Paris 40. Éditions du Muséum, Paris, pp 1–159, 66 pls

Hottinger L (2009) The Paleocene and earliest Eocene foraminiferal family Miscellaneidae: neither nummulites nor rotaliids. Carnets Géol Article 2009/06, CG2009_A06

Hottinger L, Halicz E, Reiss Z (1991) The foraminiferal genera *Pararotalia*, *Neorotalia* and *Calcarina*: taxonomic revision. J Paleontol 69:1–33

Ismail AA, Boukhary M (2008) Larger foraminifera from the Early Eocene of Shabwa area, Southeastern Yemen. Rev Paléobiol 27:89–97

Leppig U (1988) Structural analysis and taxonomic revision of Miscellanea, Paleocene, larger Foraminifera. Eclog geol Helv 81(3):689–721

Nicora A, Garzanti E, Fois E (1987) Evolution of the Tethys Himalaya continental shelf during Maastrichtian to Paleocene. Riv Ital Paleontol Strat 92:439–496

Nuttall WL (1926) The larger foraminifera of the upper Ranikot Series (Lower Eocene) of Sind (India). Geol Mag 63(2):112–121

Pignatti J, Matteucci R, Parlow T, Fantozzi L (1998) Larger foraminiferal biostratigraphy of the Maastrichtian-Ypresian Wadi Mashib succession (South Hadramawt Arch, SE Yemen). Z Geol Wiss 26(5–6):609–635

Rahaghi A (1983) Stratigraphy and faunal assemblage of Paleocene-Lower Eocene in Iran. Nat Iran Oil Comp Geol Lab 10:73 pp, 49 pls

Sander NJ (1962) Aperçu paléontologique et stratigraphique du Paléogène en Arabie Séoudite orientale. Rev Micropaléontol 5:3–40

Schwager C (1863) Die Foraminiferen aus den Eocaenablagerungen der lybischen Wüste und Aegyptens. Palaeontographica 30:79–154, 6 pls

Smout AH (1954) Lower Tertiary foraminifera of the Qatar Peninsula. Brit Mus (Nat Hist), 96 pp, 44 figs, 15 pls

Smout AH, Haque M (1956) A note of the larger foraminifera and ostracoda of the Ranikot from the Nammal Gorge, Salt Range, Pakistan. Rec Geol Surv Pak 8(2):49–60, 3 pls

Tambareau Y (1967) Sur une nouvelle espèce d'Operculine, *Operculina tenuis* n. sp. Bull Soc Hist Nat Toulouse 103(3/4):425–430

Tambareau Y (1972) Thanétien supérieur et Ilerdien inférieur des Petites Pyrénées, du Plantaurel et des Chaînons audois. Trav Lab Géol Pétrol Univ P Sabatier, 383 pp, 20 pls

Visser AM (1951) Monograph on the foraminifera of the type-locality of the Maastrichtian (south-Limburg, Netherlands). Leidse Geol Meded 14:197–359

Wan X (1991) Paleocene larger foraminifera from Southern Tibet. Rev Esp Micropaleontol 23:7–28

Weiss W (1993) Age assignments of larger foraminiferal assemblages of Maastrichtian to Eocene age in northern Pakistan. Zitteliana 20:223–252

Some Taxa That Are or Remain Excluded from the Family Rotaliidae

8

Abstract

Four groups of rotaliid foraminifera are distinguished according to their basic structural shell characteristics. These groups are or remain excluded from the family Rotaliidae. The new subfamily Laffitteininae is here assessed. *Neorotalia*, *Paralockhartia* n. gen., *Laffitteina*, *Cuvillierina*, *Smoutina*, *Storrsella* with eleven species (*N. alicantina*, *N. litothamnica*, *N. viennoti*, *P. eos* n. sp., *L. marsicana*, *L. mengaudi*, *L. monodi*, *L. bibensis*, *C. sireli*, *S. cruysi*, *S. haastersi*) are described and illustrated.

Four groups of rotaliid foraminifera are distinguished according to their most basic structural characteristics:

1. Taxa with single, areal foramina, umbilical flaps complemented by "toothplates", with umbilical fills such as plugs, umbos, piles or columellar structures, but lacking folia and dense enveloping canal systems. The chamber lumen is separated from the interlocular space by an umbilical flap coating the surface of previous whorls. Toothplates and umbilical flaps may merge into a common element. This group corresponds to the classic family Pararotaliidae with the genera *Pararotalia* with an open umbilical fissure and *Neorotalia* where the umbilical space around the umbilical fill is covered by secondary outer lamellas and transformed into a spiral canal. The new genus *Paralockhartia* is included here taking into account its corresponding foraminal characteristics.
2. Taxa with a single areal foramen and a dense enveloping canal system covering the complete surface of the shell and incorporating the numerous funnels piercing the dorsal and ventral umbos. *Laffitteina* and *Cuvillierina* are classified in this group. This group merits to be ranked as a particular subfamily Cuvillierininae Loeblich and Tappan (1964). Their position in the Rotaliidae family is excluded here but a transfer to the Pararotaliidae is questionable.
3. Taxa with a single, areal foramen lacking umbilical fills of any kind. Costae confined on each chamber surface form a diagnostic pattern of ornamentation for the genera *Thalmannita* and *Civrieuxia*. Presently, the two genera must be considered as of *incertae sedis*.
4. *Pararotalia* Le Calvez, 1949 and *Neorotalia* Bermúdez, 1952: both genera, closely related by their areal foramen combined with a "toothplate" (Hansen and Reiss 1971, p. 335, pl. 9–10; here in fig. 21), are characterized by a heavy ornamentation of the trochospiral test that is evolute on the dorsal and involute on the ventral side. The ornate aspect of the shell

L. Hottinger, *Paleogene larger rotaliid foraminifera from the western and central Neotethys*,
DOI 10.1007/978-3-319-02853-8_8,

has often been the only argument for identification of the group. This leads to errors in generic or specific identification of the ornate shells. A taxonomic reassessement of three species of *Neorotalia* (i.e., *N. alicantina*, *N. litothamnica*, *N. viennoti*) might be of some use in biostratigraphy.

8.1 *Neorotalia* Bermúdez, 1952

Neorotalia alicantina Colom 1954; Plate 8.1, Figs. 1–27.

8.2 *Paralockhartia* n. gen.

Type species: *Paralockhartia eos* n. sp.

Description: The test is bilamellar and perforate. The chambers are arranged in a low trochospiral, involute on their umbilical side, evolute on their dorsal side. The periphery is angular to faintly keel. Both sides of the shell are covered with a dense assemblage of pustules formed by the local thickening of the secondary outer lamellae. As usual in such cases, the pustules on both shell sides tend to become larger in diameter and higher in relief in adaxial direction. In late whorls, when the chambers and consequently the number of secondary outer lamellas reach a sufficient number, the gaps between the pustules may be bridged by outer lamellas that produce a loose and irregular enveloping canal system fed by the intraseptal interlocular space.

The intercameral foramen is a single, long, narrow slit in interiomarginal-areal position, directed in ventral direction at low angle to the base of the apertual face. No aperture was observed; it may be masked. The dorsal rim of the foramen is bordered by a strong lip. Ventrally, the foramen is delimited by the protruding end of a "toothplate", obliquely extending in proximal direction and glued to the previous whorl at the proximal end of the chamber bottom. There is no folium. The chamber alae delimit with their umbilical walls a deep spiral furrow around the structured umbilical fill. In proximal direction, the furrow blindly ends against the toothplates and the ventral walls of the chambers of the previous whorl. Its walls receive secondary lamellation until the furrow is covered by the next whorl and transformed into a spiral interlocular space. The latter communicates with the intraseptal space restricted to the ventral side of the test. The dorsal side of the septal face is covered by a septal flap.

The wide umbilicus is filled with a complex umbilical structure. It is crowded with stout umbilical piles shooting off from the ventral tip of each chamber. Seen from the exterior, the heads of these piles constitute the umbilical pustules. The umbilical area is covered in part by extensions of the lateral-ventral chamber wall (lw on Pls. 74–75) leaving open umbilical spaces in between them. The latter are communicating with each other by vertical gaps between the protruding piles and open to the ambient environment by orifices distributed all over the surface of the umbilicus. This umbilical structure is similar to the one of *Lockhartia* but lacks any foliar elements and in particular an umbilical plate. Each chamber covers a considerable part of the umbilical area with a bilamellar extension of its ventral lateral wall. These lateral walls cover the tops of the umbilical piles arising from previous walls. They are attached obliquely with their far end to the base of the opposite chambers in a whorl and produce a kind of propeller structure where the blades are inclined in respect to the equatorial plane.

The strong ornamentation on both sides of the shell, covering the walls and bordering the interlocular spaces with numerous pustules and piles, belongs to the architectural style of all pararotaliinae and is probably more than just a specific feature.

Differential diagnosis: *Paralockhartia* differs from *Pararotalia* (see Hottinger et al. 1991) by its umbilical cavities. It differs from *Lockhartia* by exhibiting an areal intercameral foramen

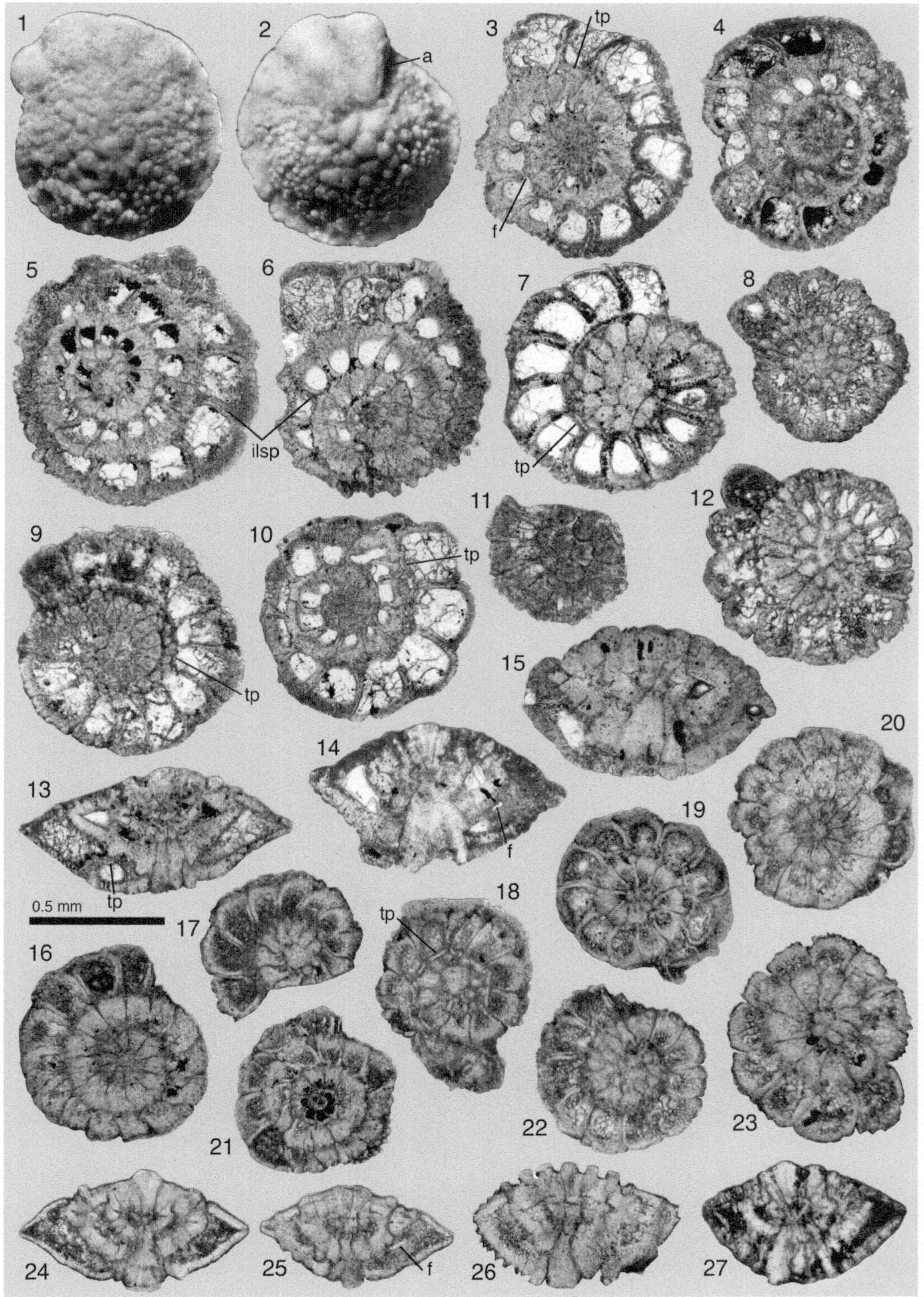

Plate 8.1 *Neorotalia alicantina* Colom, 1954. (**1–2**) External view, dorsal (**1**) and ventral (**2**). (**3–6**) Microspheric specimens (?), sections perpendicular to coiling axis. (**7–12**, **16–23**) Sections not centered, perpendicular to coiling axis. (**13–15**) Axial sections. (**24–27**) Subaxial sections. Abbreviations: *f* foramen, *a* aperture, *tp* toothplate, *ilsp* intraseptal interlocular space

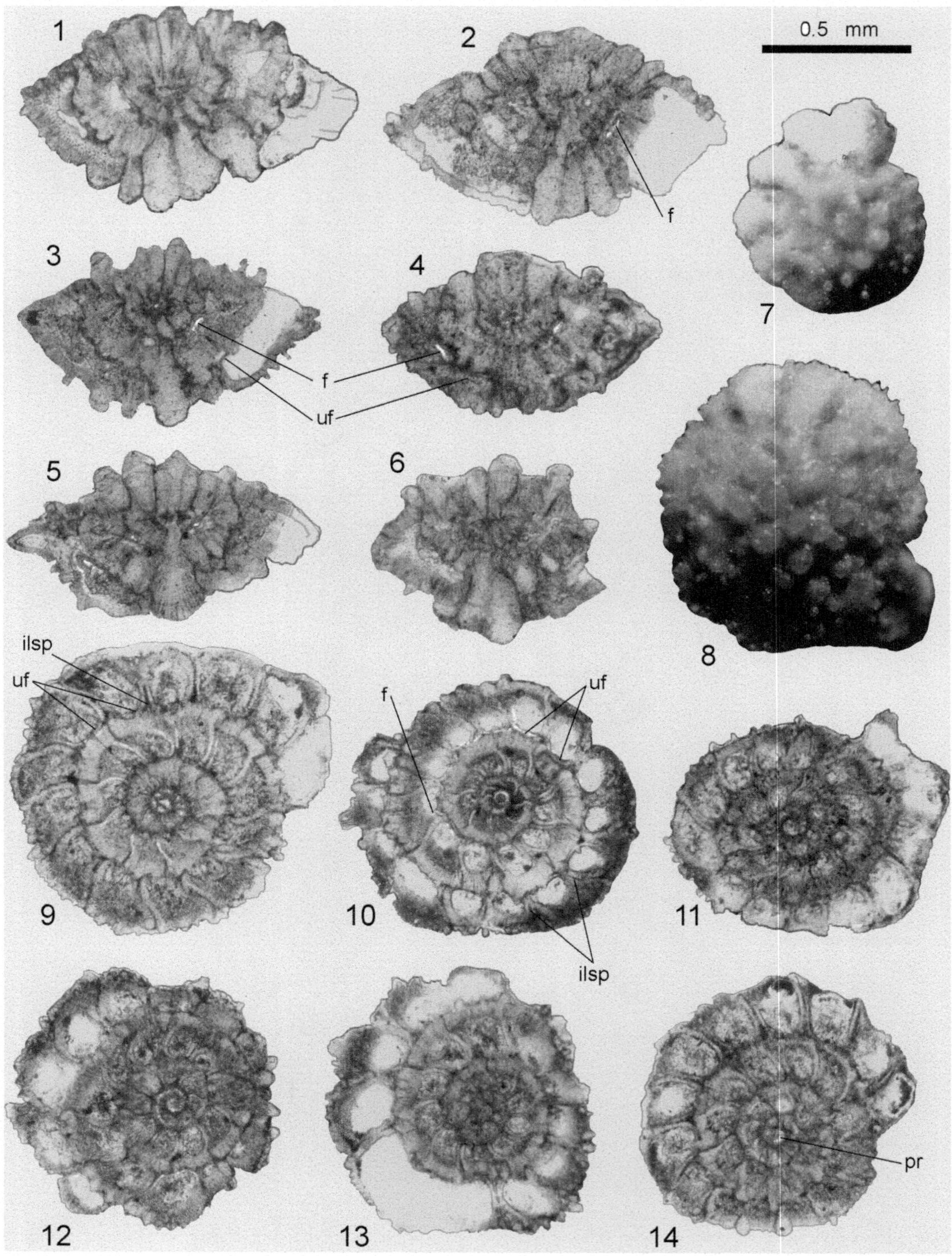

Plate 8.2 *Neorotalia litothamnica* (Uhlig, 1886); sample AV 10 Verona (Italy); Lutetian, *Alveolina munieri* Zone (SBZ 14). (**1–6**) Axial sections. (**7–8**) External views, dorsal (**7**) and ventral (**8**). (**9–14**) Centered sections, perpendicular to coiling axis. Abbreviations: *f* foramen, *uf* umbilical flap, *pr* proloculus, *ilsp* intraseptal interlocular space

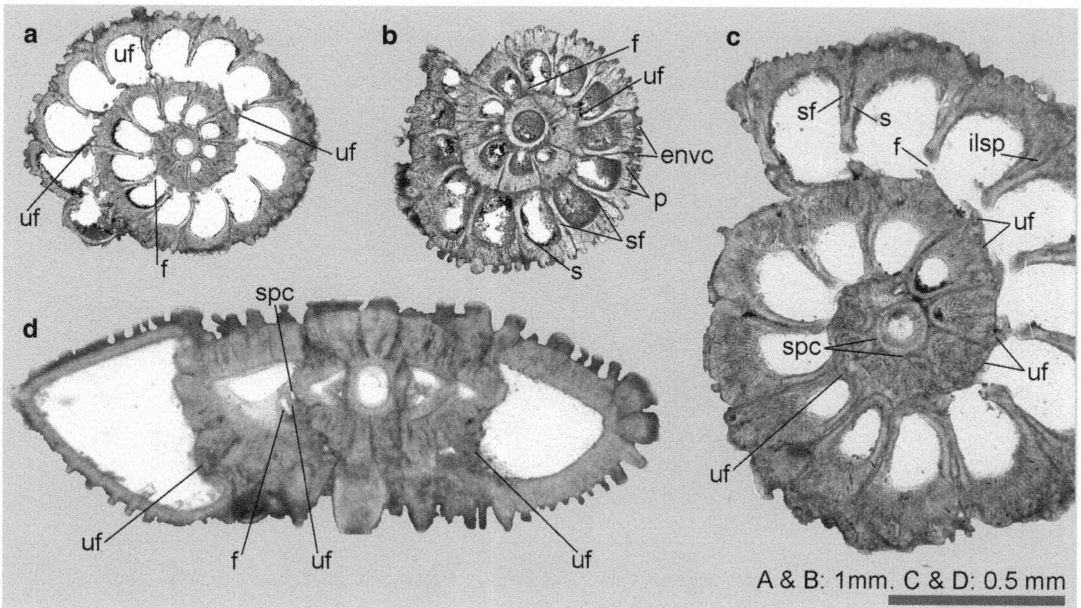

Fig. 8.1 *Neorotalia viennoti* (Greig, 1935). (**A–C**) Sections perpendicular to coiling axis. (**D**) Axial section. Abbreviations: *f* foramen, *uf* umbilical flap, *envc* enveloping canals, *p* pores, *sf* septal flap, *s* septum, *ilsp* intraseptal interlocular space, *spc* spiral canal

bordered by a heavy lip and a protruding toothplate whereas *Lockhartia* (see Müller-Merz 1980) has an umbilical plate, a folium extending over the umbilical region, and foliar apertures. *Daviesina*, although heavily ornamented on both sides of the shell, differs from all pararotaliids by its umbilical plate and from *Lockhartia* by its lack of umbilical spaces. *Calcarina*, although exhibiting a supplemental skeletal fill of the umbilicus with an umbilical canal system similar to the one of *Paralockhartia*, has multiple apertures and umbilical flaps separating the chamber lumina from the umbilical interlocular space.

The presence of a toothplate in *Paralockhartia* is the reason to assign this new genus to the subfamily Pararotaliinae Reiss, 1963.

***Paralockhartia eos* n. sp.**; Plate 8.3, Figs. 1–14; Plate 8.4, Figs. 1–15.

1972 ?*Rotalia* sp. 4 pars—Samuel et al., pl. 40, figs. 1–2.

Type specimens: Syntypes from sample 94702, axial section and section perpendicular to coiling axis figured in Plate 8.4, Figs. 1–2; deposited in the Museum of Natural History of Basel, Switzerland.

Type locality and type level: Gorges du Cabaret, bed 94702, Petites Pyrénées, south-western France, SBZ 2.

Derivation of name: a genus combining features from *Pararotalia* (toothplate) with features of *Lockhartia* (umbilical cavities) with a type species occurring very early in the Paleocene.

Diagnosis: Specimens with the characteristics of the genus *Paralockhartia*, as defined above, having strikingly radial septa which are slightly bent backward only in the dorsal, peripheral part of the chamber. 16–17 chambers in the last whorl of adult specimens reach 1 mm in equatorial diameter. The shell is thickly lenticular; dorsal and ventral surface are approximately equal as to their convexity. The ratio of equatorial to axial diameter ranges from 1.25 to 1.75. The megalosphere diameter is 0.04–0.06 mm. Possibly, there is an inconspicuous dimorphism. The genus is monospecific.

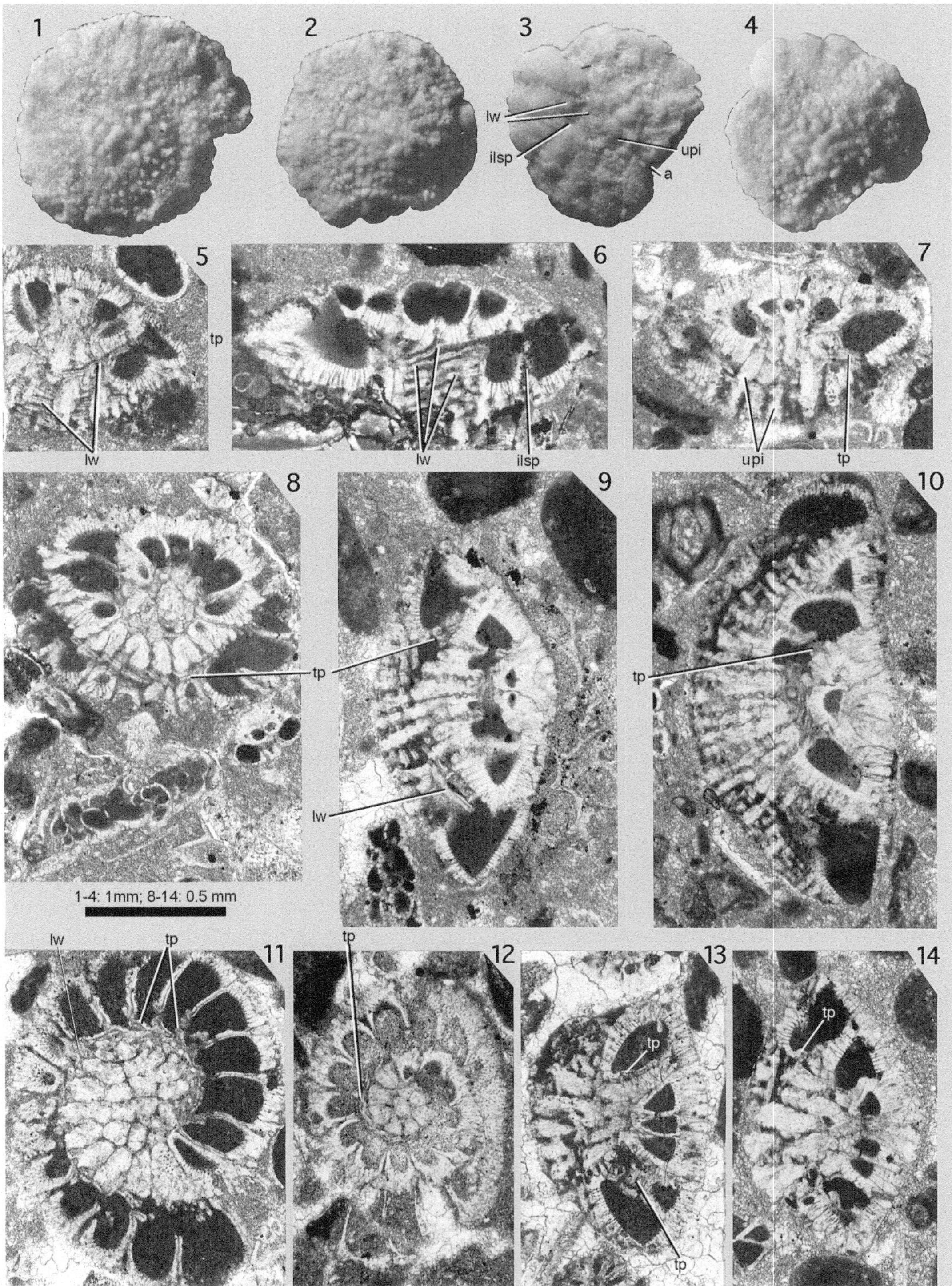

Plate 8.3 *Paralockhartia eos* n. sp.; Gorges du Cabaret, bed 94702, Petites Pyrénées, southwestern France; Paleocene (SBZ 2). (**1–4**) External views, dorsal (**1–2**) and ventral (**3–4**). (**5**, **7**) Axial sections. (**6**, **9–10**, **13–14**) Subaxial and adaxial sections. (**8**, **11–12**) Oblique sections respect to coiling axis. Abbreviations: *a* aperture, *lw* lateral-ventral chamber wall, *upi* umbilical piles, *ilsp* intraseptal interlocular space, *tp* toothplate

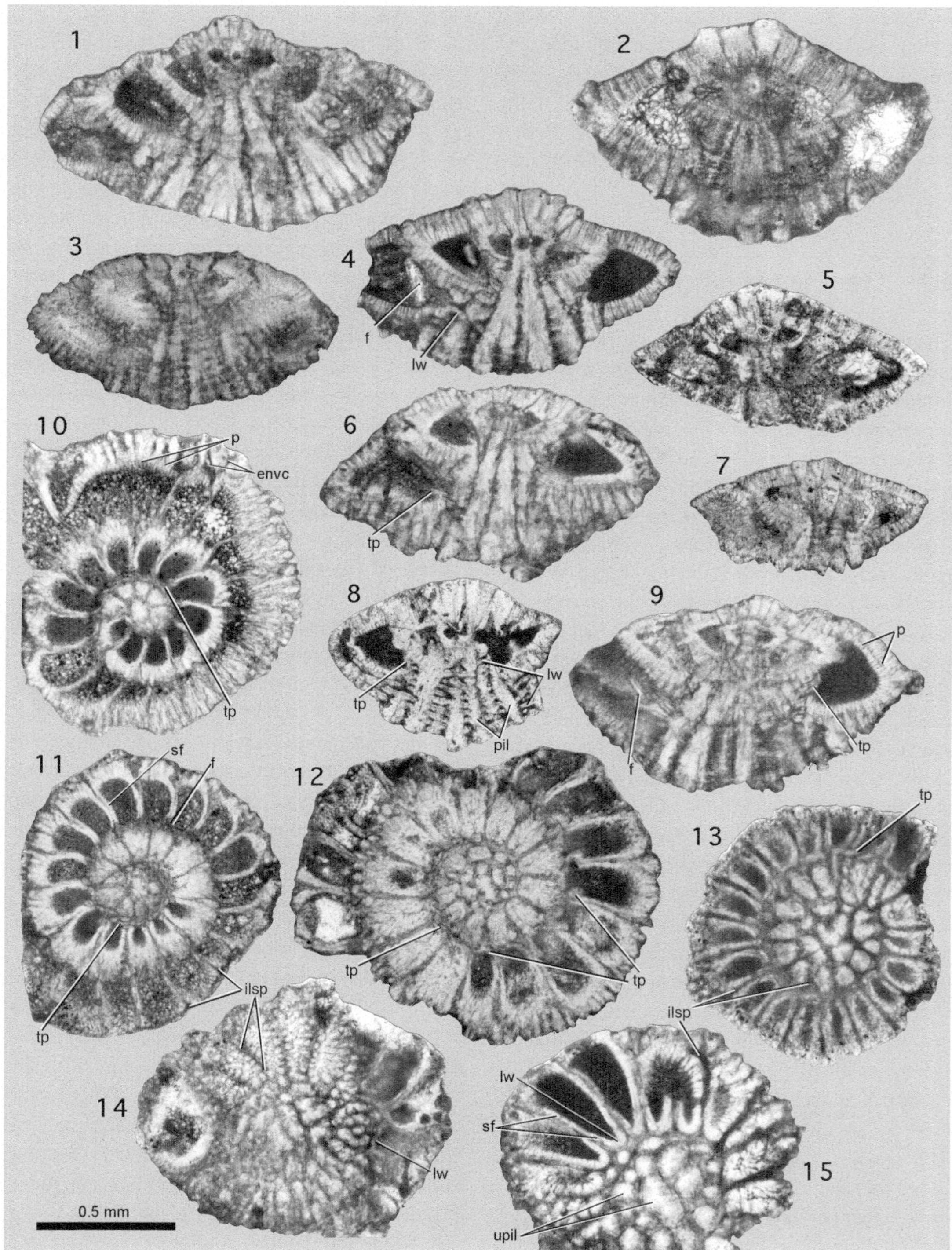

Plate 8.4 *Paralockhartia eos* n. sp.; Gorges du Cabaret, bed 94702, Petites Pyrenees, southwestern France; Paleocene (SBZ 2). (**1–9**) **A**xial sections. (**10–13**) Centered sections perpendicular to coiling axis. (**13–15**) Not centered sections perpendicular to coiling axis. Abbreviations: *f* foramen, *lw* lateral-ventral chamber wall, *p* pores, *ilsp* intraseptal interlocular space, *tp* toothplate, *envc* enveloping canals, *pil* pile, *sf* septal flap, *upil* umbilical pile

Material and distribution: about 20 isolated specimens have been sectioned, many appear in random sections in cemented limestones.

Localities: Gorges du Cabaret, Volp, Vaquemorte, Cassagne (all from the Petites Pyrénées, southwestern France), see Tambareau et al. (1997).

8.3 New subfamily Laffitteininae

The members of this subfamily have in common a single foramen, slit-shaped, straight and areal, extending from a ventral interiomarginal position in oblique direction into the dorsal parts of the septal face. They exhibit no foliar chamberlets but a plate of which the nature and its relation to the areal foramen is unclear. In our material of *Laffitteina marsicana* Farinacci, 1965, the plate is particularly well preserved (Fig. 8.2D). It may consist of only a single lamella, the inner lining and forms an adaxial paries proximus that is glued to the previous whorl. Where the adaxial part of the septum bridges canal orifices of the previous whorl, a narrow triangular space develops that is the site of funnel generation. Thus, the chamber lumen is separated from an extremely restricted, narrow space in the ventral umbilicus that may constitute a kind of interrupted or continuous spiral canal. The somewhat larger ventral and the narrower dorsal umbilicus are filled with a solid, umbonal mass of secondary lamellas, perforated in axial direction by parallel funnels, as in the ventral side of the members of the Kathininae. The continuity of the funnels in successive whorls are guaranteed by blueprinting processes similar to the one of the pores in successive lamellas.

Dorsal and ventral faces of the shells are covered by an intensive enveloping canal system. The canal system derives from an undivided, broad intraseptal interlocular space that opens in the chamber sutures to a chevron-shaped system of grooves giving the latest chambers their feathering. Supplemental outer lamellas overgrow the grooves to produce the enveloping canal system, covering dorsal and ventral surfaces of the chambers (see Hottinger and Leutenegger 1980). The evenly and densely distributed canal orifices have often been mistaken for regular pore mouths. Their alignment in radial rows along the septa has motivated other authors to include the *Laffitteina* in the Elphidiidae by considering the *Laffitteina* sutural canal orifices as fossettes (Hottinger 2001, fig. 2). The latter, however, are generated by the backward folds of the septal walls, the so-called retral processes that are missing in *Laffitteina*.

There is a certain analogy in canal orifice distribution with the Neogene *Pseudorotalia indopacifica* (Thalmann 1935) see Billman et al. (1980, pl. 30). Pseudorotalias of this type occur in deltas of tropical rivers where nutrients from land are carried together with fine-grained clay minerals into the shallow water environments. We have interpreted the similar distribution of the orifices on both, ventral and dorsal faces of the shell as reflecting a shallow-infaunal lifestyle in a meso- to eutrophic environment. In the same way we interpret the enveloping canal system of the Laffitteininae as reflecting meso- to eutrophic environments. This is supported by the stratigraphic distribution of the Laffitteinas from Late Cretaceous to the end of the Paleocene. They are the only group of "larger" foraminifera surviving the trophic changes in shallow water at the K-T boundary.

The subfamily Laffitteininae has comparatively numerous members (see Sirel 1998; Rahaghi 1992). However, they need a revision in a separate monograph in order to clarify the nature of the foramen and of the umbilical walls of the spiral chambers. Here we can only redescribe some key species of *Laffitteina* from the Late Cretaceous based on topotypes. Whether the Lower Eocene *Cuvillierina* species attributed by Sirel (1998, p. 85) to the Miscellaneidae (see also Müller-Merz 1980, pl. 7, fig. 3) must be included in the Laffitteininae or not can be decided only with additional detailed studies.

We also include Drooger's (1960a, b) two Caribbean Paleogene genera in this subfamily:

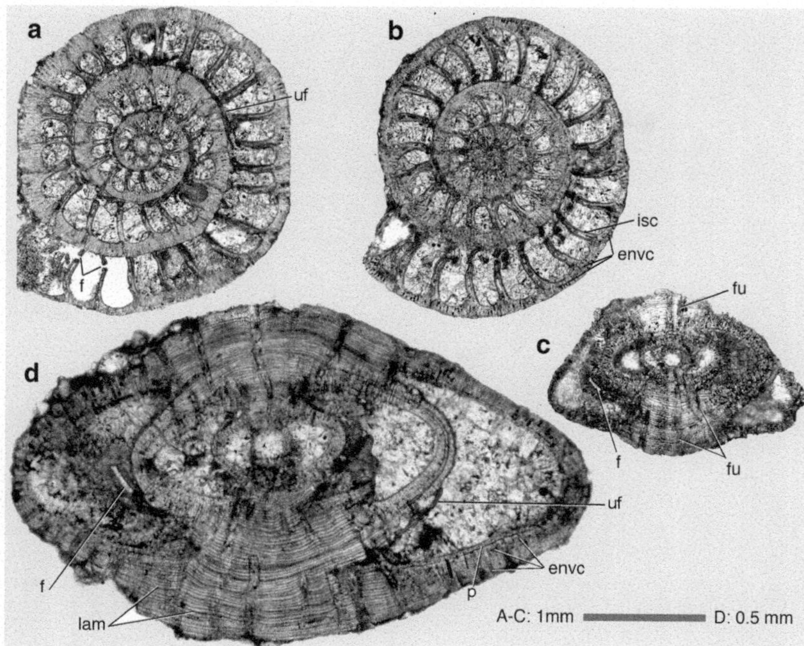

Fig. 8.2 *Laffitteina marsicana* Farinacci, 1965; megalospheric forms. (**A**, **B**) Equatorial sections. (**C**, **D**) Axial sections. Abbreviations: *f* foramen, *uf* umbilical flap, *fu* funnels, *p* pores, *isc* intraseptal canal system, *envc* enveloping canals, *lam*, lamellae

Storrsella and *Smoutina*. Both have a single areal slit as foramen, a minute plate of uncertain nature and an umbilical fill with numerous funnels. *Smoutina* is clearly trochospiral and evolute on the dorsal side. *Storrsella* is involute on both sides of the shell and presents a biumbilical shell. Both umbilical fills are pierced by numerous parallel funnels.

The name *Smoutina* has occasionally been used before for what is called in this paper *Plumokathina lenticula* n. sp. The two Caribbean genera are briefly redescribed below, based on free topotypes.

"*Laffitteina*" *melona* Rahaghi, 1976 and "*Laffitteina*" *khorassanica* Rahaghi, 1976 have to be excluded from the Laffitteininae because they have multiple foramina.

8.3.1 *Laffitteina* Marie, 1946

Type species: *Laffitteina bibensis* Marie, 1946

Description: Low-trochospiral, involute shells covered on both sides with an enveloping canal system. Septa with an intraseptal interlocular space that communicates in the umbonal areas with the ambient environment by funnels, over the rest of the shell surface and in particular the periphery by the enveloping canals. These are arranged in double rows per underlying chamber suture. The periphery is rounded and unkeeled, lacking a marginal cord. The foramen is areal; it forms a single oblique slit in the septum. There is no folium but a short umbilical flap of unknown lamellar nature that closes off the chamber lumen from the interlocular space around the ventral umbo and in the chamber septa of the previous whorls. There is only a minute passage between the intraseptal spaces of succeeding chambers forming a kind of spiral canal. Numerous species are in need of revision.

Remarks: The Laffitteinas are one of the few groups of larger, complex foraminifera surviving the Cretaceous–Cenozoic boundary event in shallow water. Among the Laffitteinas, *Laffitteina bibensis* Marie, 1946 (Fig. 8.4A–F) is the only one recorded in SBZ 1 age, in which this species represents a link between Maastrichtian and Paleocene species. The distribution of the orifices of the canal system over the entire surface of the shell recalls similar pattern in the genetically unrelated, Plio–Pleistocene to Recent *Pseudorotalia indopacifica* (Thalmann, 1935). Today, these live

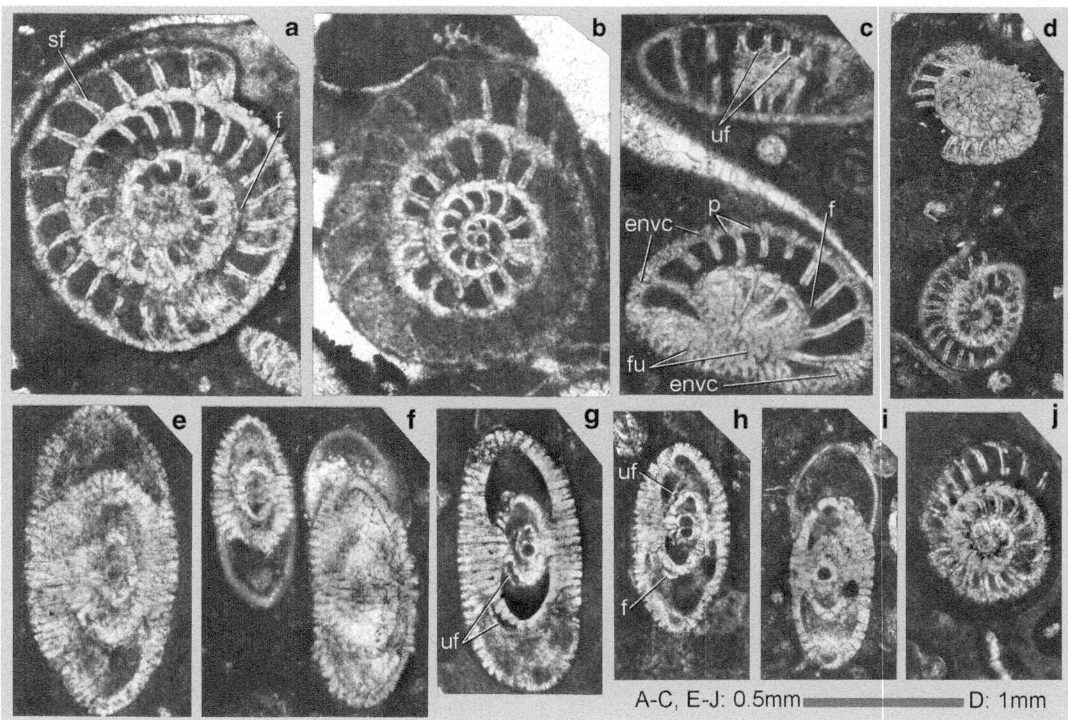

Fig. 8.3 *Laffitteina mengaudi* (Astre, 1923), topotypes, Marnes à Huitres, Aquitaine, France; Maastrichtian. (**A**, **B**) Equatorial sections. (**C**) Oblique sections. (**D**, **J**) Equatorial sections slightly obliques. (**E–I**) Axial sections. Abbreviations: *f* foramen, *uf* umbilical flap, *fu* funnels, *p* pores, *envc* enveloping canals, *sf* septal flap

on or in soft substrates in deltas of tropical rivers (see Billman et al. 1980), where the oligotrophic realm in shallow water, documented by coral reefs, is interrupted by the affluents from the continent carrying nutrients into the sea. This might be a hint to an interpretation of the Cretaceous–Tertiary boundary event as a widespread eutrophication of the tropical shallow waters.

Laffitteina marsicana Farinacci, 1965; Fig. 8.2A–D.

Description: Thickly-lenticular *Laffitteina* with a rounded periphery. The equatorial to axial diameter ratio ranges from 1.5 to 1.7. Dorsal side more convex than ventral side that may show, however, a slight umbonal swelling. Twenty two to twenty four chambers are counted in adult whorls. The megalosphere measures about 0.04 mm. The density of funnels and enveloping canals is lower than in all other species.

The Pyrenean material available of this species exhibits the lamellar construction of the shell (Fig. 8.2D) by the mineralisation of the organic layers, separating the lamellas of the shell from each other.

Laffitteina mengaudi (Astre, 1923); Fig. 8.3A–J; Plate 8.5, Figs. 1–29.

1923 *Nummulites mengaudi*—Astre, p. 360, pl. 1–2.

1975 *Laffitteina mengaudi* (Astre)—Blanc, p. 61–68, pl. 1, figs. 1–8; pl. 2, figs. 1–8.

1998 *Laffitteina mengaudi* (Astre)—Sirel, p. 88, pl. 46, figs. 1–23; non pl. 47, figs. 1–22.

Description: Lenticular *Laffitteina* with a rounded periphery. The ventral and dorsal umbilical filling are flattened or slightly convex, protruding. The equatorial to axial diameter ratio varies from 2.0 to 2.7. There are 20–22 chambers in late, adult whorls. The megalosphere measures

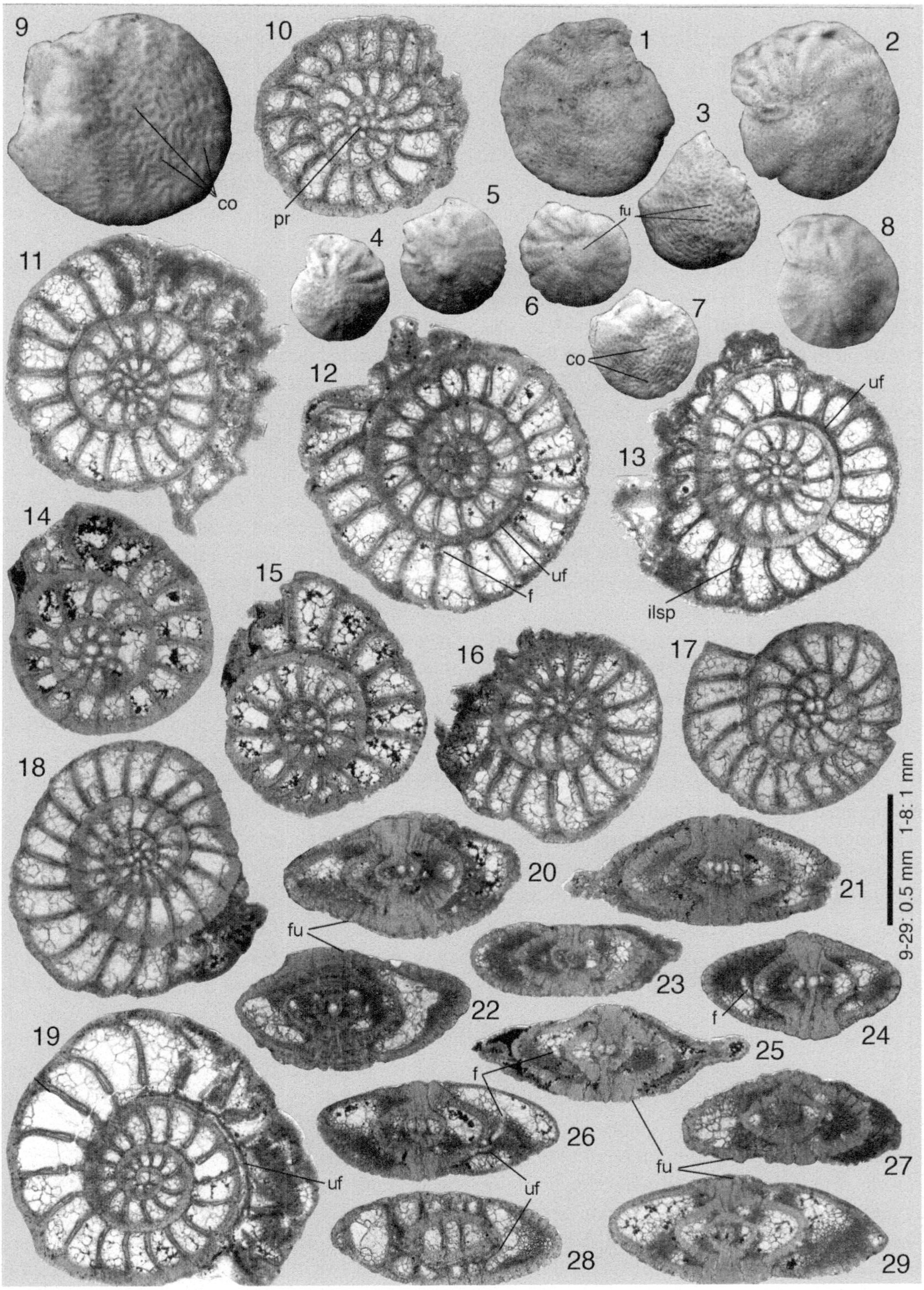

Plate 8.5 *Laffitteina mengaudi* (Astre, 1923); Marnes à Huitres, Aquitaine, France; Maastrichtian. (**1–9**): external views, dorsal and ventral sides. (**10–19**) Equatorial sections. (**12**) Probably microspheric specimen. (**20–29**) Axial sections. Abbreviations: *f* foramen, *co* canal orifices, *pr* proloculus, *fu* funnel, *uf* umbilical flap, *ilsp* intraseptal interlocular space

0.03–0.04 mm in diameter. If there is a dimorphism of generations at all, it is restricted to the first few nepiontic whorls (microspheric specimen, Plate 8.5, Fig. 12).

Laffitteina monodi Marie, 1946; Plate 8.6, Figs. 1–17.

1998 *Laffitteina* aff. *mengaudi* (Astre)—Sirel, pars, p. 90, pl. 48, fig.16.

Description: Lenticular *Laffitteina* with a rounded periphery, compressed, with flattened umbonal areas on both sides, semi-evolute. The equatorial to axial diameter ratio ranges from 2.7 to 5.0. The adult last whorls are composed of 24–28 chambers. The megalosphere diameter is around 0.08 mm. *Laffitteina monodi* is distinguished from *L. mengaudi* by its larger, more evolute and more compressed shells.

Laffitteina bibensis Marie, 1946; Fig. 8.4A–I.

1987 *Laffitteina bibensis* Marie—Loeblich and Tappan, p. 661, pl. 759, figs. 1–2, 9.

Description: Lenticular, low-trochspiral *Laffitteina* with chambers that have more involute dorsal and more evolute ventral alar prolongations. The umbo on the dorsal side is restricted in diameter, often slightly inflated. On the ventral side, the umbo has a flattened surface or is slightly concave. The equatorial to axial shell diameter ratio varies from 2.4 to 3.0. 27–30 chambers in late, adult whorls, are present. The megalosphere diameter is 0.03–0.06 mm.

8.3.2 *Cuvillierina* Debourle, 1955

Type species: *Laffitteina vallensis* Ruiz de Gaona 1948 (= *Laffitteina vanbelleni* Grimsdale, 1952 = *Cuvillierina eocaenica* Debourle, 1955).

Description: Almost planispiral-involute shells with a single areal foramen next to an umbilical flap. Both adaxial areas are occupied by umbos perforated by a dense group of funnels. The chamber sutures are dominated by a deep intraseptal interlocular space that gets subdivided by parallel ridges of the chevron-shaped feathering. The ridges bridge the outer part of the intraseptal interlocular space and give rise to a uniformly dense enveloping canal system. The orifices of the radial canals in the enveloping system have the same diameter as the orifices of the funnels. There is no folium but the umbilical flap detaches itself from the shell surface of the previous whorl over a short distance below the foramen and creates a thin spiral canal, connecting the interlocular spaces of successive chamber sutures.

Remarks: Reiss and Merling (1958), Reiss (1963, pl. 3, fig. 7; pl. 5, fig. 22; pl. 6, fig. 9; pl. 7, fig. 8) and Müller-Merz (1980) have produced detailed descriptions of the type species of the genus *Cuvillierina*, *C. vallensis* (Ruiz de Gaona, 1948). A redescription of this Lower Eocene (Cuisian) species seems to be superfluous. However, a copy of some of Reiss's illustrations is given here in Fig. 8.5 A–B at standard enlargement and with designations of the structural elements, in order to emphasise the similarity of architecture between the genera *Cuvillerina* and *Laffitteina*. Moreover, a redescription of the Paleocene *Cuvillierina sireli* Inan, 1988 might be useful and is given below.

Cuvillierina sireli Inan, 1988; Plate 8.7, Figs. 1–11; Plate 8.8, Figs. 1–12.

1976 *Cuvillierina* n. sp.—Sirel and Gündüz, p. 33, fig. 2.

1986 *Cuvillerina* sp.—Sirel et al., p. 387, fig. 2.

1988 *Cuvillerina sireli*—Inan, p. 121, pl. 1, figs. 1–9; pl. 2, figs. 1–8.

1998 *Pseudocuvillerina sireli* (Inan)—Sirel, p. 86, pl. 44, figs. 1–20.

2000 "*Cuvillerina*" *sireli* Inan—Peybernès et al., p. 46, fig. 6/1–3.

Description: *Cuvillierina*, easy to identify by its enveloping canal system in random sections, was treated so far as index fossil for the Cuisian (upper part of Lower Eocene). The Paleocene species *C. sireli*, was treated by Sirel (1998) as a separate genus, *Pseudocuvillerina*, based on "the lack of septal and umbilical flaps". This is obviously a mistake because all foraminifera with a so-called double septum, that is with an intraseptal interlocular space, have *per definitionem* a septal flap forming the posterior septal wall of the respective chamber. Therefore,

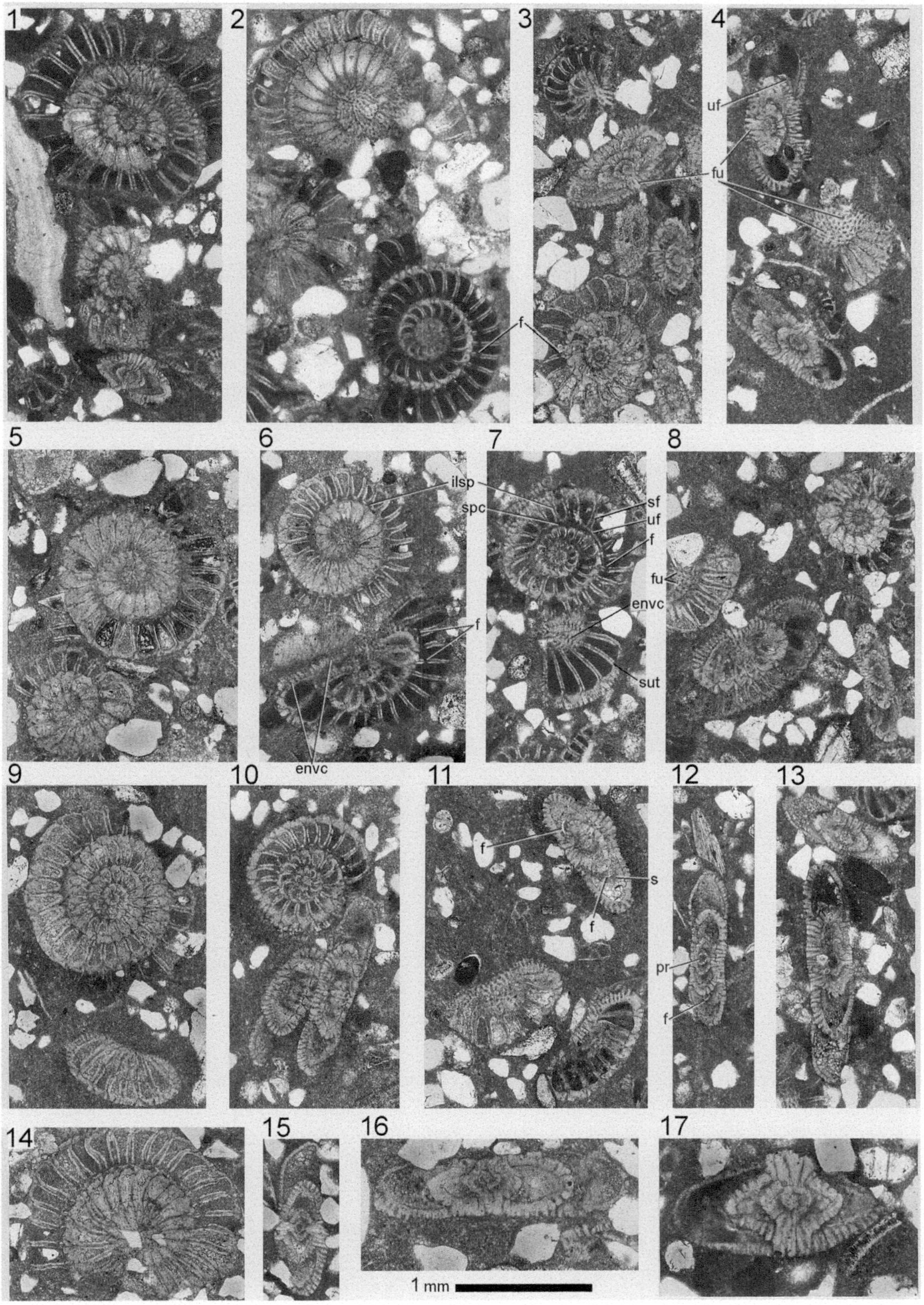

Plate 8.6 *Laffitteina monodi* Marie, 1946. (**1**) (*top*), (**2**) (*bottom*), (**3**) (*bottom*), (**5–8**) (*top, right*), (**9–10, 14**) Equatorial sections. (**1**) (*bottom*), (**12, 13, 15–17**): axial sections. (**4, 8**) (*bottom, right*), (**10, 11**) (*top, right*): subaxial and adaxial sections. (**4**) (*center*), (**7**) (*bottom*), (**9**) (*bottom*), (**11**) (*bottom left*) Tangential sections. (**3**) (*center*), (**4**), (*top*), (**6**) (*bottom*), (**8**) (*left*) Oblique sections. Abbreviations: *f* foramen, *pr* proloculus, *s* septum, *fu* funnel, *uf* umbilical flap, *ilsp* intraseptal interlocular space, *spc* spiral canal, *envc* enveloping canals, *sut* suture, *sf* septal flap

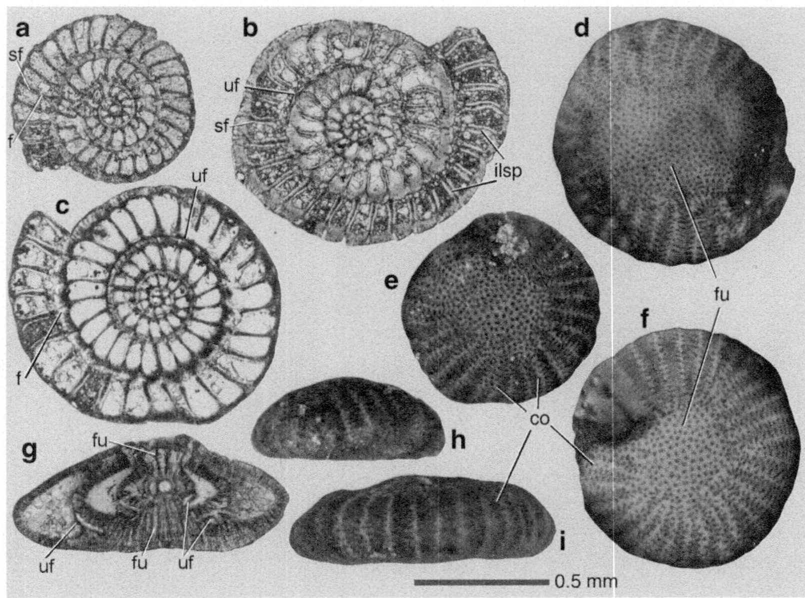

Fig. 8.4 *Laffitteina bibensis* Marie, 1946; megalospheric forms; specimens from the Campo section, sample Stop 1 in Robador et al. (1991). (**A–C**) Sections perpendicular to coiling axis. (**C**) Axial section. (**D**, **E**) Dorsal view. (**F**) Ventral view. (**G**) Axial section. (**H**, **I**) Lateral view. Abbreviations: *f* foramen, *uf* umbilical flap, *co* canal orifices, *sf* septal flap, *ilsp* intraseptal interlocular space, *fu* funnels

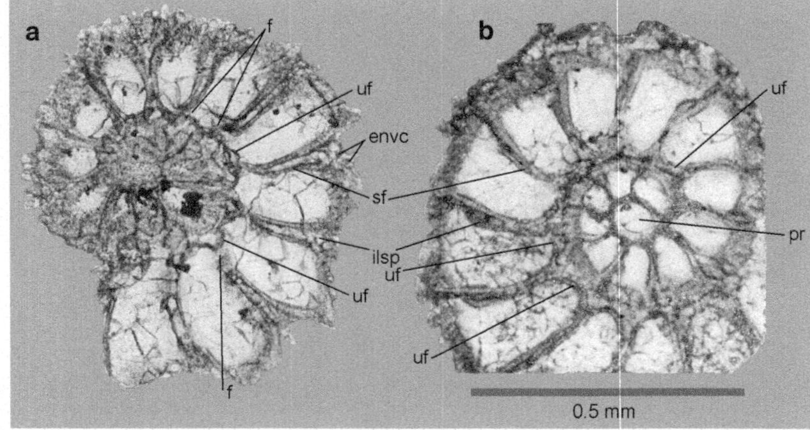

Fig. 8.5 *Cuvillierina vallensis* (Ruiz de Gaona, 1948), illustrations after Reiss (1963). (**A**) Subequatorial section. (**B**) Equatorial section. Abbreviations: *f* foramen, *pr* proloculus, *uf* umbilical flap, *sf* septal flap, *envc* enveloping canals, *ilsp* intraseptal interlocular space

the genus *Pseudocuvillerina* is not considered here as valid.

Remarks: The species *C. sireli* and *C. vallensis* with their identical architecture are complemented by a third species, *Cuvillierina yarzai* (Ruiz de Gaona, 1948) in the Early Eocene (Ilerdian) of the Pyrenees redescribed by Müller-Merz (1980; text-figs. 14–15 and pl. 7, figs. 1–2; see also Ferrer et al. 1973, pl. 10). In this species the chambers of the last whorl are covered only by feathered ridges without a complete enveloping canal system. Whether this *yarzai* species belongs to *Cuvillierina* or to *Ornatanomalina* needs additional investigations about the latter's architecture.

8.3.3 *Smoutina* Drooger, 1960

Type species: *Smoutina cruysi* Drooger, 1960

Description: Trochospiral, lenticular shells with an angular periphery, lacking any keels. The dorsal side of the shell is evolute, flush.

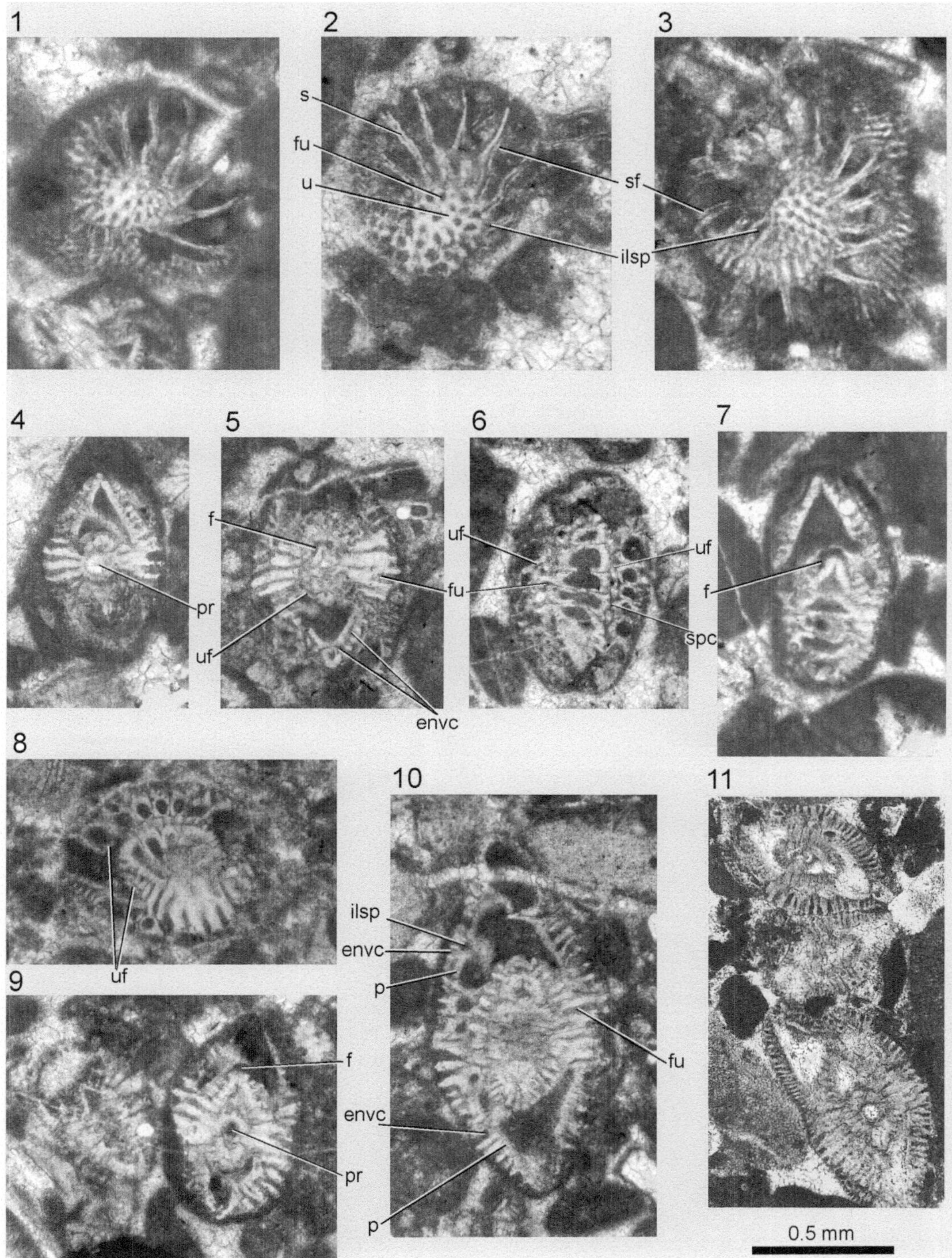

Plate 8.7 *Cuvillierina sireli* Inan, 1988, topotypes; eastern Pontides, Paleocene. (**1–3**) Tangential sections. (**4–5**, **9–11**) Axial sections. (**6–7**) Subaxial sections. (**8**) Oblique section. Abbreviations: *f* foramen, *pr* proloculus, *s* septum, *fu* funnel, *uf* umbilical flap, *ilsp* intraseptal interlocular space, *spc* spiral canal, *envc* enveloping canals, *sf* septal flap, *u* umbiliculus, *p* pores

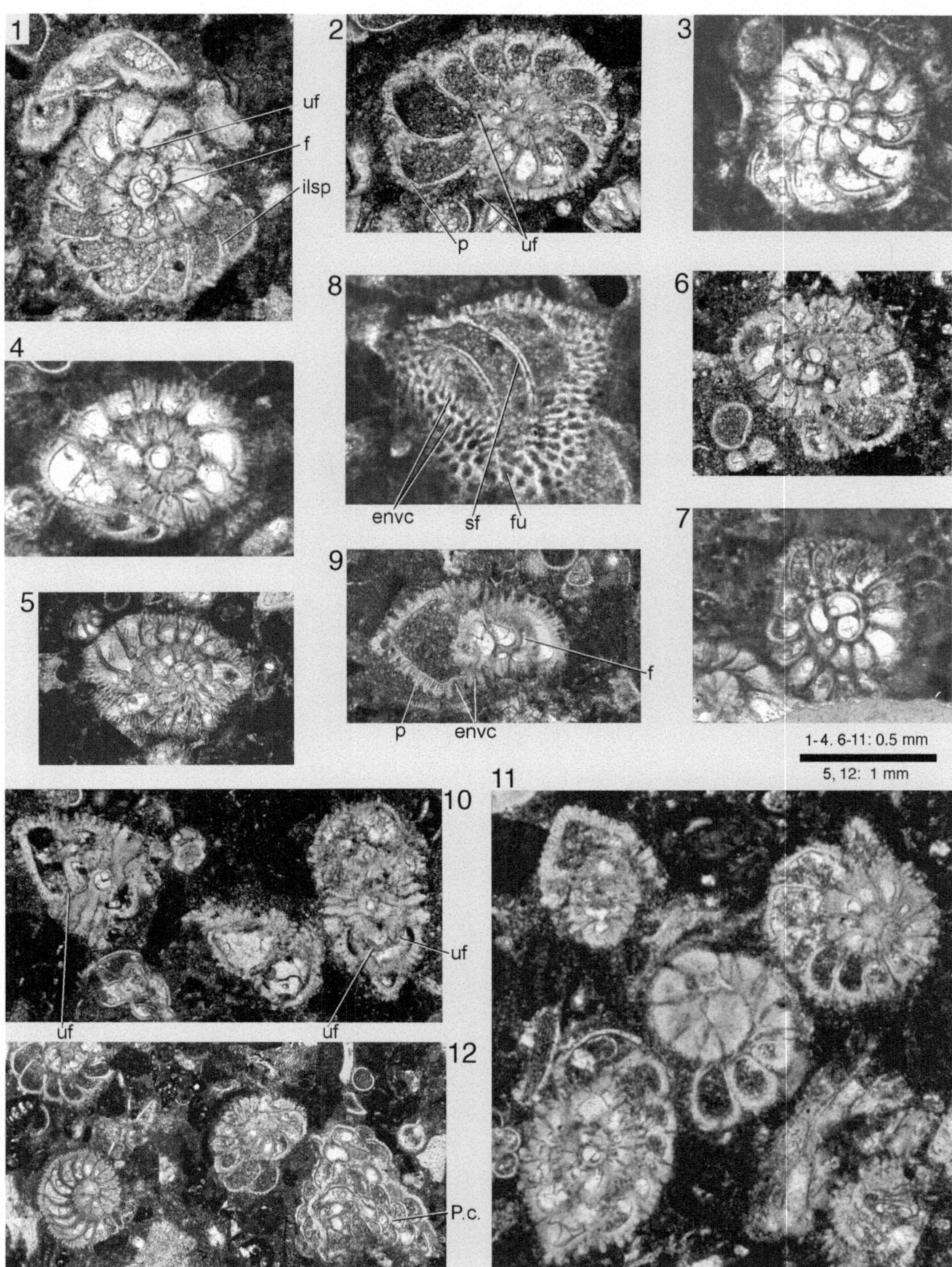

Plate 8.8 *Cuvillierina sireli* Inan, 1988; topotypes; eastern Pontides, Paleocene. (**1**, **3–7**) Centered equatorial sections. (**2**, **11**) Uncentered equatorial sections. (**8**) Tangential section. (**9–10**) Axial sections. (**11**) (*bottom*, *left*) Oblique section. (**12**) Uncentered equatorial section with *Planorbulina cretae* (Marsson, 1878), *bottom right* (P. c.). Abbreviations: *f* foramen, *fu* funnel, *uf* umbilical flap, *ilsp* intraseptal interlocular space, *envc* enveloping canals, *sf* septal face, *p* pores

The ventral side is involute but with a wide umbonal area with numerous funnels. Spiral and septal sutures are without any relief. The septa are dorsally arcuate, inclined backwards; they are ventrally radial, with sutural grooves marking narrow intraseptal interlocular spaces. The apex of the spiral on the dorsal side of the test has no lamellar thickening and forms not more than a tiny, sharp peak in the axis of the shell. The foramen is a straight areal slit, starting from its dorsal end in interiomarginal position to reach, in ventral direction, the middle line of the apertural face. Compared with *Laffitteina*, the foramen is inclined in the opposite direction. The "umbilical" architecture is comparatively simple. There is no folium and, consequently, no true umbilical plate. The ventral umbo, consisting of numerous superposed lamellae, is perforated by funnels that spring from a triangular ventral enlargement of the interlocular space over the periphery of the previous whorl. The adaxial chamber wall is glued to the previous whorl in the shell and consists probably of a single lamella, the inner lining. Drooger has drawn a spiral canal (see Fig. 8.6L) that I have not found again in Drooger's topotypes.

Remarks: Sirel (1998) has used this generic name for rotaliid species with a similar, lenticular habit that are frequent and widespread all over the Paleogene Neotethys. These are described here as *Plumokathina subsphaerica* (Sirel, 1972) and *P. lenticula* n. sp. They show a heavy feathering of the chamber sutures that lack in Drooger's (1960a, b) Caribbean species. These have areal foramina, whereas *Plumokathina* exhibits a wide interiomarginal arch combined with a heavy umbilical plate and small, narrow foliar cavities fused to a spiral canal.

Smoutina cruysi Drooger, 1960; Fig. 8.6A–L; Plate 5.18, Figs. 1–7.

1960a *Smoutina cruysi*—Drooger, p. 307, pl. 4, figs. 1–13.

1960b *Smoutina cruysi* Drooger—Drooger, p. 459, pl. 2, figs. 7–8.

1987 *Smoutina cruysi* Drooger—Loeblich and Tappan, p. 663, pl. 760, figs. 11–17.

Description: Lenticular shells with almost identical convexity of both sides. The equatorial to axial diameter ratio varies from 1.6 to 2.0. 20–22 chambers are counted in adult whorls. Over a radial distance of 0.25 mm, there are about 7 funnels in the ventral umbo. The megalosphere diameter is 0.06 mm.

Remarks: *Rahaghia khorassanica* (Rahaghi 1976) has a similar, lenticular habit with a similar, strong umbonal area perforated by numerous funnels. However, the latter species has a ventral septal vortex of multiple apertures with peristomes and a restricted dorsal umbonal area that permits easy distinction.

8.3.4 *Storrsella* Drooger 1960

Type species: *Cibicides haastersi* Van den Bold, 1946

Description: Trochospiral, almost planispiral-involute shells with falciform septa. The periphery is angular, sharp, but not keeled. The foramen is a single areal slit in the ventral part of the septum. A larger, ventral and a smaller dorsal umbo are perforated by funnels running in parallel directions with the coiling axis of the shell. They take their origin in the enlargement of the interlocular space between the adaxial tips of the alar prolongations. The adaxial interlocular space extends in the septum towards the periphery of the shell to the point where the septum crosses the periphery of the previous whorl. The lateral chamber walls on both sides of the test do not bear any enveloping canal system.

Storrsella haastersi (Van den Bold, 1946); Fig. 8.7A–K; Plate 5.18, Figs. 8–15.

1946 *Cibicides haastersi*—Van den Bold, p. 125, pl. 18, fig. 9.

1960b *Storrsella haastersi* (Van den Bold)—Drooger, p. 296–301, pl. 1, fig. 1; pl. 11, figs. 1–13.

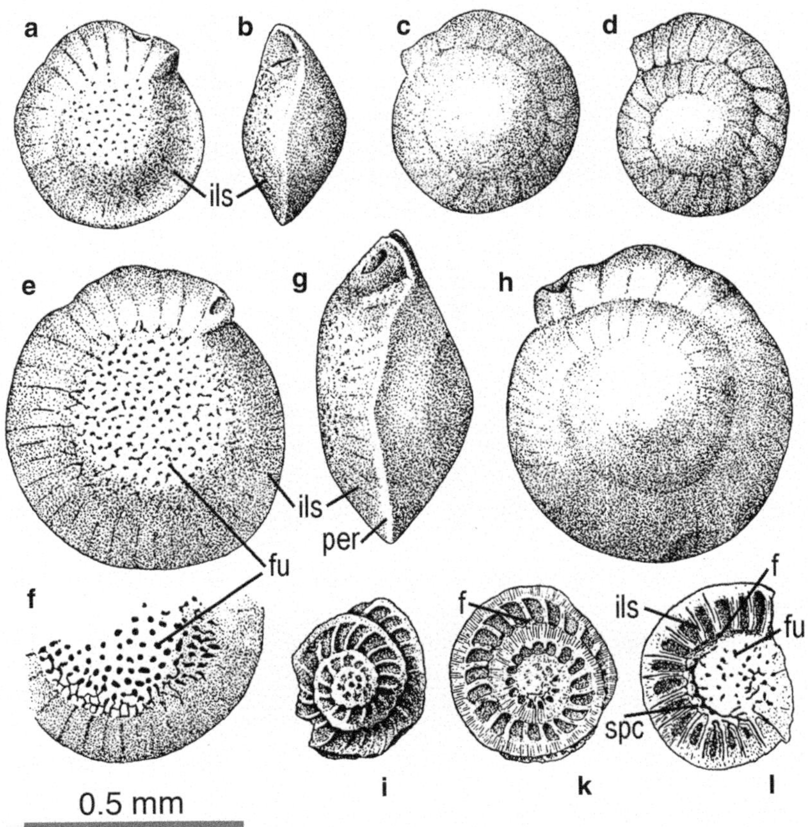

Fig. 8.6 *Smoutina cruysi* Drooger, 1960, illustrations after Drooger (1960a, b). (**A**, **E**, **F**) Ventral view. (**B**, **G**) Lateral views. (**C**, **D**, **H**) Dorsal views. (**I**, **K**) Equatorial sections. (**L**) Subequatorial section. Abbreviations: *f* foramen, *ils* intraseptal interlocular space, *per* periphery, *fu* funnels, *spc* spiral canal

1976 *Storrsella haastersi* (Van den Bold)—Rahaghi, p. 35, pl. 1, figs. 1–7 (part only? see Remarks in the text).

1983 *Storrsella haastersi* (Van den Bold)—Rahaghi, no text; pl. 38, figs. 12–19.

Remarks: The genus is monotypic. The generic description is therefore also valid for the type species. The equatorial to axial diameter ratio varies from 1.7 to 1.8. There are 19–21 chambers in adult whorls. The megalosphere diameter is about 0.04 mm.

Rahaghi (1976) describes the Caribbean species *Storrsella haastersi* (Van den Bold, 1946) from the Kopet Dagh near the ex-USSR border with Iran, in association with "*Laffitteina*" *khorassanica* Rahaghi, 1976 and "*L.*" *melona* Rahaghi, 1976. Rahaghi (1976, p. 35) mentions the topotypes from French Guinea but does not indicate which of the figured specimens are from French Guinea and which from Iran. Rahaghi's (1976) material, documenting the associate species "*L.*" *khorassanica* and "*L.*" *melona* deposited in the Natural History Museum (London), exclusively consists of random sections of cemented rock. Consequently, Rahaghi's free specimens (1976, pl 1, figs. 1–4) are unlikely to originate from Iran. As the rest of the documentation is poor in quality and without detail, we must be cautious to use the presence of *Storrsella haastersi* in Iran as indication for a cosmopolitan distribution of these Laffitteininae.

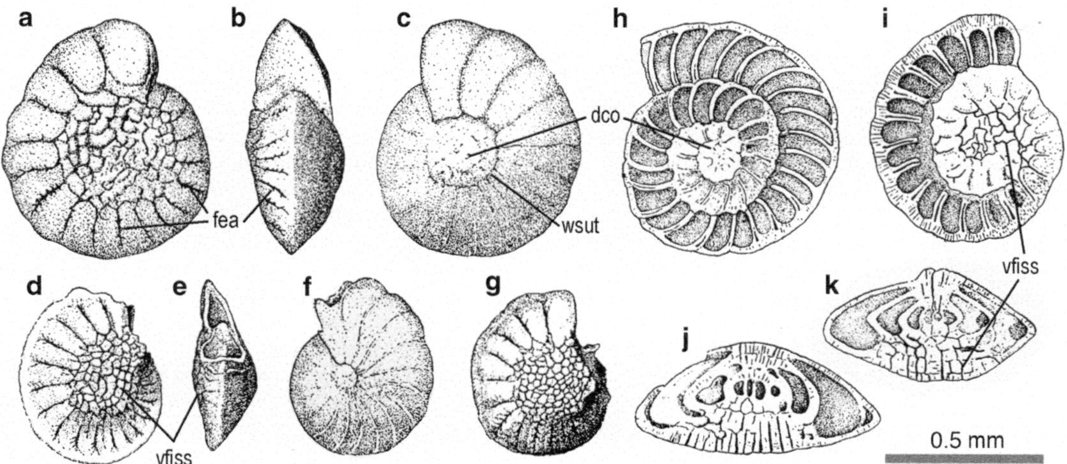

Fig. 8.7 *Storrsella haastersi* (Van den Bold, 1946), illustrations after Drooger (1960a, b). (**A**, **D**, **G**) Ventral view. (**B**, **E**) Lateral views. (**C**, **F**) Dorsal views. (**H**) Equatorial section. (**I**) Subequatorial section. (**J**, **K**) Axial sections. Abbreviations: *fea* feathering of the intraseptal interlocular space, *wsut* whorl suture, *dco* dorsal canal orifice, *vfiss* ventral fissures

References

Astre G (1923) Étude paléontologique des *Nummulites* du Crétacé supérieur de Cezan Laverdens (Gers). Bull Soc Géol Fr 4(23):360–368

Bermúdez PJ (1952) Estudio sistematico de los foraminíferos rotaliformes. Bol Geol Caracas 2 (4):230 pp, 35 pls

Billman H, Hottinger L, Oesterle H (1980) Neogene to Recent Rotaliid foraminifera from the Indopacific Ocean; their canal system, their classification and their stratigraphic use. Schweiz Paläontol Abh 101:71–113, 27 text figs, 39 pls

Blanc PL (1975) Contribution à l'étude du genre *Laffitteina*, Elphidiidae du Crétacé Terminal. Rev Micropaléontol 18(2):61–68

Colom G (1954) Estudios de las biozonas con foraminiferos del Terciario de Alicante. Bol Inst Geol Min Esp 66:101–279, 135 pls

Debourle A (1955) *Cuvillierina eocaenica*, nouveau genre et nouvelle espèce de foraminifère de l'Yprésien d'Aquitaine. Bul Soc Géol Fr (sér 6) 5:55–57

Drooger CW (1960a) Microfauna and age of the Basses Plaines Formation of French Guyana I & II. Proc K Ned Akad Wet Amsterdam, Ser B 63(4):449–468

Drooger CW (1960b) Some Early Rotaliid Foraminifera I-III. Proc K Ned Akad Wet Amsterdam, Ser B 63 (2):287–334

Farinacci A (1965) "Laffitteina marsicana", nuova specie di Rotalide nel calcare maastrichtiano a "Rhapydionina liburnica" di M. Turchio (Marsica). Riv Ital Paleont 71(4):115–111

Ferrer J, Le Calvez Y, Luterbacher H-P, Premoli-Silva I (1973) Contribution à l'étude des foraminifères Ilerdiens de la region de Tremp (Catalogne), Mém Mus Nat Hist Nat Paris, Sér C, 29. Éditions du Muséum, Paris, pp 3–107

Greig DA (1935) Rotalia viennoti, an important foraminiferal species from Asia Minor and Western Asia. J Paleontol 9(6):523–526

Grimsdale TF (1952) Cretaceous and Tertiary foraminifera from the Middle East. Bull Brit Mus (Nat Hist) 1:223–247

Hansen H-J, Reiss Z (1971) Electron microscopy of Rotaliacean wall structures. Bull Geol Soc Denmark 20:329–346

Hottinger L (2001) The shell cavity systems in Elphidiid and Pellatispirine bilamellar foraminifera. Micropaleontology 47(suppl 2):1–4

Hottinger L, Leutenegger S (1980) The structure of calcarinid Foraminifera. Schweiz Paläontol Abh 101:115–151

Hottinger L, Halicz E, Reiss Z (1991) The foraminiferal genera *Pararotalia*, *Neorotalia* and *Calcarina*: taxonomic revision. J Paleontol 69:1–33

Inan N (1988) Sur la presence de la nouvelle espèce *Cuvillerina sireli* dans le Thanétien de la Montagne de Tecer (Anatolie centrale, Turquie). Rev Paléobiol 7:121–127

Le Calvez Y (1949) Révision des Foraminifères lutétiens du Bassin de Paris. 2. Rotaliidae et families affines. Mém Carte géol France, 54 pp, 1tab, 6 pls. Imprimarie Nationale, Paris

Loeblich AR, Tappan H (1964) Sarcodina, chiefly "Thecamoebians" and Foraminiferida. In: Moore RC (ed) Part C, Protista 2. The University of Kansas Press,

Lawrence, and The Geological Society of America, Boulder, 1, 510 pp; 2, pp 511–900

Loeblich AR, Tappan H (1987) Foraminiferal genera and their classification. Van Nostrand Reinhold New York, 1:970 pp, 2:212 pp, 847 pls

Marie P (1946) Sur Laffitteina bibensis et Laffitteina Monodi, nouveau genre et nouvelles espéces de Foraminifères du Montien. Bull Soc géol France (5), 15 (1945):419–434

Marsson T (1878) Die Foraminiferen der Weissen Schreibkeide der Inseln Rüggen. Mitteeilungen des Naturwissenschaftlichen Vereins für Neu-Vorpommern und Rugen in Greifswald 10:115–196

Müller-Merz E (1980) Strukturanalyse ausgewählter rotaloider Foraminiferen. Schweiz Paläontol Abh 101:5–68, 15 pls

Peybernès B, Fondecave-Wallez M-J, Hottinger L, Eichène P, Segonzac G (2000) Limite Crétacé-Tertiaire et biozonation micropaléontologique du Danien-Sélandien dans le Béarn occidental et la Haute-Soule (Pyrénées Atlantiques). Geobios 33:35–48

Rahaghi A (1976) Contribution a l'étude de quelques grands foraminifères de l'Iran. Nat Iranian Oil Corp Geol Lab 6:1–68, Teheran

Rahaghi A (1983) Stratigraphy and faunal assemblage of Paleocene-Lower Eocene in Iran. Nat Iranian Oil Comp Geol Lab 10:73 pp, 49 pls

Rahaghi A (1992) The geographic and stratigraphic range of the genus *Laffitteina* Marie, 1946 in Iran, with description of *Laffitteina jaskii* n. sp. Rev Esp Micropaleontol 24(3):5–11

Reiss Z (1963) Reclassification of Perforate Foraminifera. Bull Geol Surv Israel 36:1–111

Reiss Z, Merling P (1958) Structure of some Rotaliidea. Bull Geol Surv Israel 21:1–19

Robador A, Samsó JM, Serra-Kiel J, Tosquella J (1991) Field Guide. In: Introduction to the Early Paleogene of the south Pyrenean Basin. Early Paleogene Benthos 1st Meeting IGCP nº 286, Jaca 1990, pp 131–159

Ruiz de Gaona M (1948) La fauna principalmente nummulitica de la serie terciaria guipuzcoana. Estud Geol, Inst Invest Geol "Lucas Mallada", 9:133–158, Madrid

Samuel O, Borza K, Köhler E (1972) Microfauna and Lithostratigraphy of the Paleogene and adjacent Cretaceous of the Middle Vah valley (West Carpathians). Geol Ustav Dionysza Stura, 246 pp, 153 pls

Sirel E (1972) Systematic study of new species of the genera *Fabularia* and *Kathina* from Paleocene. Türk Jeol Kur Bül 15:277–249, 8 pls

Sirel E (1998) Foraminiferal description and biostratigraphy of the Paleocene-Eocene shallow water limestones and discussion of the Cretaceous-Tertiary boundary in Turkey. Gen Dir Min Res Expl (MTA), Monograph ser 2:117 pp, 68 pls

Sirel E, Gündüz H (1976) Description and stratigraphical distribution of the some species of the genera *Nummulites*, *Assilina* and *Alveolina* from the Ilerdian, Cuisian and Lutetian of Haymana region (S Ankara). Bull Geol Soc Turkey 19(1):31–44, 15 pls

Sirel E, Dager Z, Sozeri B (1986) Some biostratigraphic and paleogeographic observations on the Cretaceous/Tertiary boundary in the Haymana polatli region (Central Turkey). In: Walliser O (ed) Global bio-events, vol 8, Lecture note in Earth Sciences. Springer, Berlin, pp 385–396

Tambareau Y, Hottinger L, Rodriguez-Lazaro J, Vilatte J, Babinot IF, Colin I-P, Garcia Zaraga E, Rocchia R, Guerrero N (1997) Communités benthiques fossiles aux alentours de la limite Crétacé–Tertiaire dans les Pyrénées. Bull Soc Geol Fr 168(6):795–804

Thalmann HE (1935) Mitteilungen über Foraminiferen. 2. 7. Liste der Foraminiferen von der Typus-Lokalitaet der miozänen Tuxpan-Stufe (Ciudad de Tuxpan, Veracruz, Mexico). Eclog geol Helv 28(2):602–605

Uhlig V (1886) Über eine Mikrofauna aus dem Alttertiär der westgalizischen Karpathen. Jb k-k geol Reichsanst 36:141–214, Wien

Van den Bold WA (1946) Contribution to the study of Ostracoda with special reference to the Tertiary and Cretaceous microfauna of the Caribbean region, 167 pp, 18 pls. Proefschrift Rijks-Univ. Utrecht

Rotaliid Taxa of Uncertain Affinities

9

Abstract

Some rotaliid taxa with single foramina and heavy ornamentation have no umbilicus but umbilical depressions: *Thalmannita* and *Civrieuxia*. Rotaliid taxa with multiple foramina are rare in the Paleogene: *Scarificatina*, *Cincoriola* and *Rahaghia* n. gen. The new subfamily Cincoriolinae is here assessed. These are grouped here into different subfamilies according to their foraminal features and are considered as of *incertae sedis* on all higher systematic levels. In the family Victoriellidae, a considerable number of genera show an attachment surface on the spiral side of their trochospiral test: *Gyroidinella laevis*, *G. eocaenica*, *G. magna* are described and illustrated.

9.1 Rotalid Taxa Without Umbilical Fills

There is a number of low-trochospiral to planispiral smaller benthic foraminiferal forms that have no umbilicus but umbilical depressions. Their ornamentation is always heavy. Their foramina are single. More, much extended studies on such benthics will be necessary in order to decide whether such a group of taxa should be classified together in a particular subfamily or family and where to place it in the system. Loeblich and Tappan (1987), however, classify the two genera *Thalmannita* and *Civrieuxia*, described here, in the same subfamily Cuvillerininae.

9.1.1 *Thalmannita* Bermúdez, 1952

Type species: *Rotalia madrugaensis* Cushman and Bermúdez, 1947

Thalmannita madrugaensis (Cushman and Bermúdez, 1947); Plate 9.1, Figs. 1–21.

1947 *Rotalia madrugaensis*—Cushman and Bermúdez, p. 24, pl. 5, Fig. 4.

1951 *Rotalia madrugaensis* Cushman and Bermúdez—Cushman, p. 55, pl. 15, Fig. 12.

1952 *Thalmannita madrugaensis* (Cushman and Bermúdez)—Bermúdez, p. 76.

1962 *Thalmannita madrugaensis* (Cushman and Bermúdez)—Hofker, text-Fig. 24D.

Description: Small-sized, heavily ornate, almost planispiral, evolute shells with a dorsal

L. Hottinger, *Paleogene larger rotaliid foraminifera from the western and central Neotethys*,
DOI 10.1007/978-3-319-02853-8_9,

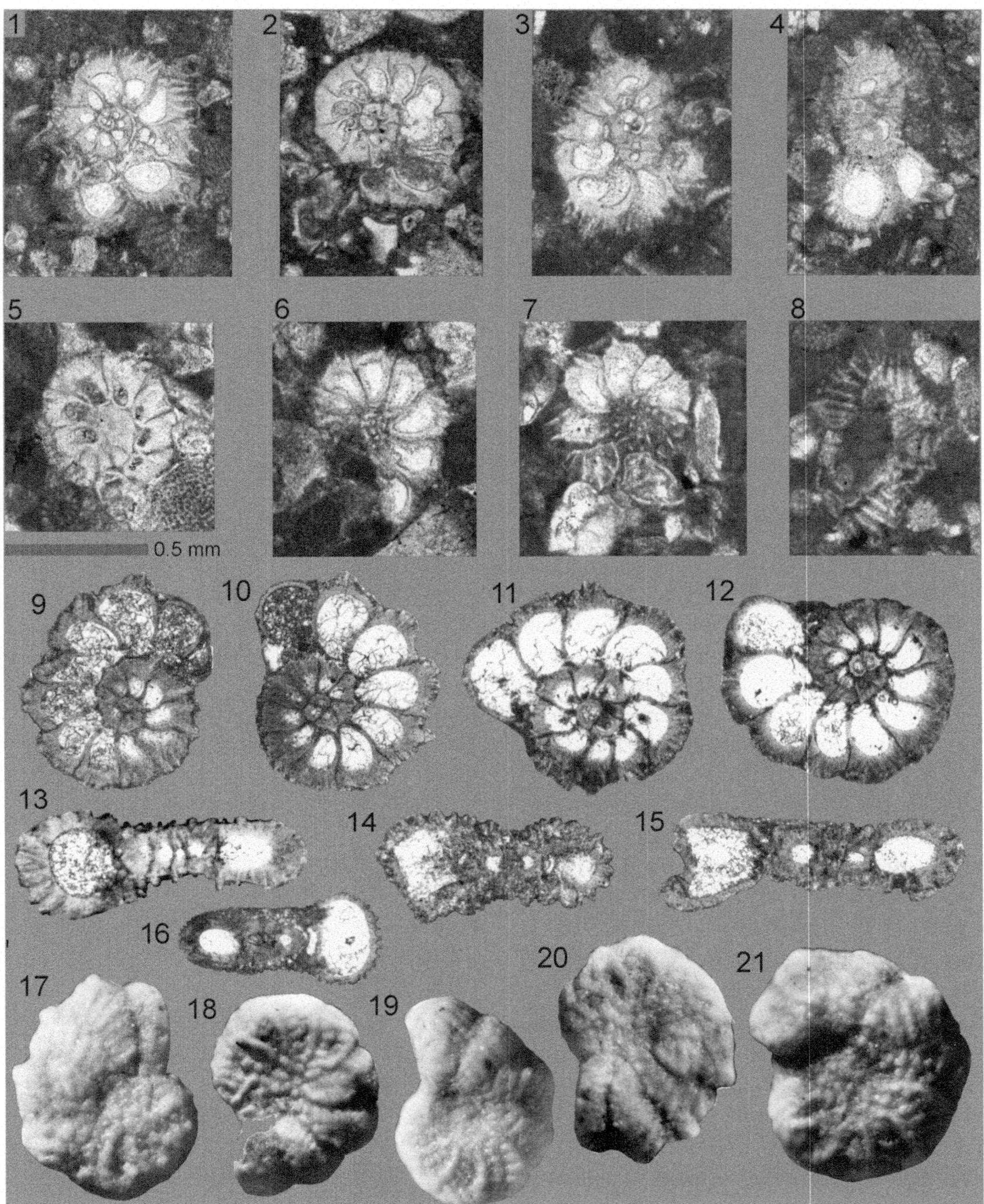

Plate 9.1 *Thalmannita madrugaensis* (Cushman and Bermúdez, 1947). (**1–8**) Exceptionally well preserved specimens in cemented rock, from sample Kar 8, Kuh-e-Kargan, Kermanshah, Iranian Zagros, collected by J. Braud; SBZ 4. (**1–2**) Equatorial sections. (**3–4**) Oblique sections respectively inclined for about 60° and 30° in respect to the coiling axis. (**5–7**) Sections more or less perpendicular to the coiling axis. (**8**) Tangential section showing the course of the costae on subsequent chamber walls. (**9–21**) Isolated specimens from Narp, Petites Pyrénées; Aquitaine, southwestern France; Paleocene (SBZ 3). (**9–12**) Equatorial sections. (**13–16**) Sections parallel and close to the coiling axis. (**17–21**) External, lateral view of free shells showing the ornamentation

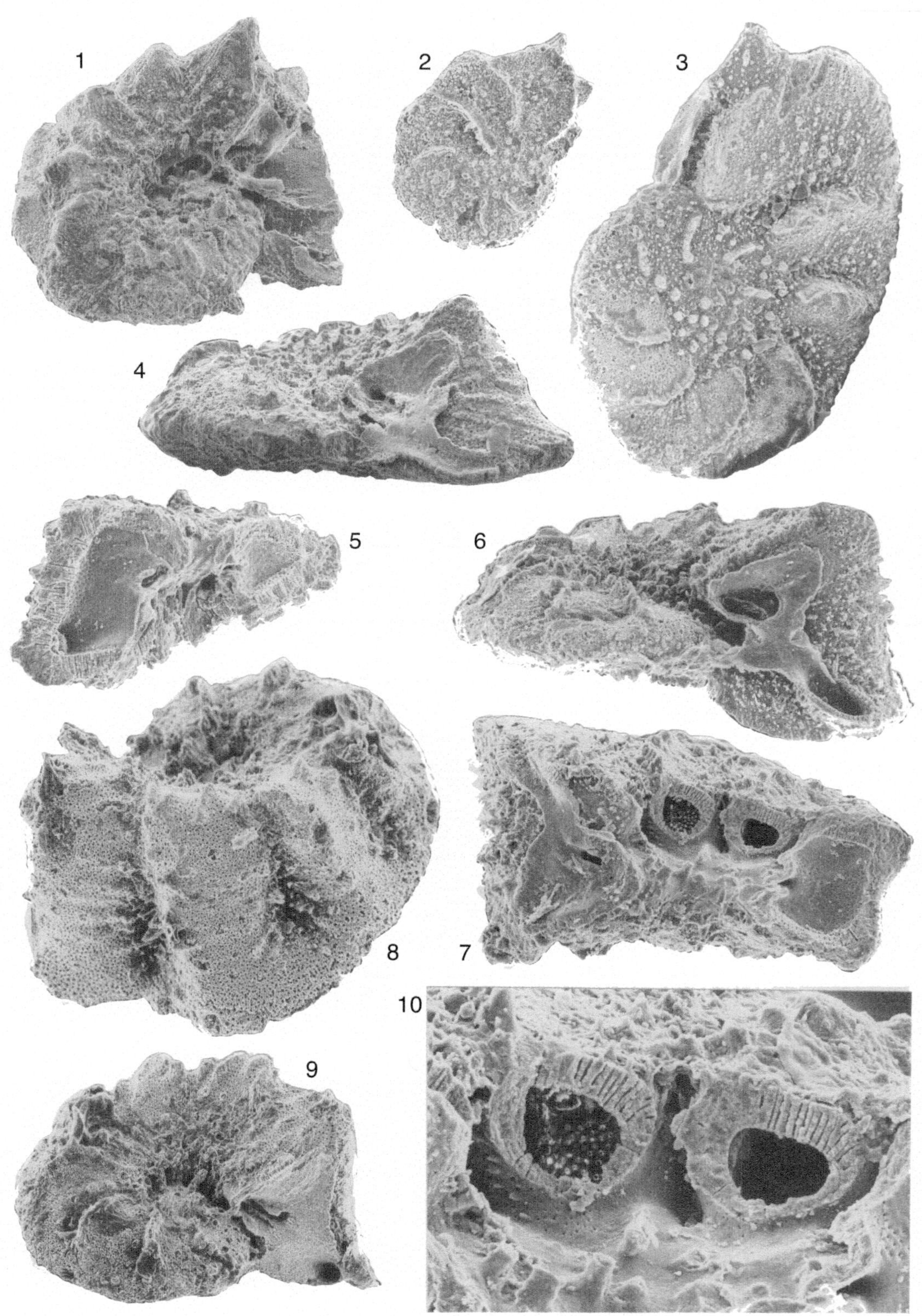

Plate 9.2 *Civrieuxia bicarinata* (Colom, 1954); SEM photographs; Tuilerie de Gan near Pau, Aquitaine occidental, southern France; Cuisian, Early Eocene (SBZ 10). (**1**) Oblique ventral view. (**2–3**) Dorsal view. (**4–5**) Peripheral views, ventral side up. (**6**) Fragment broken in proximally axial direction, ventral side up, showing

and a ventral umbilical depression. The chambers in adult whorls are covered by costae that continue the ponticuli subdividing the intraseptal interlocular space in the septal suture. The costae are confined to each slightly inflated chamber and run in oblique direction in respect to the course of the periphery. The latter is rounded, without keels, and sometimes gets doubled (Plate 9.1, Fig. 15) as in miscellaneids or *Daviesina langhami* (Plate 7.8, Figs. 1, 5–6). The bottom of the umbilical depression is ornate with papillae. They may obscure the orifices of the interlocular space.

There are 9–11 chambers in adult whorls.

9.1.2 Civrieuxia *Bermúdez, 1978*

Type species *Rotalia*? *palmerae* Cushman and Bermúdez, 1947

Civrieuxia bicarinata Colom (1954); Plate 9.2, Figs. 1–10.

Description: Comparatively small-sized benthics with a low-trochospiral shell. The dorsal side is flat, almost completely evolute, orned with limbate, curved chamber sutures and parallel costae or rows of papillae, aligned in oblique direction in respect to the course of the periphery. The latter is truncate, doubled in adult stages of growth. The angles forming the two peripheries are thickened but regularly perforate without keels (Plate 9.2, Fig. 5). The interlocular space opens widely in the chamber sutures in between the two peripheries with a single orifice protected by spikes (Plate 9.2, Fig. 7). These protect the orifices, larger than the foramina, from getting blocked by indigestible hardparts of the food, by diatom frustules for example. The spikes in the fossettes of Elphidiids have the same function.

The ventral side of the shell is dominated by an umbilical depression. On its bottom, the adaxial walls of the chambers partially remain open providing for all spiral chambers a permanent, direct passage into the ambient environment. Subsequent chambers are connected with a foramen of oval outline in areal position. The foramen has an abaxial lip and an adaxial "toothplate" that needs further study.

9.2 Rotaliid Taxa with Multiple Foramina

Multiple foramina are a rare feature in benthic foraminifera of the Paleogene. We do not know at what taxonomic level this feature has to be rated as diagnostic. The higher level systematics (on family level and higher) is uncertain; most genera are described from external features only, without understanding their significance in comparative anatomy and/or in functional morphology. Therefore we restrain from taxonomic emendations above subfamily level for all genera that we place out of the Rotaliidae. They are grouped here into different subfamilies according to their foraminal features and are considered as of *incertae sedis* on all higher systematic levels. *Scarificatina* is a minute species from the Lower Paleocene (SBZ 1). Y. Tambareau had discovered exceptionally well preserved, empty shells in borehole material from the region of Mons. SEM pictures are published here as first contribution to this group, in order to demonstrate the existence of multiple foramina often poorly preserved and difficult to see in thin sections. In most cases, they have not been recognized in their systematic descriptions. The new subfamily Cincoriolinae is here assessed and groups the genera *Cincoriola* Haque, 1958 and *Rahaghia* n. gen. (see 2.6 *Identification Key for Some Rotaliid Shells with Multiple Areal Foramina*).

Plate 9.2 (Continued) foramen in frontal view and perforation of the angular peripheries. (**7**) Oblique peripheral view, ventral side up, shows main orifice of the intraseptal interlocular space. (**8**) Oblique umbilical view, ventral side up. (**9**) Shell broken in approximately axial direction, off center, ventral side down, showing foramina in oblique view. (**10**) Detail of Fig. (**9**) showing the perforation of the chamber walls and the spiking of the interlocular space

Plate 9.3 *Scarificatina reinholdi* (Pozaryska and Szczechura, 1970), from Mons, Belgium, Paleocene, showing linear, single or multiple, parallel umbilical crests and multiple apertures with peristomes. The exceptionally well preserved material was collected and placed at my disposal for study by Yvette Tambareau (Toulouse). Abbreviations: *f* foramen, *p* pores, *up* umbilical plate, *ucr* umbilical crest, *lh?* loop hole, *pst* peristome, *sf* septal face, *p* pores, *rp* radial pustules, *s* septum

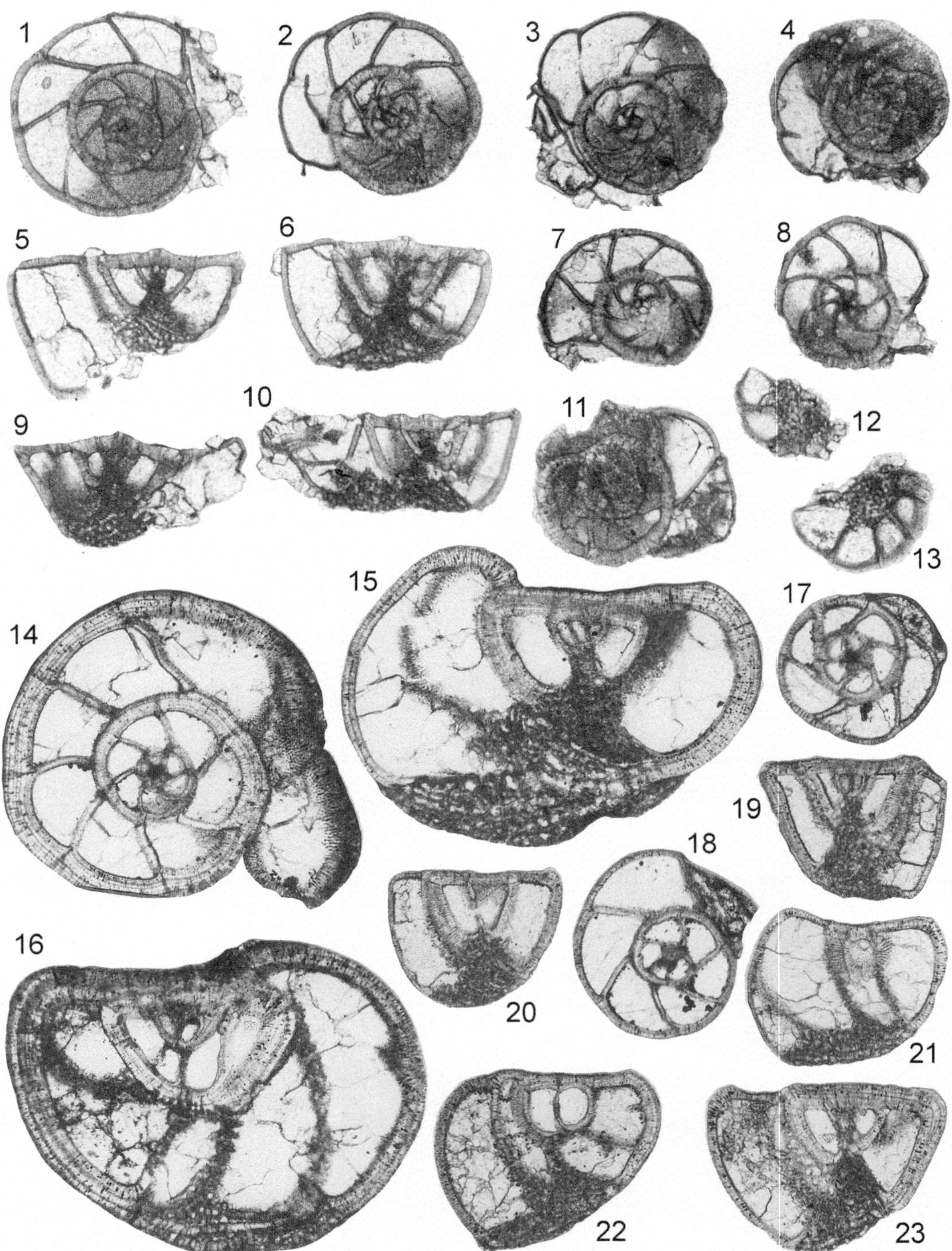

Plate 9.4 (**1–13**) *Cincoriola patalensis* Haque, 1956; sample 6, Lafarge Quarry, Aquitaine, France; Paleocene (SBZ 3). (**1–4**, **7–8**, **11**) Sections perpendicular to coiling axis. (**5–6**, **9–10**) Axial sections. (**12–13**) Tangential sections. (**14–23**) *Cincoriola ovoidea* (Haque, 1958); sample 6, Lafarge Quarry, Aquitaine, France; Paleocene (SBZ 3). (**14**, **17–18**) Sections perpendicular to coiling axis, microspheric specimen (**14**) and probably megalospheric forms (**17–18**). (**15–16**) Microspheric specimens, axial sections. (**19–21**, **23**) Adaxials sections. (**22**) Megalospheric specimen, axial centered section

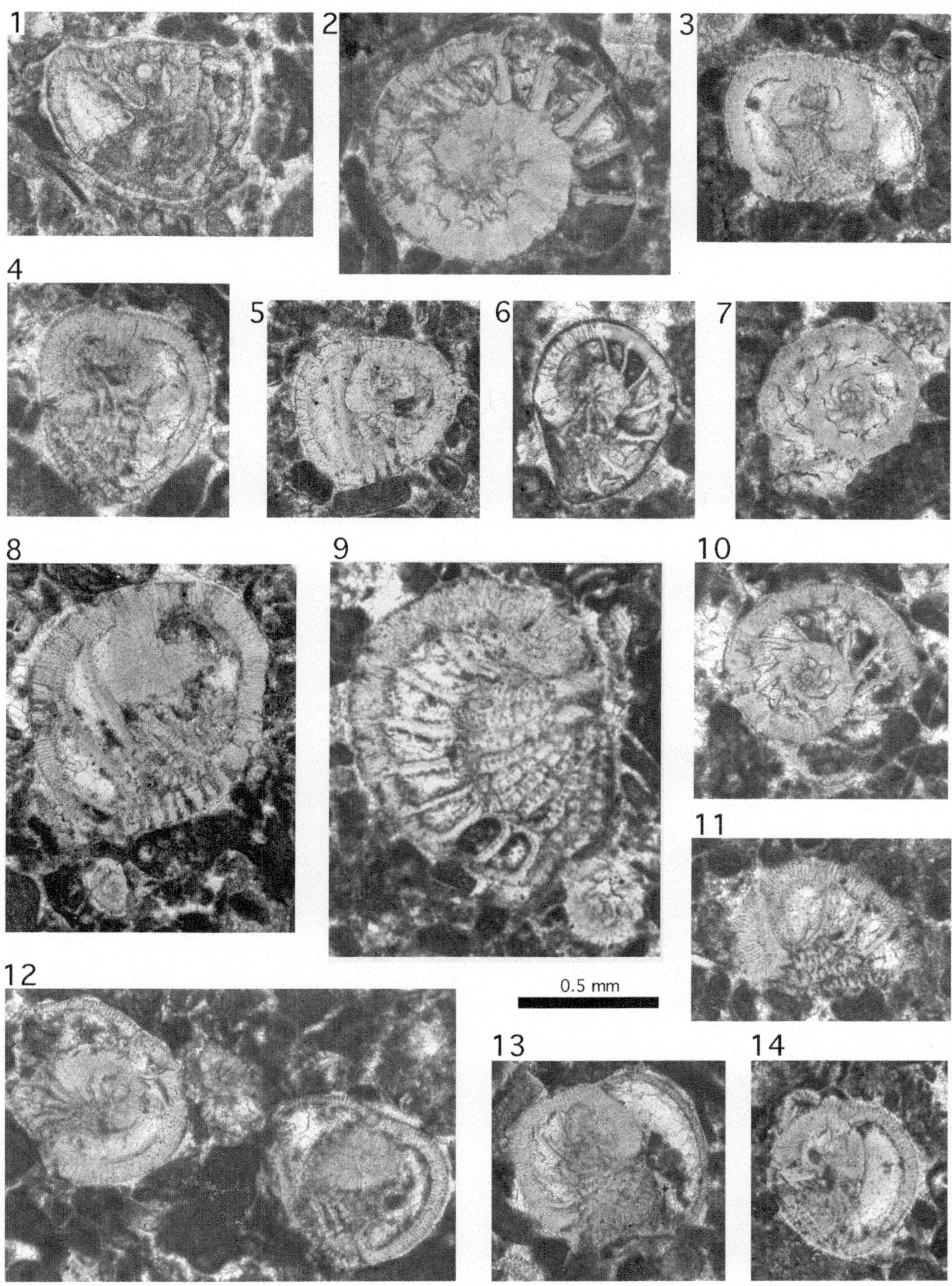

Plate 9.5 *Cincoriola ovoidea* (Haque, 1958); sample 6, Lafarge Quarry, Aquitaine, France; Paleocene (SBZ 3). (**1**, **3**, **5**) Axial sections. (**2**, **6**) Uncentered section perpendicular to coiling axis. (**4**, **8–9**, **11–14**) Oblique sections. (**7**, **10**) Centered section perpendicular to coiling axis

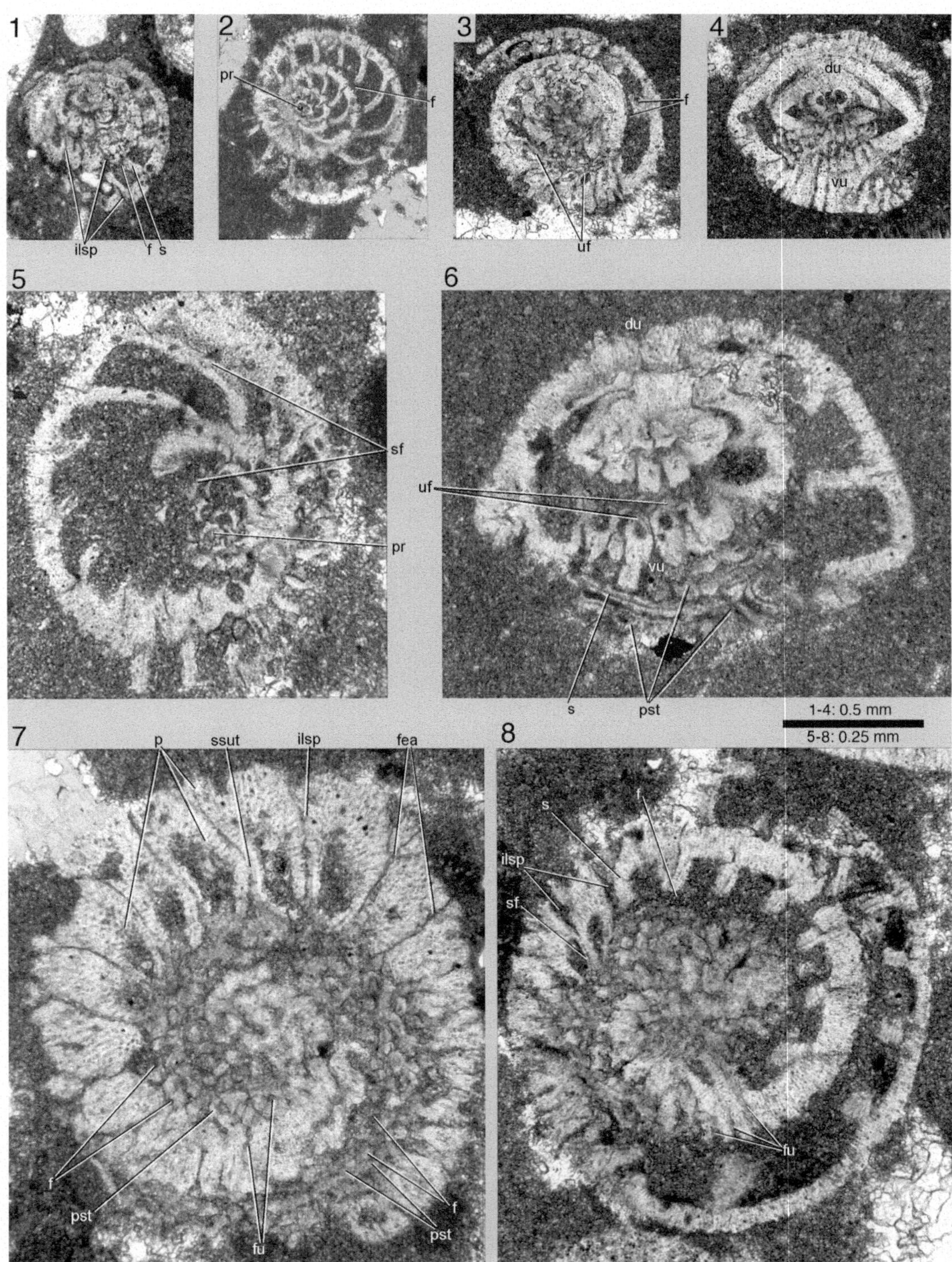

Plate 9.6 (**1**, **3**, **8**) *Rahaghia melona* (Rahaghi, 1976). (**1**) Almost centered, oblique section with an inclination of about 80° in respect to the coiling axis. (**3**) Oblique section with an inclination of about 45° in respect to the coiling axis. (**8**) Section perpendicular to the coiling axis and tangential to the ventral umbilicus of the penultimate whorl; note the spiking of the funnels (*fu*) and the oblique sections of the peristomes in the septum. (**2**, **4–7**) *Rahaghia khorassanica* (Rahaghi, 1976). (**2**) Slightly oblique, equatorial section; note long, arcuate septa. (**4**) Section parallel to nearby coiling axis showing dorsal and ventral umbilicus. (**5**) Microspheric specimen, oblique

9.2.1 *Scarificatina* Moorkens, 1982

Type species: *Boldia reinholdi* Pozaryska and Szczechura, 1970

Remarks: Almost planispiral, lamellar-perforate shells of small size but with a coarse perforation. The dorsal surface is flat or concave; the ventral surface is convex with a flattened "umbilical" cover. The periphery is angular or rounded, corresponding to the moderately inflated spiral chambers. These have septa that are inclined slightly backwards. There is no sign of a septal flap. In the area adjacent to the previous whorl, the septum bears a group of numerous foramina that are irregularly disposed and bear each a circular peristome. In addition, there seems to be a slit in the spiral suture of the last few chambers that may represent a sutural aperture similar to species of the genera *Cibicides* or *Lobatula* (Cibicididae).

The peculiar umbilical architecture is most unusual by its parallel umbilical crests that run in parallel lines over the entire umbilical extension, without any apparent relationship to the radial septa of the spiral chambers. To my knowledge there is only one more group with a similar, linear umbilical architecture: *Soriella* Haque, 1960 included by Loeblich and Tappan (1987) in the Rotaliinae. Their architecture however needs to be analysed in detail.

Each umbilical crest of *Scarificatina* has a median wall composed of several lamellae. The median wall is folded in alternating sites on both sides of the crest forming halfpipes in alternating position (Plate 9.3, Fig. 5). These evidently protect protoplasmic strands extruding at their base from the umbilical cavity below through orifices in alternating position. The space in between the crests are covered by pustules in radial rows as in *Glabratellina* (Hottinger 2006, Fig. 7a). The pustule cover of the shell's face provides mechanical support for the pseudopods that have to keep two shells face to face during a plastogamic process of reproduction. The umbilical crests may canalize the movement of the gametes towards their partner in the narrow space between the two shell faces.

Scarificatina reinholdi (Pozaryska and Szczechura, 1970); Plate 9.3.

1960 *Boldia carinata* Cushman and Bérmudez, 1947—Drooger, partim, p. 454, pl. 1, figs. 12–13, non figs. 10, 11.

1970 *Boldia reinhold*—Pozaryska and Szczechura, p. 99, pl. 2, figs. 3–4; pl. 3, figs. 1–4.

1977 *Scarificatina reinholdi* (Pozaryska and Szczechura)—Guillevin, pl. 11, figs. 3–4; pl. 12, figs. 1–3.

1982 *Scarificatina reinholdi* (Pozaryska and Szczechura)—Moorkens, pl. 9, figs. 5–7.

1987 *Scarificatina reinholdi* (Pozaryska and Szczechura)—Loeblich and Tappan, p. 653, pl. 713, figs. 10–14.

2000 *Scarificatina* sp.—Peybernès et al., p. 46, fig. 6/9.

Remarks: the genus *Scarificatina* is so far monospecific. Therefore, the generic description is valid also for the type species. There are 9 chambers in the last whorl, about 20 foramina in late septa, each with a peristome. The number of parallel umbilical crests varies from one to five. The proloculus has not been observed.

9.2.2 *Cincoriola* Haque, 1958

Type species: *Punjabia ovoidea* Haque, 1956

Plate 9.6 (Continued) centered section inclined for more than 45° in respect to coiling axis. (**6**) Oblique section inclined for less than 45° in respect to coiling axis. (**7**) Section almost perpendicular to coiling axis, tangential to the ventral umbilicus. All specimens of this plate are topotypes from the Paleocene Chehel-Kaman Formation, Kopeh-Dagh. North-eastern Iran collected by A. Rahaghi and deposited in the Natural History Museum (London). Abbreviations: *f* foramen, *p* pores, *pst* peristome, *sf* septal flap, *uf* umbilical flap, *fea* feathering of interlocular space, *pr* proloculus, *ilsp* intraseptal interlocular space, *fu* funnel, *ssut* septal suture, *s* septum, *du* dorsal umbilicus, *vu* ventral umbilicus

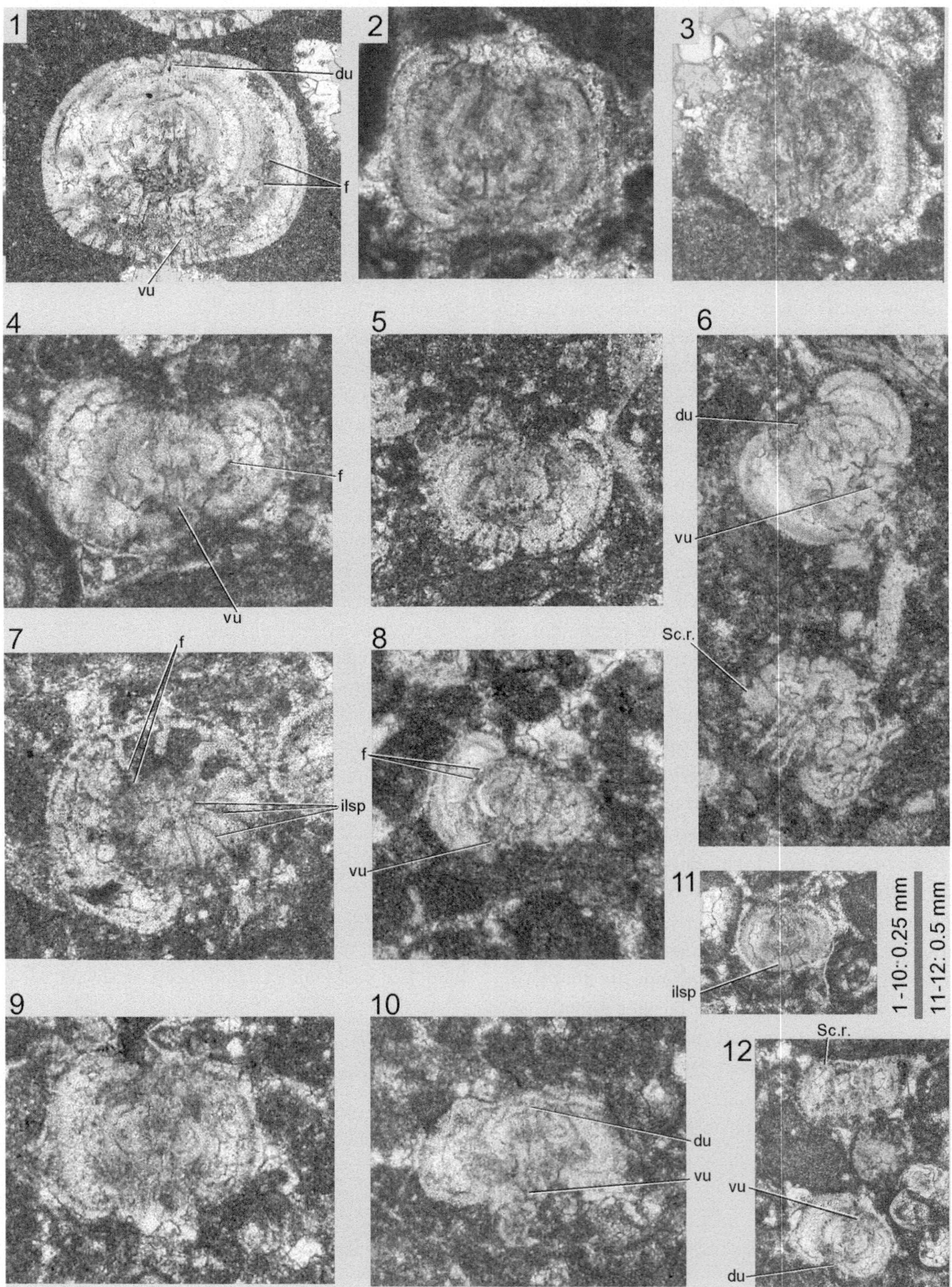

Plate 9.7 (**1**) *Rahaghia melona* (Rahaghi, 1976); topotype from Chehel-Kaman Formation, Kopeh-Dagh, north-eastern Iran, collected by A. Rahaghi and deposited in the Natural History Museum (London); this occurrence is dated according to Rahaghi (1976) as Paleocene. (**1**) Section parallel to coiling axis. (**2–12**) *Rahaghia biumbilicata* n. sp.; specimens collected by Y. Tambareau in the western Aquitaine, southern France; Paleocene

Description: Lamellar-perforate shells with chamber septa lacking any septal flaps or intraseptal interlocular spaces. In a dense vortex, the septa cover the entire adaxial ("umbilical") area and bear numerous, closely spaced foramina, each with a circular peristome (Plate 9.4, Figs. 2–3). The foramina of subsequent septa are aligned with their axes in loose, parallel spirals. The alignment is visible only in oblique sections of an appropriate inclination (Plate 9.5, Fig. 9). The most interiomarginal row of foramina may have a larger diameter in respect to foramina in more areal position or be transformed into a long interiomarginal slit.

Cincoriola ovoidea (Haque, 1958); Plate 9.4, Figs. 14–23; Plate 9.5, Figs. 1–14.

1956 *Punjabia ovoidea*—Haque, p. 153, pl. 30, figs. 1a–c.
1958 *Cincoriola ovoidea* (Haque)—Haque, p. 103.
1998 *Soriella bitlisica*—Sirel, (pars), p. 76; pl. 35, figs. 5, 7, 9–14.
1998 *Cincoriola* ? sp.—Sirel, p. 77; pl. 35, figs. 15–17.
1998 ?*Kathina melona* Rahaghi—Accordi et al., p. 174, pl. 3, fig. c.
1998 *Soriella* cf. *bitlisica* Sirel—Accordi et al., p. 190, pl. 11, fig. 1.
1998 ?*Kathina melona* (Rahaghi)—Accordi et al., p. 196, pl. 14, figs. 1–3.

Cincoriola patalensis Haque, 1956; Plate 9.4, figs. 1–13.

1956 *Cincoriola patalensi*—Haque, p. 154, pl. 34, figs. 1a–c.
1998 *Soriella bitlisica*—Sirel, (pars), p. 76, pl. 35, figs. 2–4, 6.

Description: Trochospiral *Cincoriola* with a flat dorsal and a highly convex ventral side. The equatorial to axial diameter ratio is 1.3 to 1.7. The periphery is angular, keeled. The septa are often limbate, dorsally straight, much inclined backwards, and ventrally radial. There is no intraseptal interlocular space. The umbilicus is narrow, covered by a vortex of septa with numerous foramina, bearing comparatively small peristomes.

9.2.3 *Rahaghia* n. gen.

Type species: *Laffitteina melona* Rahaghi, 1976

Description: Spherical or lenticular shells composed of almost planispiral chambers. The poles of the shell are narrow areas around the axis of coiling, smaller on the dorsal, a little wider on the ventral side. They do not produce any polar depressions. The septa are radial or moderately backwards inclined. Around the ventral umbonal area they are transformed by an abrupt torsion into a vortex that covers the ventral adaxial area. In that area, the septa are perforated by a dense group of foramina with peristomes. The peristomes of successive chambers may fuse to an umbonal mass, perforated, however, by numerous parallel funnels. There are neither folia nor umbilical plates but, similar to *Laffitteina*, there is a closure of the chamber lumen against the previous septum in the vortex. The radial part of the septa exhibits an undivided interlocular space that communicates with the ambient environment by narrow slits

Plate 9.7 (Continued) (SBZ 2–3). (**2–5**) Sections parallel or slightly oblique in respect to the nearby coiling axis. (**6**) Holotype, oblique section inclined for about 30° in respect to the coiling axis. Note the ventral tips of the totally involute chambers in the ventral umbilicus; the holotype is associated with an oblique section of (Sc. r.) *Scarificatina reinholdi* (Pozaryska and Szczechura, 1970). (**7**) Section perpendicular to the coiling axis showing the stout, almost radial septa, apparently without preservation of the intraseptal space or of the suture between primary bilamellar chamber wall and septal flap. (**8–12**) Sections more or less parallel to the coiling axis, associated in 12 with *Scarificatina reinholdi* (Sc. r.). Abbreviations: *f* foramen, *vu* ventral umbiliculus, *du* dorsal umbiliculus, *ilsp* intraseptal interlocular space

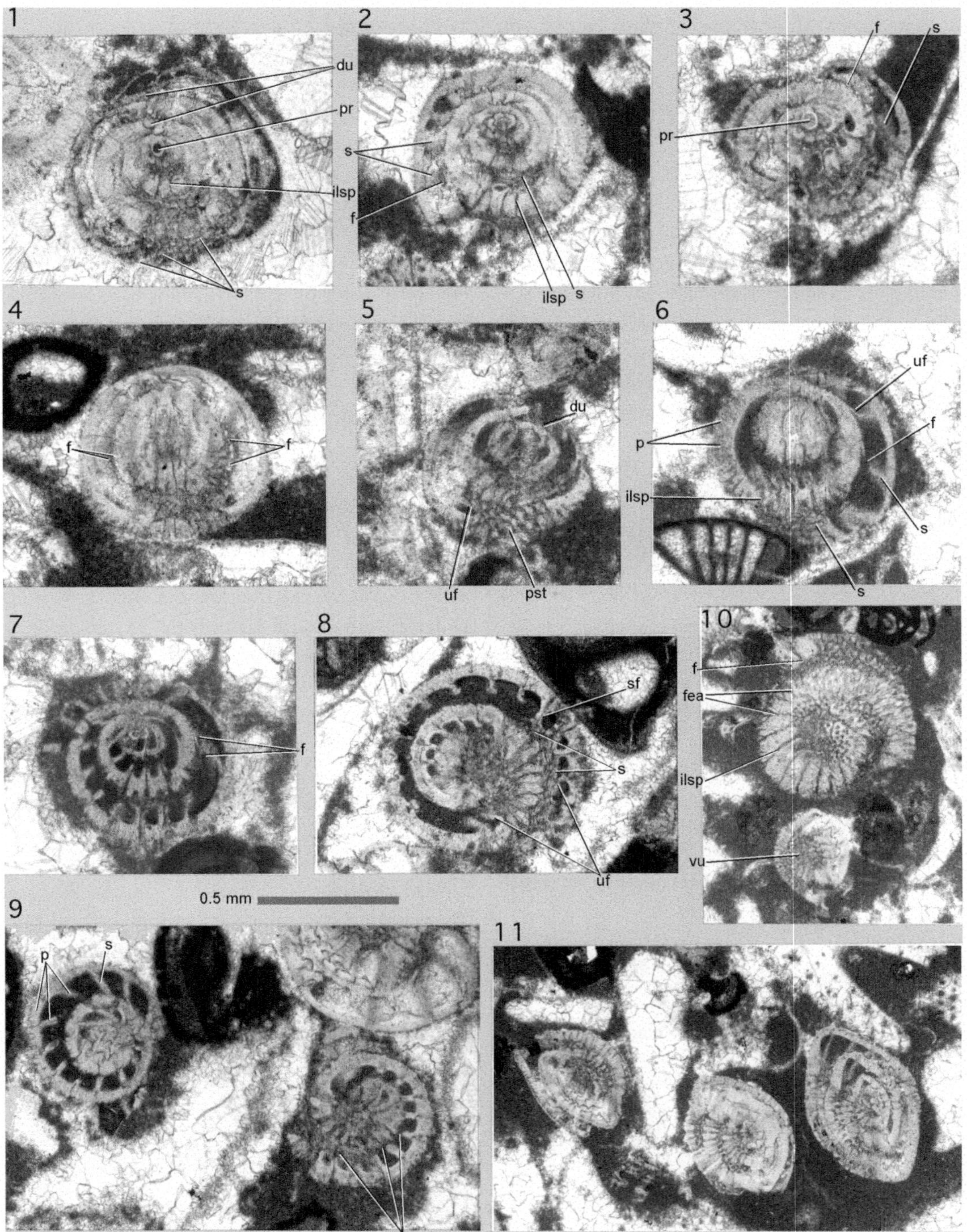

Plate 9.8 (**1–4**, **6–9**) *Rahaghia melona* (Rahaghi, 1976); topotypes from Chehel-Kaman Formation, Kopet Dagh, north-eastern Iran, collected by A. Rahaghi and deposited in the Natural History Museum (London); this occurrence is dated according to Rahaghi (1976) as Paleocene. (**1–3**, **7**) Oblique, centered sections with an inclination of less than 45° in respect to the coiling axis. (**4**) Section parallel to nearby coiling axis. (**6**) Oblique section far off center, inclined for less than 45° in respect to the coiling axis. (**8–9**) Oblique sections off center inclined for 60°–80° in respect to the coiling axis. Specimen (**8**) may be the holotype as far as the quality of the photograph in Rahaghi 1976 allows identifying an individual. (**10–11**) *Rahaghia khorassanica* (Rahaghi, 1976); topotypes from

in the otherwise smooth shell surface. A dimorphism of generations is observed only in a single case (Plate 9.7, Fig. 5), restricted to the first two nepionic whorls.

Remarks: *Rahaghia melona* and *R. khorassanica* have been found in material from Kopet Dagh, North-Eastern Iran, in a region that may represent the shores of the Asiatic continent. These shallow water deposits represent a marginal area in respect to the Central Neotethys. They may be an endemic group related to *Cincoriola* but sufficiently different to justify a new generic name. In Rahaghi's original material, deposited in the British Museum of Natural History, there are no diagnostic associates that may prove directly the Paleocene age except some sections of *Rotorbinella hermi* von Hillebrandt, 1962.

***Rahaghia biumbilicata* n. sp.**; Plate 9.7, Figs. 2–12.

1998 "*Protelphidiid*"—Accordi et al., p. 188, pl. 10, fig. 12.

Holotype: specimen illustrated in Plate 9.7, Fig. 6.

Type locality and type level: western Aquitaine, southern France; Paleocene (SBZ 2–SBZ 3).

Derivation of name: double, dorsal and ventral umbilicus.

Diagnosis: Lamellar-perforate, involute, almost planispiral shells that exhibit similar structural elements as found in the type species of the genus, but with an additional umbilical depression on the dorsal side of the shell. Both umbilici may be filled by some free, unfused piles that are growing on the shoulders of the adaxial chamber walls. These are thickened and admit between adjacent chambers a short, slitlike interlocular space. There are 11 chambers in the last whorl. The very small shells are poorly preserved; their description can only be rather approximate.

Rahaghia melona (Rahaghi, 1976); Plate 9.6, Figs. 1, 3, 8; Plate 9.7, Fig. 1; Plate 9.8, Figs. 1–9.

1976 *Laffitteina melona*—Rahaghi, p. 40; pl. 2, figs. 6–12.

1983 *Laffitteina melona* Rahaghi—Rahaghi, no text; pl. 38, fig. 11.

1998 ?*Kathina melona* (Rahaghi)—Accordi et al., pl. 14, figs. 1–3.

Description: Smooth, spherical *Rahaghia* with involute spiral chambers in the shape of orange wedges. The polar area is without adaxial depression. The equatorial to axial diameter ratio varies from 1.2 to 1.4. 12–17 chambers are counted in late adult whorls.

Rahaghia khorassanica (Rahaghi, 1976); Plate 9.6, Figs. 2, 4–7; Plate 9.8, Figs. 10–11; Plate 9.9, Figs. 1–11.

1976 *Laffitteina khorassanica*—Rahaghi, p. 37, pl. 1, figs. 16–19.

1978 *Laffitteina khorassanica* Rahaghi—Rahaghi, pl. 38, fig. 1.

Description: Lenticular, clearly trochospiral-involute *Rahaghia* with a narrowly rounded to angular periphery. The dorsal side is less convex than ventral side. The equatorial to axial diameter ratio varies from 1.3 to 1.5. There are 25–28 chambers in the last adult whorl.

Plate 9.8 (Continued) the Paleocene Chehel-Kaman Formation, Kopet Dagh; north-eastern Iran collected by A. Rahaghi and deposited in the Natural History Museum (London). (**10**) Sections tangential to ventral umbilicus, inclined for about 70° in respect to the coiling axis. (**11**) Several oblique sections. Abbreviations: *f* foramen, *p* pores, *pst* peristome, *uf* umbilical flap, *pr* proloculus, *ilsp* intraseptal interlocular space, *du* dorsal umbo, *s* septum, *sf* septal flap, *fea* feathering of interlocular space, *vu* ventral umbo

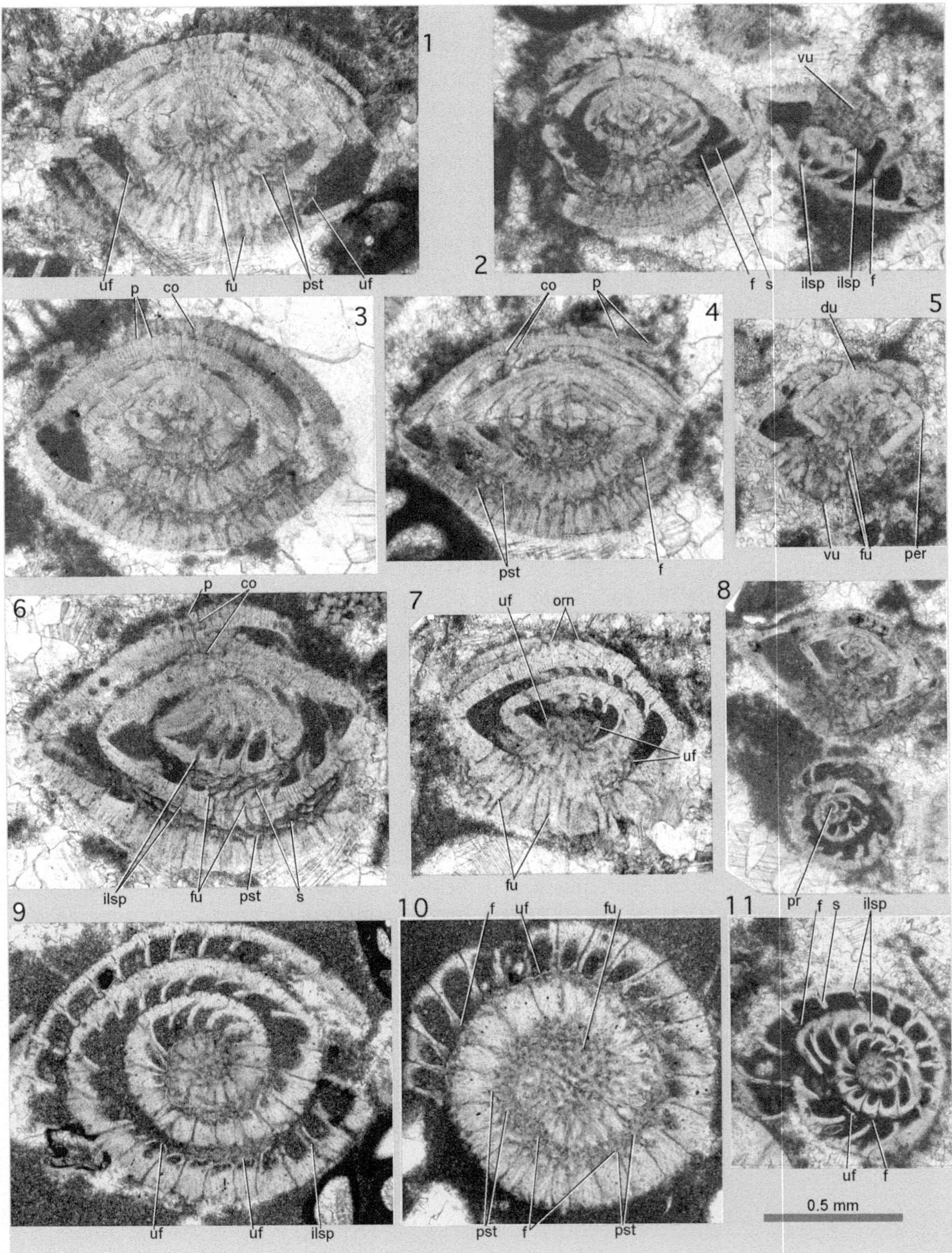

Plate 9.9 *Rahaghia khorassanica* (Rahaghi, 1976); topotypes from the Paleocene Chehel-Kaman Formation, Kopet Dagh, north-eastern Iran collected by A. Rahaghi and deposited in the Natural History Museum (London). (**1**) Approximately axial section. (**2**) (*left*) Oblique section inclined for less than 45° in respect to the coiling axis. (**2**) (*right*) Oblique section far from center, almost parallel to the coiling axis; note the upside down orientation of the section on the Plate. (**3**) Oblique section inclined for less than 45° in respect to the coiling axis. (**4**) Transverse section parallel to the coiling axis. (**5**) Near-axial section. (**6**) Oblique section far from center, inclined for about 20°

9.3 Some Rotaliids of the Family Victoriellidae Chapman and Crespin, 1930

In the family Victoriellidae, a considerable number of genera show an attachment surface on the spiral side of their trochospiral test. Lamellar-perforate foraminifera develop their surface of attachment always on the spiral, dorsal side of the shell. Usually, small orifices of an interlocular canal system are distributed over the attachment surface, probably in order to provide the glue for fixing the shell to its substrate. Thus, the ventral side of the test functioning as face of the shell, keeps the main apertures free during growth. Many of them are adapted to fishing food particles in the open water in contrast to so many benthics living face down to collect the food from the bottom of the sea. A permanent attachment would block the access to their apertures. Therefore, the specimens illustrated here are mounted face downward on the plates.

With regard to their shell architecture, numerous questions remain unsolved because the internal morphology of most genera is inadequately described. A suitable terminology of the structural elements compatible with better known benthic families remains therefore out of reach. In this paper we limit the discussion to two species of the genus *Gyroidinella* in order to demonstrate the difference from the true rotaliids.

9.3.1 *Gyroidinella* Le Calvez, 1949

Type species: *Gyroidinella magna* Le Calvez, 1949

Description: Trochospiral shell with a flattened to concave dorsal side that in several species obviously serves as surface of permanent attachment. The septa are inclined backwards on the dorsal side of the chamber, radial on the ventral side. The foramen is an interiomarginal, short, low arch restricted to an area shortly below the periphery of the shell. It has a rim directed backwards. The septum is double and admits a broad triangular interlocular space in the adaxial ventral chamber suture. Below the ventral end of the interiomarginal foramen the successive interlocular spaces fuse to form a large spiral canal. This communicates with the dorsal extension of the interlocular space, extending in the adaxial chamber sutures to open on the dorsal shell surface. Likewise, the adaxial sutural interlocular space runs into an open "umbilicus" (pseudumbilicus in the sense of Reiss, 1963, p. 40 and 83, pl. 7, Fig. 9) filled with a coarse pillared construction of unknown lamellar composition. The umbilical pillars may fuse to a grit covering the umbilical cavity. There are neither folia nor foliar extensions as in sakesarias of the same habit.

Gyroidinella laevis (Grimsdale, 1952); Plate 9.10, Figs 1–13.

1952 *Eorupertia incrassata* (Uhlig) var. *laevis*—Grimsdale, p. 239, pl. 20, figs. 15–21.

1971 *Eorupertia cristata laevis* (Grimsdale)—Boulanger and Poignant, pl. 3, figs. 5–6.

Remarks: This species is very close to *Gyroidinella*'s type species *G. magna* but also close to *G. eocaenica* discussed below. It is characterized by an architecture typical for the genus and by a smooth surface all over the subspherical shell with a rounded periphery. The equatorial to axial diameter ratio varies from 1.2 to 1.4. 11–12 chambers are counted in adult whorls. Microspheric specimens have not been observed.

Plate 9.9 (Continued) in respect to the coiling axis. (**7**) Section parallel to the coiling axis. (**8**) (*top*) Oblique section. (**8**) (*bottom*) Centered section with an inclination of about 75° in respect to the coiling axis. (**9**) Oblique section near to center, inclined for about 60° in respect to the coiling axis. (**10**) Section perpendicular to coiling axis, tangential to the ventral umbilicus of penultimate whorl. (**11**) Almost centered section of megalospheric specimen inclined for about 80° in respect to the coiling axis. Abbreviations: *f* foramen, *p* pores, *pst* peristome, *orn* ornamentation, *uf* umbilical flap, *fu* funnel. *pr* proloculus, *ilsp* intraseptal interlocular space, *vu* ventral umbilicus, *du* dorsal umbilicus, *co* canal orifice, *s* septum, *per* periphery

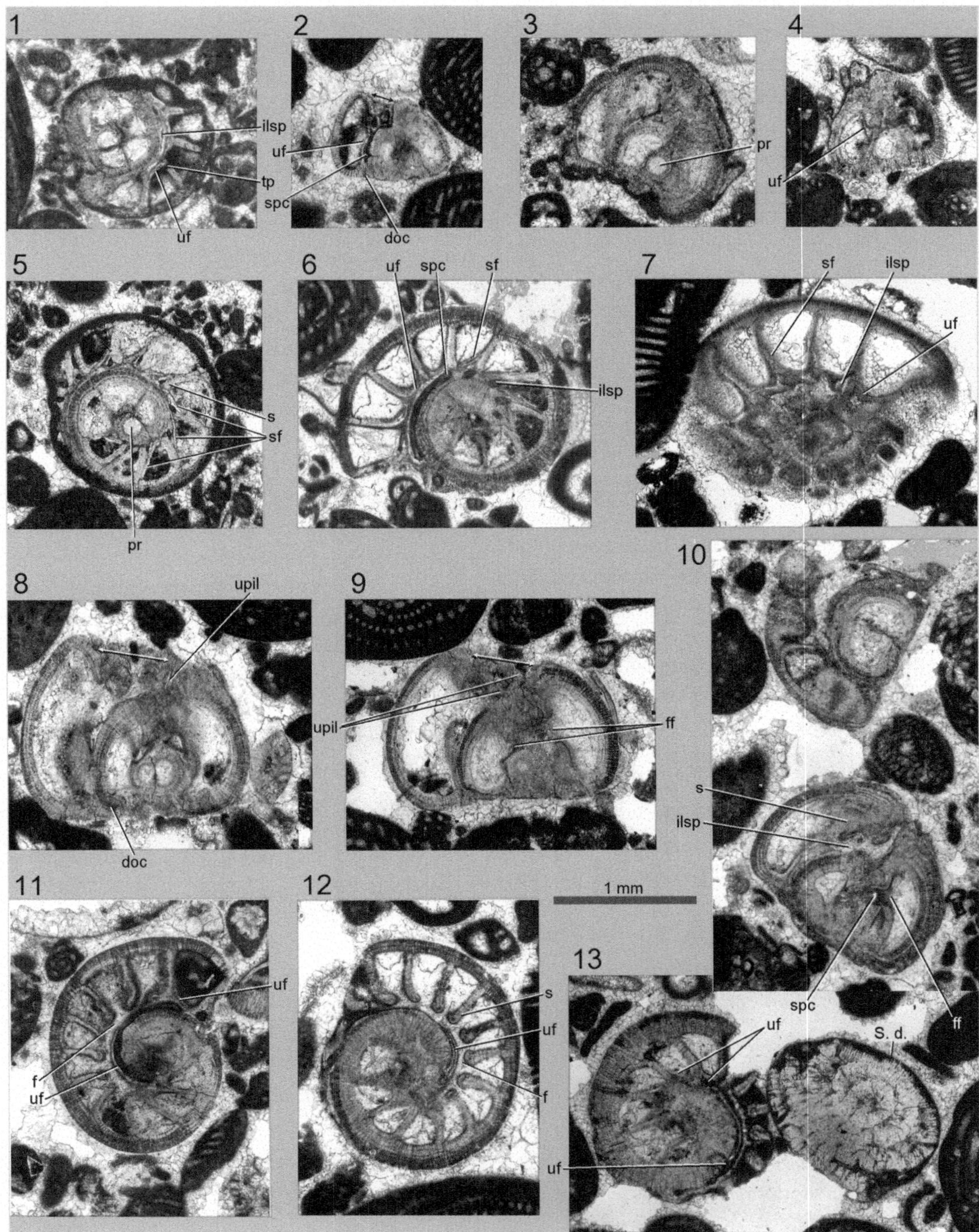

Plate 9.10 *Gyroidinella laevis* (Grimsdale, 1952); samples AS4340–AS4344, Monte Gargano, Puglia, Italy, collected by P. De Castro and deposited in the collections of the Dipartimento di Scienze della Terra, University of Naples "Federico II", Italy; Eocene (SBZ 10–13). (**1**) Centered section perpendicular to the coiling axis. (**2**) Section parallel to the nearby coiling axis. (**3**) Oblique section inclined for less than 45° in respect to the coiling axis. (**4**) Section parallel to the nearby coiling axis. (**5–6**) Centered sections perpendicular to the coiling axis, close to the spiral (dorsal) shell surface. (**7**) Section inclined for about 70° in respect to the coiling axis, near to the umbilical end of the last whorl's chambers. (**8–9**) Axial sections. (**10**) (*top*) Oblique section. (**10**) (*bottom*) Axial section. (**11–12, 13**) (*left*) Sections perpendicular to coiling axis, proximally at the level of the shell's periphery. (**13**) (*right*) Oblique section of the associated (S. d.) *Slovenites decastroi* n. sp. Abbreviations: *f* foramen, *s* septum, *uf* umbilical flap, *ilsp* intraseptal interlocular space, *spc* spiral canal, *sf* septal flap, *doc* dorsal canal orifice, *tp* toothplate, *pr* proloculus, *upil* umbilical pile

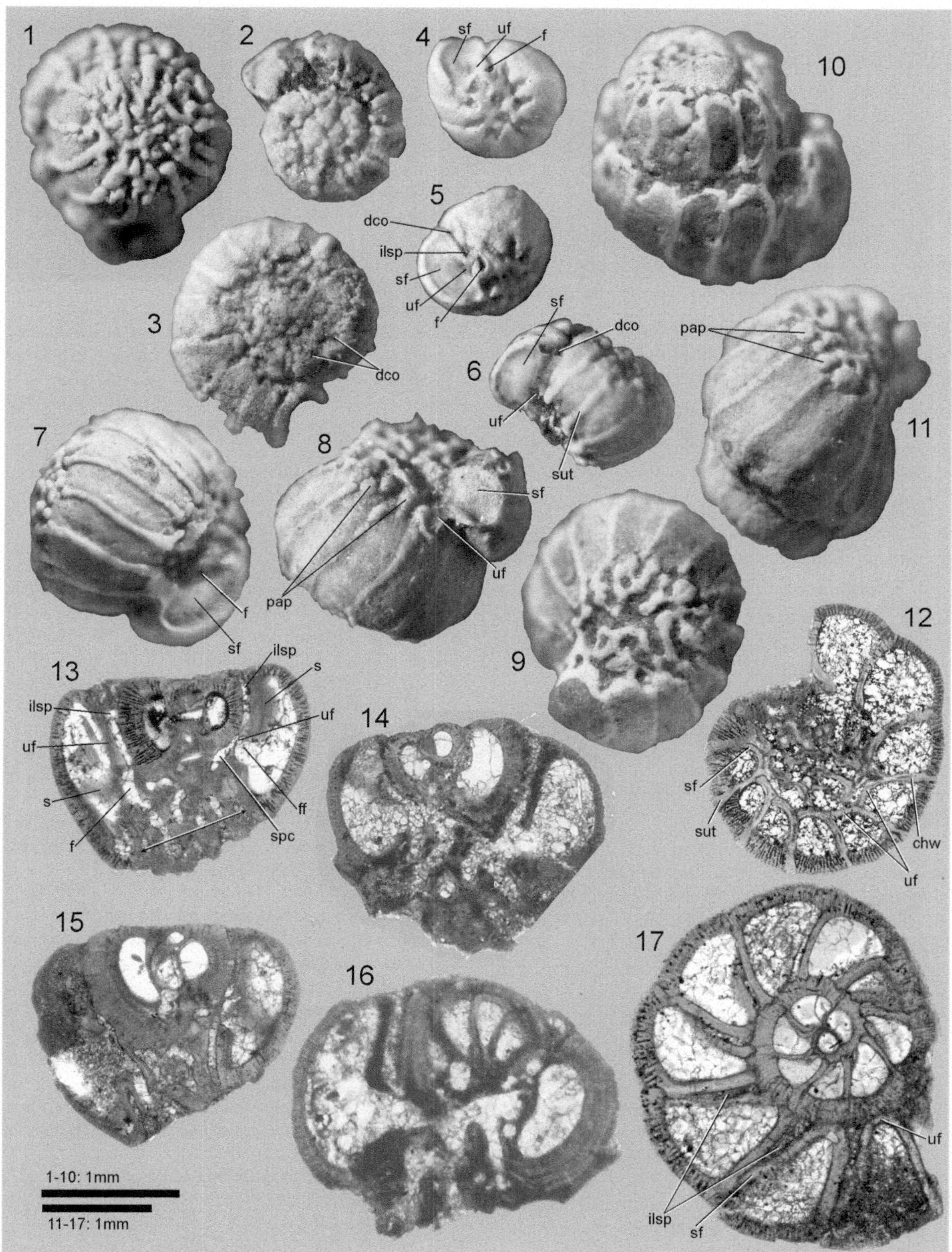

Plate 9.11 (**1–16**) *Gyroidinella eocaenica* (Sacal and Debourle, 1957). (**1–11**) External views showing the pear-shaped to subspherical, high-trochospiral shells. (**12**) Nearly equatorial section. (**13–16**) Oblique to tangential sections. (**17**) *Gyroidinella magna* Le Calvez, 1949; equatorial section

Gyroidinella eocaenica (Sacal and Debourle, 1957); Plate 9.11, Figs. 1–16.

1957 *Asanoina eocaenica*—Sacal and Debourle, p. 40, pl. 17, Fig. 11.

2000 *Sakesaria eocaenica* (Sacal and Debourle)—Sztràkos, p. 114, pl. 7, fig. 7; pl. 18, figs. 5–8.

Description: Pear-shaped to subspherical, high-trochospiral shells with strongly limbate whorl and cameral sutures. The shells are free, lacking any attachment surface. The spiral side of the shell is convex or flattened, ventral side involute, pseudoumbilicate. The umbilical opening is inclined in respect to the coiling axis of the shell. The inclination corresponds to the location of the last chamber in respect to the coiling axis. The adaxial dorsal tips of the chambers support each a single, scarcely or non-perforate, large papilla flanked by beads on the dorsal chamber and whorl sutures. The triangular interlocular space forms a dorsal series of canal orifices between the beads of the dorsal ornamentation. The pseudumbilicus is covered by a grid consisting of fused papilla heads. There are 12–14 chambers in adult whorls.

References

Accordi G, Carbone F, Pignatti J (1998) Depositional history of a Paleogene ramp (Western Cephalonia, Ionian islands, Greece). Geol Romana 34:131–205

Bermúdez PJ (1952) Estudio sistematico de los foraminíferos rotaliformes. Bol Geol Caracas 2(4): 230 pp, 35 pls

Bermúdez PJ (1978) Un género nuevo de foraminifèro de la familia Rotaliidae y otros géneros relacionados en la Region Caribe Antillana. Rev Espan Micropaleontol 10(2):191–204

Boulanger D, Poignant A (1971) Sur quelques foraminifères fixés de l'Eocène supérieur et de l'Oligocène en Aquitaine méridionale. Rev Micropaléontol 14(3): 167–176

Chapman F, Crespin I (1930) Rare foraminifera from deep borings in the Victorian Tertiaries. Part 2. R Soc Victoria, Proc, Melbourne 43(1):110–114

Colom G (1954) Estudios de las biozonas con foraminiferos del Terciario de Alicante. Bol Inst Geol Min Esp 66:101–279, 135 pls

Cushman JA (1951) Paleocene foraminifera of the Gulf coastal region of the United States and adjacents areas. USGS Professional Paper, 232,75 pp, Washington

Cushman JA, Bermúdez PJ (1947) Some Cuban Foraminifera of the genus *Rotalia*. Cushman Lab Foram Res, Sharon, Massachusetts 23(2):23–29, 5–10 pls

Drooger CW (1960) Some Early Rotaliid Foraminifera I-III. Proc K Ned Akad Wet Amst B 63(2):287–334

Grimsdale TF (1952) Cretaceous and Tertiary Foraminifera from the Middle East. Bull Brit Mus (Nat Hist) 1: 223–247

Guillevin Y (1977) Contribution à l'étude des Foraminifères du Montien du Bassin de Paris. Cah Micropaléontol, 79 pp

Haque A (1956) The smaller foraminifera of the Ranikot and the Laki of the Nammal Gorge, Salt Range. Palaeontol pakistanica, Mem Geol Sur Pak 1:1–228, 35 pls

Haque A (1958) *Cincoriola*, a new generic name for *Punjabia* Haque, 1956. Contr Cushman Lab Foram Res 9(4):103

Haque AFMM (1960) Some Middle- to Late Eocene smaller foraminfiera from the Sor Range, Quetta District, West Pakistan. Mem Geol Surv Pakistan 2 (1959):1–79, Karachi

Hofker J senior (1962) Correlation of the Chalk of Maestricht (Type Maestrichtian) with the Danske Kalk of Denmark (Type Danian), the Stratigraphic position of the Type Montian, and the planktonic Foraminiferal Faunal Break. J Paleontol 36:1091–1089

Hottinger L (2006) The "face" of benthic foraminifera. Boll Soc Paleontol Ital 45:75–89

Le Calvez Y (1949) Révision des Foraminiféres lutétiens du Bassin de Paris. 2. Rotaliidae et families affines. Mém Carte géol France, 54 pp, 1 tab, 6 pls, Imprimerie Nationale, Paris

Loeblich AR, Tappan H (1987) Foraminiferal genera and their classification. Van Nostrand Reinhold, New York, 1, 970 pp; 2, 212 pp, 847 pls

Moorkens TL (1982) Foraminifera of the Montian stratotype and of subjacent strata in the "Mons well 1969" with a review of Belgian Paleocene stratigraphy. Mém Serv géol Belgique 2(2): 185 pp

Peybernès B, Fondecave-Wallez M-J, Hottinger L, Eichène P, Segonzac G (2000) Limite Crétacé-Tertiaire et biozonation micropaléontologique du Danien-Sélandien dans le Béarn occidental et la Haute-Soule (Pyrénées Atlantiques). Geobios 33: 35–48

Pozaryska K, Szczechura J (1970) On some warm-water foraminifers from the Polish Montian. Acta Palaeontol Pol 15:95–113

Rahaghi A (1976) Contribution a l'étude de quelques grands foraminifères de l'Iran, Nat Iranian Oil Corp Geol Lab 6:1–68

Rahaghi A (1978) Paleogene biostratigraphy of some parts of Iran. Nat Iran Oil Comp Geol Lab 7:82 pp, 41 pls

Rahaghi A (1983) Stratigraphy and faunal assemblage of Paleocene-Lower Eocene in Iran. Nat Iranian Oil Comp Geol Lab 10:73 pp, 49 pls

Reiss Z (1963) Reclassification of Perforate Foraminifera. Bull Geol Surv Israel 36:1–111

Sacal V, Debourle A (1957) Foraminifères d'Aquitaine. 2e partie - Peneroplidae à Victoriellidae. Mém Soc Géol Fr (n s) 78(36):1–87, 35 pls

Sirel E (1998) Foraminiferal description and biostratigraphy of the Paleocene-Eocene shallow water limestones and discussion of the Cretaceous-Tertiary boundary in Turkey. Gen Dir Min Res Expl (MTA), Monograph ser 2, 117 pp, 68 pls

Sztràkos K (2000) Les foraminifères de l'Eocène du Bassin de l'Adour (Aquitaine, France): Biostratigraphie et taxinomie. Rev Micropaléontol 43(1–2): 71–172, 23 pls

von Hillebrandt A (1962) Das Paleozän und seine Foraminiferenfauna im Becken von Reichenhall und Salzburg. Bayer Akad Wiss Math-Naturwiss Kl 108: 9–182

References

Abramovich S, Keller G, Adatte T, Stinnesbeck W, Hottinger L, Stueben D, Berner Z, Ramanivosoa B, Randriamanantenasoa A (2002) Age and paleoenvironment of the Maestrichtian to Paleocene of the Mahajanga Basin, Madagascar: a multidisciplinary approach. Mar Micropaleontol 47:17–70

Hottinger L (ed) (1980) Rotaliid Foraminifera. Schweiz Paläontol Abh 101, 154 pp, Basel

Singh NP (2007) Cenozoic Lithostratigraphy of the Jaisalmer Basin, Rajasthan. J Palaeontol Soc India 52 (2):129–154, 12 pls

L. Hottinger, *Paleogene larger rotaliid foraminifera from the western and central Neotethys*,
DOI 10.1007/978-3-319-02853-8,

Index

L. Hottinger, *Paleogene larger rotaliid foraminifera from the western and central Neotethys*,
DOI 10.1007/978-3-319-02853-8,

The manufacturer's authorised representative in the EU is Springer Nature Customer Service Centre GmbH, Europaplatz 3, 69115 Heidelberg, Germany. If you have any concerns regarding our products, please contact ProductSafety@springernature.com

Printed and bound by CPI Group (UK) Ltd, Croydon, CR0 4YY
15/07/2026
02167649-0005